NONLINEAR ANALYSIS

A Collection of Papers in Honor of Erich H. Rothe

Erich H. Rothe

NONLINEAR ANALYSIS

A Collection of Papers in Honor of Erich H. Rothe

Edited by *LAMBERTO CESARI*
Department of Mathematics
University of Michigan
Ann Arbor, Michigan

RANGACHARI KANNAN
Mathematics Department
The University of Texas at Arlington
Arlington, Texas

HANS F. WEINBERGER
School of Mathematics
University of Minnesota
Minneapolis, Minnesota

ACADEMIC PRESS New York San Francisco London 1978
A Subsidiary of Harcourt Brace Jovanovich, Publishers

ACADEMIC PRESS, INC.
111 Fifth Avenue, New York, New York 10003

United Kingdom Edition published by
ACADEMIC PRESS, INC. (LONDON) LTD.
24/28 Oval Road, London NW1 7DX

Library of Congress Cataloging in Publication Data

Main entry under title:

Nonlinear analysis.

"Some of the papers in this book were presented at an international conference held in January 1969 at the University of the West Indies in Kingston, Jamaica."

Includes bibliographies and index.

CONTENTS: Amann, H. Periodic solutions of semilinear parabolic equations. --Brézis, H., and Browder, F. E. Linear maximal monotone operators and singular nonlinear integral equations of Hammerstein type. --Cesari, L. Nonlinear problems across a point of resonance for nonselfadjoint systems. [etc.]

1. Mathematical analysis--Addresses, essays, lectures. 2. Nonlinear theories--Addresses, essays, lectures. 3. Rothe, Erich H. I. Rothe, Erich H. II. Cesari, Lamberto. III. Kannan, Rangachary. IV. Weinberger, Hans F. V. Mona, Jamaica. University of the West Indies.

QA300.5.N66 515 77-6599
ISBN 0-12-165550-4

PRINTED IN THE UNITED STATES OF AMERICA

To ERICH H. ROTHE, scholar and friend

We offer this volume in appreciation of his lifework as a mathematician. We hope that the spirit of our seminar discussions and the more tranquil mood of our *Waldwanderungen* is reflected here—and projected into the future.

CONTENTS

LIST OF CONTRIBUTORS

Numbers in parentheses indicate the pages on which the authors' contributions begin.

HERBERT AMANN (1), Institute of Mathematics, Ruhr-University, Bochum, Federal Republic of Germany

H. BRÉZIS (31), Department of Mathematics, Universite P. et M. Curie, Paris, France

F. E. BROWDER (31), Department of Mathematics, University of Chicago, Chicago, Illinois 60637

LAMBERTO CESARI (43), Department of Mathematics, University of Michigan, Ann Arbor, Michigan 48104

JANE CRONIN (69), Department of Mathematics, Rutgers—The State University, New Brunswick, New Jersey 08903

JACK K. HALE (83), Lefschetz Center for Dynamical Systems, Division of Applied Mathematics, Brown University, Providence, Rhode Island

PETER HESS (99), Mathematics Institute, University of Zurich, Zurich, Switzerland

R. KANNAN (109), Mathematics Department, The University of Texas, Arlington, Texas 76019

M. KUČERA (125), Mathematical Institute of the Czechoslovak Academy of Sciences, Prague, Czechoslovakia

JEAN MAWHIN (145), Mathematics Institute, Université de Louvain, Louvain-la-Neuve, Belgium

J. NEČAS (125), Mathematical Institute of the Czechoslovak Academy of Sciences, Prague, Czechoslovakia

PAUL H. RABINOWITZ (161), Department of Mathematics, University of Wisconsin, Madison, Wisconsin 53706

D. SATHER (179), University of Colorado, Boulder, Colorado 80302

D. H. SATTINGER (193), School of Mathematics, University of Minnesota, Minneapolis, Minnesota 55455

J. SOUČEK (125), Mathematical Institute of the Czechoslovak Academy of Sciences, Prague, Czechoslovakia

M. M. VAINBERG (211), Ped. Institute, Moscow, U.S.S.R.

H. F. WEINBERGER (219), School of Mathematics, University of Minnesota, Minneapolis, Minnesota 55455

PREFACE

Professor Erich Rothe has made significant contributions to various aspects of nonlinear functional analysis. His early interests were in the field of parabolic and elliptic partial differential equations. Since then he has made fundamental contributions to the theory of nonlinear integral equations, gradient mappings, and degree theory. His more recent interests have been in critical point theory and the calculus of variations.

This volume is a collection of articles on nonlinear functional analysis dedicated to Professor Rothe on the occasion of his eightieth birthday. The intent of this collection is not to present a complete exposition of any particular branch of nonlinear analysis, but to provide an overview of some recent advances in the field.

Periodic Solutions of Semilinear Parabolic Equations

Herbert Amann
Ruhr-University

Dedicated to Professor Erich H. Rothe on the occasion of his 80th birthday.

Introduction

In this paper we use some methods of nonlinear functional analysis, namely fixed-point theorems in ordered Banach spaces, to prove existence and multiplicity result for periodic solutions of semilinear parabolic differential equations of the second order.

The most natural and oldest method for the study of periodic solutions of differential equations is to find fixed points of the Poincaré operator, that is, the translation operator along the trajectories, which assigns to every initial value the value of the solution after one period (e.g., Krasnosel'skii [16]). In the case of parabolic equations it turns out that the Poincaré operator is compact in suitable function spaces. Moreover, by involving the strong maximum principle for linear parabolic equations, it can be shown that it is strongly increasing in some closed subspace of $C^{2+\nu}(\bar{\Omega})$, $0 < \nu < 1$.

This paper is motivated by some papers of Kolesov [12–14], who has used essentially the same approach. However, he considered the Poincaré operator in the space of continuous functions, and he did not realize that, even in the case of the general semilinear parabolic equations, this operator is strongly increasing. This latter fact is the basis for nontrivial existence and multiplicity results. For simplicity we present only one multiplicity result, namely we establish the existence of at least three periodic solutions, given certain conditions. But having shown that the Poincaré operator is strongly increasing, it is clear that we can put the problem in the general framework of nonlinear equations in ordered Banach spaces. Hence, by applying other general fixed-point theorems for equations of this type (e.g., [3,4,5,15]), it is possible to obtain further existence and multiplicity results.

We refer to the papers of Kolesov for further references. In addition we mention a paper by Fife [8], who, by different methods, obtained some existence theorems for periodic solutions of linear and quasi-linear parabolic equations. More recently, the work of Fife has been used by Bange [6] and Gaines and Walter [11] to obtain existence theorems in the case of one space variable. For further results on periodic solutions of nonlinear parabolic equations we refer to the references [7,10,20,23,26]. These authors use the theory of monotone operators to deduce existence theorems. However, none of these papers contains multiplicity results.

In the following section we introduce our hypotheses and present the main results. In Section 2 we collect some facts on abstract evolution equations in Banach spaces. Section 3 presents semilinear abstract evolution equations. It contains the basic a priori estimates, which, for further uses, are presented in somewhat greater generality than needed in this paper.

In Section 4 we study initial boundary value problems for semilinear parabolic differential equations. In particular we prove a global existence theorem (Theorem 4.5), which is of independent interest. In the last paragraph we establish the basic properties of the Poincaré operator and prove the existence and multiplicity results of Section 1.

1. Definitions and Main Results

Throughout this paper all functions are real-valued.

Let X and Y be nonempty sets with $X \subset Y$, and let $u: X \to \mathbb{R}$ and $v: Y \to \mathbb{R}$. Then we write $u \leq v$ if $u(x) \leq v(x)$ for every $x \in X$. If $u \leq v$ and $u \neq v \mid X$, then we write $u < v$. If $u > 0$, we say that u is positive, and if $u \geq 0$, it is called nonnegative.

We denote by Ω a bounded domain in $\mathbb{R}^N$, whose boundary Γ is an $(N-1)$-dimensional $C^{2+\mu}$-manifold for some $\mu \in (0, 1)$, such that Ω lies locally on one side of Γ.

We let

$$A(x, t, D)u := -\sum_{i,k=1}^{N} a_{ik}(x, t)D_i D_k u + \sum_{i=1}^{N} a_i(x, t)D_i u + a_0(x, t)u,$$

where (x, t) denotes a generic point of $\bar{\Omega} \times \mathbb{R}$. The coefficients a_{ik}, a_i, a_0 are supposed to be μ-Hölder continuous functions on $\bar{\Omega} \times \mathbb{R}$, where we use the metric $d((x, t), (y, s)) := (|x - y|^2 + |s - t|)^{1/2}$ for the computation of the Hölder constant (that is, a_{ik}, a_i, $a_0 \in C^{\mu, \mu/2}(\bar{\Omega} \times \mathbb{R})$). We assume that the coefficients of $A(x, t, D)$ are ω-periodic in t, for some $\omega > 0$, that $a_{ik} = a_{ki}$, and that there exists a positive constant μ_0 such that

$$\sum_{i,k=1}^{N} a_{ik}(x, t)\xi^i\xi^k \geq \mu_0 |\xi|^2,$$

for $(x, t) \in \bar{\Omega} \times \mathbb{R}$ and $\xi \in \mathbb{R}^N$. Hence

$$Lu := \frac{\partial u}{\partial t} + A(x, t, D)u$$

is a uniformly parabolic differential operator in $\bar{\Omega} \times \mathbb{R}$.

We denote by $\beta \in C^{1+\mu}(\Gamma, \mathbb{R}^N)$ an outward pointing, nowhere tangent vector field on Γ. Then we let $B = B(x, D)$ be a boundary operator on $\Gamma \times \mathbb{R}$ of the form

$$Bu := b_0 u + \delta \frac{\partial u}{\partial \beta},$$

where either $\delta = 0$ and $b_0 = 1$ (Dirichlet boundary operator), or $\delta = 1$ and $b_0 \in C^{1+\mu}(\Gamma)$ with $b_0 \geq 0$ (regular oblique derivative boundary operator). Observe that B is independent of t.

Let (x, t, ξ, η) be a generic point of $\bar{\Omega} \times \mathbb{R}^{N+2}$ with $x \in \bar{\Omega}$ and $\eta = (\eta^1, \ldots, \eta^N) \in \mathbb{R}^N$. Then we denote by $f\colon \bar{\Omega} \times \mathbb{R}^{N+2} \to \mathbb{R}$ a continuous function which is ω-periodic in t, such that $f(\cdot, \cdot, \xi, \eta)\colon \bar{\Omega} \times \mathbb{R} \to \mathbb{R}$ is μ-Hölder continuous, uniformly for (ξ, η) in bounded subsets of $\mathbb{R} \times \mathbb{R}^N$, and such that $\partial f/\partial \xi$ and $\partial f/\partial \eta^i$, $i = 1, \ldots, N$, exist and are continuous on $\bar{\Omega} \times \mathbb{R}^{N+2}$. Lastly, we suppose that there exist functions $c\colon \mathbb{R}_+ \to \mathbb{R}_+ := [0, \infty)$ and $\varepsilon\colon \mathbb{R}_+ \to (0, 1)$ such that

$$|f(x, t, \xi, \eta)| \leq c(\rho)(1 + |\eta|^{2-\varepsilon(\rho)}), \tag{1.1}$$

for every $\rho \geq 0$ and $(x, t, \xi, \eta) \in \bar{\Omega} \times \mathbb{R} \times [-\rho, \rho] \times \mathbb{R}^N$.

Under the above assumptions we study the existence of ω-periodic solutions of the semilinear parabolic boundary value problem (BVP)

$$\begin{aligned} Lu &= f(x, t, u, \nabla u) \quad &&\text{in} \quad \Omega \times \mathbb{R}, \\ Bu &= 0 \quad &&\text{on} \quad \Gamma \times \mathbb{R}. \end{aligned} \tag{1.2}$$

By an *ω-periodic solution* of the BVP (1.2) we mean a function $u \in C^{2,1}(\bar{\Omega} \times \mathbb{R})$, which is ω-periodic in t, such that $Lu(x, t) = f(x, t, u(x, t), \nabla u(x, t))$ for $(x, t) \in \Omega \times \mathbb{R}$ and $Bu(x, t) = 0$ for $(x, t) \in \Gamma \times \mathbb{R}$, where $\nabla u = (D_1 u, \ldots, D_N u)$ denotes the gradient of u with respect to x. Of course, $u \in C^{2,1}$ means that u is continuously differentiable, twice with respect to x and once with respect to t. (In fact, it will be shown, that every ω-periodic solution of (1.2) belongs to $C^{2+\mu, 1+\mu/2}(\bar{\Omega} \times \mathbb{R})$.)

In the special case that the coefficients of $A(x, t, D)$ are independent of t (in which case we write $A(x, D)$), we can consider the linear elliptic eigenvalue problem (EVP)

$$\begin{aligned} A(x, D)u &= \lambda u && \text{in } \Omega, \\ Bu &= 0 && \text{on } \Gamma. \end{aligned} \tag{1.3}$$

It is known (cf. Amann [3, Theorem 1.16]) that this EVP possesses a smallest eigenvalue λ_0. Moreover, $\lambda_0 > 0$ if $a_0 \geq 0$ and if, in the case that $\delta = 1$, $a_0 > 0$ for $b_0 = 0$.

After these preparations we can state an *existence and uniqueness theorem* for the linear case.

1.1 THEOREM Let one of the following hypotheses be satisfied:

(i) $a_0 \geq 0$. Moreover, if $\delta = 1$, then $a_0 > 0$ if $b_0 = 0$.
(ii) The coefficients of $A(x, t, D)$ are independent of t, and the smallest eigenvalue of the EVP (1.3) is positive.

Then, for every Hölder continuous function w on $\bar{\Omega} \times \mathbb{R}$, which is ω-periodic in t, the linear BVP

$$\begin{aligned} Lu &= w && \text{in } \Omega \times \mathbb{R}, \\ Bu &= 0 && \text{on } \Gamma \times \mathbb{R}, \end{aligned}$$

has exactly one ω-periodic solution u, and $u > 0$ if $w > 0$.

A function u is called an *ω-subsolution* for the BVP (1.2) if there exists a number $T = T(u) > \omega$ such that $u \in C^{2,1}(\bar{\Omega} \times [0, T])$ and

$$\begin{aligned} Lu &\leq f(\cdot, \cdot, u, \nabla u) && \text{in } \Omega \times (0, T], \\ Bu &\leq 0 && \text{on } \Gamma \times (0, T], \\ u(\cdot, 0) &\leq u(\cdot, \omega) && \text{on } \bar{\Omega}. \end{aligned} \tag{1.4}$$

It is called a *strict ω-subsolution* if either $u(\cdot, 0) < u(\cdot, \omega)$ or $B(u(\cdot, 0)) < 0$. The notions of *ω-supersolutions* and *strict ω-supersolutions* are defined by reversing the above inequalities.

An ω-subsolution $\bar{v}$ and an ω-supersolution $\hat{v}$ are said to be *B-related*, if there exists a function $u \in C^{2+\mu}(\bar{\Omega})$ with $Bu = 0$, such that $\bar{v}(\cdot, 0) \leq u \leq \hat{v}(\cdot, 0)$. It can be shown (cf. Remark 5.5) that in the case of the first BVP (that is, if $\delta = 0$) every pair of ω-sub- and supersolutions $\bar{v}$, $\hat{v}$ satisfying $\bar{v}(\cdot, 0) \leq \hat{v}(\cdot, 0)$ are B-related.

After these preparations we are ready for the statement of the main *existence theorem* for ω-periodic solutions of the semilinear parabolic BVP (1.2).

1.2 THEOREM Suppose that $\bar{v}$ is an ω-subsolution and $\hat{v}$ is an ω-supersolution for the BVP (1.2) such that $\bar{v}$ and $\hat{v}$ are B-related. Then there exists at least one ω-periodic solution u such that $\bar{v} \leq u \leq \hat{v}$.

More precisely, there exist a minimal ω-periodic solution $\bar{u}$ and a maximal ω-periodic solution $\hat{u}$ with $\bar{v} \leq \bar{u} \leq \hat{u} \leq \hat{v}$, in the sense that $\bar{u} \leq u \leq \hat{u}$ for every ω-periodic solution u satisfying $\bar{v} \leq u \leq \hat{v}$.

In the case of the first BVP the above theorem is due to Kolesov [14] (cf. also Kolesov [12,13]). In fact, in Kolesov's theorem the coefficients of L are allowed to depend on u, and $\varepsilon(\rho)$ in (1.1) can be equal to zero (cf. Kolesov [14, Theorems 6 and 7]).

By combining Theorems 1.1 and 1.2, it is easy to give sufficient conditions for the existence of ω-periodic solutions.

1.3 THEOREM Suppose that there exist nonnegative Hölder continuous functions a and b on $\bar{\Omega} \times \mathbb{R}$, which are ω-periodic in t, such that

$$f(x, t, \xi, \eta) \leq a(x, t)\xi + b(x, t),$$

for $(x, t, \xi, \eta) \in \bar{\Omega} \times \mathbb{R} \times \mathbb{R}_+ \times \mathbb{R}^N$ and

$$f(x, t, \xi, \eta) \geq a(x, t)\xi - b(x, t), \tag{1.5}$$

for $(x, t, -\xi, \eta) \in \bar{\Omega} \times \mathbb{R} \times \mathbb{R}_+ \times \mathbb{R}^N$. Moreover, suppose that the operator $L - a$ satisfies the hypotheses of Theorem 1.1. Then the BVP (1.2) has at least one ω-periodic solution.

Proof Theorem 1.1 implies that the BVP

$$\begin{aligned} (L - a)u &= b \quad \text{in} \quad \Omega \times \mathbb{R}, \\ Bu &= 0 \quad \text{on} \quad \Gamma \times \mathbb{R}, \end{aligned} \tag{1.6$_b$}$$

has exactly one ω-periodic solution $\hat{v}$, and $\hat{v} \geq 0$. Hence $\bar{v} := -\hat{v}$ is the unique ω-periodic solution of the BVP $(1.6)_{-b}$. It is clear that $\bar{v}$ is an ω-subsolution and $\hat{v}$ is an ω-supersolution which are B-related. Hence the assertion follows from Theorem 1.2. ●

1.4 Corollary Let the hypotheses of Theorem 1.3 be satisfied, but assume, instead of inequality (1.5), that $f(x, t, 0, 0) \geq 0$ for $(x, t) \in \bar{\Omega} \times \mathbb{R}$. Then the BVP (1.2) has at least one nonnegative ω-periodic solution.

Proof It suffices to observe that 0 is an ω-subsolution and that the solution $\hat{v}$ of $(1.6)_b$ is a nonnegative ω-supersolution, which is B-related to 0. ●

Corollary 1.4 generalizes considerably the results of Kolesov [14, Theorem 8], where it had been assumed that $\delta = 0$, and that f is nonnegative and independent of η.

In addition to the above existence results we prove the following *multiplicity theorem*, which is new, even in the simplest case that f is only a function of ξ.

1.5 Theorem Suppose that $\bar{v}_1$ is an ω-subsolution, $\hat{v}_1$ is a strict ω-supersolution, $\bar{v}_2$ is a strict ω-subsolution, and $\hat{v}_2$ is an ω-supersolution, such that each one of the pairs $(\bar{v}_1, \hat{v}_1)$ and $(\bar{v}_2, \hat{v}_2)$ is B-related. Moreover, assume that $\hat{v}_1(\cdot, 0) \leq \bar{v}_2(\cdot, 0)$. Then the BVP (1.2) has at least three ω-periodic solutions u_j such that

$$\bar{v}_1 \leq u_1 < u_3 < u_2 \leq \hat{v}_2 .$$

Moreover, $\bar{v}_j \leq u_j \leq \hat{v}_j$ for $j = 1, 2$, and $\bar{v}_2 \nleq u_3 \nleq \hat{v}_1$.

It should be remarked that, instead of $\hat{v}_1(\cdot, 0) \leq \bar{v}_2(\cdot, 0)$, it suffices to assume that $\bar{v}_1(\cdot, 0) < \hat{v}_2(\cdot, 0)$ and $\bar{v}_2(\cdot, 0) \nleq \hat{v}_1(\cdot, 0)$.

We close this section with a simple example. Consider the BVP

$$\begin{aligned} \frac{\partial u}{\partial t} - \Delta u = 4\pi^2 \cos u + u^2 a(x) & \\ + e^u \sum_{k=1}^{N} a_k(x) \cos(kt)(D_k u)^{5/3} & \qquad \text{in} \quad \Omega \times \mathbb{R}, \\ \frac{\partial u}{\partial \beta} = 0 \qquad \text{on} \quad \Gamma \times \mathbb{R}, & \end{aligned}$$

where the functions a and a_k are Hölder continuous on $\bar{\Omega}$ and $|a| \leq 1$. Then it is easily verified that the constant functions $\bar{v}_1 := -2\pi$, $\bar{v}_1 := -\pi$, $\bar{v}_2 := 0$, and $\hat{v}_2 := \pi$ satisfy the hypotheses of Theorem 1.2. Hence there exist at least three 2π-periodic solutions such that $-2\pi \leq u_1 < u_2 < u_3 \leq \pi$.

2. Preliminaries on Linear Evolution Equations

Throughout this paper all vector spaces are over the reals. If A is a linear operator in some Banach space, then we denote by $R(\lambda, A)$ the resolvent of the *complexification* of A. If X and Y are Banach spaces such that X is continuously imbedded in Y, then we write $X \hookrightarrow Y$.

Let X be a Banach space and let T be a fixed positive number. Suppose that

(A1) $\{A(t) \mid 0 \le t \le T\}$ is a family of closed densely defined linear operators in X such that the domain $D(A(t))$ of $A(t)$ is independent of t.

(A2) For each $t \in [0, T]$ the resolvent $R(\lambda, A(t))$ exists for all λ with $\operatorname{Re} \lambda \le 0$, and

$$\|R(\lambda, A(t))\| \le c(1 + |\lambda|)^{-1},$$

where c is some constant that is independent of λ and t.

These assumptions imply that $A(t)$ has an inverse $A^{-1}(t) \in L(X)$, where $L(X)$ denotes the Banach algebra of bounded linear operators on X. For abbreviation we write $A := A(0)$. Then $\|x\|_1 := \|Ax\|$ defines a norm on $D(A)$, which is equivalent to the graph norm. Consequently $X_1 := (D(A), \|\cdot\|_1)$ is a Banach space, and $X_1 \hookrightarrow X$. Moreover, by the closed graph theorem, $A(s)A^{-1}(t) \in L(X)$ for every $s, t \in [0, T]$. Hence $x \to \|A(s)x\|$ defines for each s an equivalent norm on $D(A)$, and $A(s) \in L(X_1, X)$, that is, $A(t)$ is a bounded linear operator from X_1 to X.

Using these notations, we suppose

(A3) The map $A(\cdot)\colon [0, T] \to L(X_1, X)$ is Hölder continuous.

In the following we denote by $c, c(\alpha, \ldots)$ generic constants, not necessarily the same in different formulas, which depend in an increasing way on the indicated quantities.

Assumptions (A1)–(A3) imply the existence of constants $c > 0$ and $\nu \in (0, 1)$ such that

$$\|(A(s) - A(t))A^{-1}(\tau)\| \le c|s - t|^\nu \tag{2.1}$$

for $s, t, \tau \in [0, T]$. In fact, it is easily seen that the map $\tau \to B(\tau) := A(\tau)A^{-1}$ is continuous from $[0, T]$ into the group $GL(X)$ of invertible operators in $L(X)$. Since the map $B \to B^{-1}$ is continuous on $GL(X)$, there exists a constant c such that $\|B^{-1}(\tau)\| = \|AA^{-1}(\tau)\| \le c$ for $\tau \in [0, T]$. Hence the assertion follows from the inequalities

$$\|(A(s) - A(t))A^{-1}(\tau)\| \le \|A(s) - A(t)\|_{L(X_1, X)} \|AA^{-1}(\tau)\| \le c|s - t|^\nu$$

for $s, t, \tau \in [0, T]$.

Assumptions (A1) and (A2) imply that $-A(t)$ is the infinitesimal generator of a holomorphic semigroup $\{e^{-\tau A(t)} \mid 0 \le \tau < \infty\}$ in $L(X)$. Moreover, there exist positive constants c and δ_0 such that

$$\|e^{-\tau A(t)}\| \le ce^{-\delta_0 \tau} \tag{2.2}$$

and

$$\|A(t)e^{-\tau A(t)}\| \leq c\tau^{-1}e^{-\delta_0\tau} \tag{2.3}$$

for $\tau > 0$ and $t \in [0, T]$.

Then inequality (2.2) implies the existence of the integral

$$A^{-\alpha}(t) := \frac{1}{\Gamma(\alpha)} \int_0^\infty \tau^{\alpha-1} e^{-\tau A(t)}\, d\tau \tag{2.4}$$

for every $\alpha > 0$. It follows that $A^{-1}(t) = [A(t)]^{-1}$, and each $A^{-\alpha}(t)$ is an injective continuous endomorphism of X. Hence $A^\alpha(t) := [A^{-\alpha}(t)]^{-1}$ is a closed bijective linear operator in X. It can be shown that each $A^\alpha(t)$ has dense domain and that $D(A^\alpha(t)) \subset D(A^\beta(t))$ for $a \geq \beta \geq 0$. Moreover,

$$A^{\alpha+\beta}(t)x = A^\alpha(t)A^\beta(t)x = A^\beta(t)A^\alpha(t)x$$

for every $\alpha, \beta \in \mathbb{R}$ and $x \in D(A^\gamma(t))$, with $\gamma := \max\{\alpha, \beta, \alpha + \beta\}$, where $A^0(t) = id_X$. (For proofs of these facts we refer to the literature [9,17,21,24].)

It has been shown by Sobolevskii [24, inequality (1.59)] that $D(A^\beta(s)) \subset D(A^\alpha(t))$ for $0 \leq \alpha < \beta \leq 1$ and $s, t \in [0, T]$, and that

$$\|A^\alpha(s)A^{-\beta}(t)\| \leq c(\alpha, \beta) \tag{2.5}$$

for $s, t \in [0, T]$.

In the following we let $\|x\|_\alpha := \|A^\alpha x\|$ for $x \in D(A^\alpha)$ and $0 \leq \alpha \leq 1$, and we denote by X_α the Banach space $(D(A^\alpha), \|\cdot\|_\alpha)$. Then $X_\beta \hookrightarrow X_\alpha$ for $0 \leq \alpha \leq \beta \leq 1$ (with $X_0 = X$).

We consider the linear initial value problem (IVP)

$$\begin{aligned} u' + A(t)u &= g(t), \qquad 0 < t \leq T, \\ u(0) &= x, \end{aligned} \tag{2.6}$$

with $g \in C([0, T], X)$ and $x \in X$. By a *solution* u of (2.6) we mean a function $u \in C([0, T], X) \cap C^1((0, T], X)$ with $u(0) = x$, $u(t) \in D(A)$ for $t > 0$, and $u'(t) + A(t)u(t) = g(t)$ for $0 < t \leq T$.

Our assumptions (cf. inequality (2.1)) imply that the results of Sobolevskii [24] and Tanabe [25] are applicable (cf. also Friedman [10]). Hence the IVP (2.6) has a unique solution u for every Hölder continuous right-hand side g. Moreover, $u \in C^1([0, T], X)$, provided $x \in D(A)$.

Sobolevskii and Tanabe have shown that there exists a unique evolution operator $U(t, \tau) \in L(X)$, $0 \leq \tau \leq t \leq T$, such that every solution u of the IVP (2.6) can be represented in the form

$$u(t) = U(t, 0)x + \int_0^t U(t, \tau)g(\tau)\, d\tau, \qquad 0 \leq t \leq T. \tag{2.7}$$

The function U is strongly continuous on the closure of the set $\Delta := \{(t, \tau) \in [0, T]^2 \mid 0 \leq \tau < t \leq T\}$ (that is, $U(\cdot)x \in C(\bar{\Delta}, X)$ for every $x \in X$) and satisfies $U(t, t) = id_X$, $U(s, t)U(t, \tau) = U(s, \tau)$ for $0 \leq \tau \leq t \leq s \leq T$. Moreover, U has an important smoothing property, namely $U(s, t)X \subset D(A)$ for $0 \leq t < s \leq T$.

In the following lemma we collect the most important regularity properties of the evolution operator. For abbreviation we denote the norm in $L(X_\alpha, X_\beta)$ by $\|\cdot\|_{\alpha,\beta}$.

2.1 Lemma (i) Suppose that $0 \leq \alpha \leq \beta < 1$. Then

$$\|U(t, \tau)\|_{\alpha,\beta} \leq c(\alpha, \beta, \gamma)(t - \tau)^{-\gamma} \tag{2.8}$$

for $\beta - \alpha < \gamma < 1$ and $0 \leq \tau < t \leq T$. Moreover, if $0 \leq \beta < \alpha \leq 1$, then

$$\|U(t, \tau)\|_{\alpha,\beta} \leq c(\alpha, \beta). \tag{2.9}$$

(ii) Suppose that $0 \leq \alpha < \beta \leq 1$. Then

$$\|U(t, \tau) - U(s, \tau)\|_{\beta,\alpha} \leq c(\alpha, \beta, \gamma)\,|t - s|^\gamma \tag{2.10}$$

for $0 \leq \gamma < \beta - \alpha$ and $(t, \tau), (s, \tau) \in \bar{\Delta}$.

(iii) Let $0 \leq \alpha < 1$, $0 \leq \sigma < T$, and $g \in C([\sigma, T], X)$. Then

$$\left\| \int_\sigma^t U(t, \tau)g(\tau)\, d\tau - \int_\sigma^s U(s, \tau)g(\tau)\, d\tau \right\|_\alpha \leq c(\alpha, \gamma)\,|s - t|^\gamma \max_{\sigma \leq \tau \leq T} \|g(\tau)\| \tag{2.11}$$

for $0 \leq \gamma < 1 - \alpha$ and $\sigma \leq s, t \leq T$.

Proof (i) Suppose that $\alpha > 0$ and let

$$0 < \varepsilon < \min\{\alpha, 1 - \beta, (\gamma - \beta + \alpha)/2\}.$$

Then

$$\|U(t, \tau)\|_{\alpha,\beta} \leq \|A^\beta U(t, \tau)A^{-\alpha}\|$$
$$\leq \|A^\beta A^{-\beta-\varepsilon}(t)\|\; \|A^{\beta+\varepsilon}(t)U(t, \tau)A^{-\alpha+\varepsilon}(\tau)\|\; \|A^{\alpha-\varepsilon}(\tau)A^\alpha\|.$$

Hence it follows from (2.5) and Sobolevskii [24, inequality (1.65)] (cf. also Friedman [10, inequality (II.14.12)]) that

$$\|U(t, \tau)\|_{\alpha,\beta} \leq c(\alpha, \beta, \varepsilon)(t - \tau)^{\alpha-\beta-2\varepsilon}.$$

This implies (2.8).

If $\alpha = 0$, then the estimate (2.8) follows in a similar way from

$$\|U(t, \tau)\|_{0,\beta} \leq \|A^\beta A^{-\gamma}(t)\|\; \|A^\gamma(t)U(t, \tau)\|.$$

Analogously, if $0 \leq \beta < \alpha \leq 1$, let $\varepsilon := (\alpha - \beta)/2$. Then

$$\|U(t, \tau)\|_{\alpha,\beta} \leq \|A^{\beta}U(t, \tau)A^{-\alpha}\|$$
$$\leq \|A^{\beta}A^{-\alpha+\varepsilon}(t)\| \, \|A^{\alpha-\varepsilon}(t)U(t, \tau)A^{-\alpha+\varepsilon}(\tau)\| \, \|A^{\alpha-\varepsilon}(\tau)A^{\alpha}\|,$$

and the inequality (2.9) follows again from (2.5) and Sobolevskii [24, inequality (1.65)].

(ii) and (iii) are obtained by similar arguments from Sobolevskii [24, inequalities (1.68) and (2.9)] (cf. also Friedman [10, Lemma (II.14.1) and (II.14.4)]). ●

2.2 Corollary Let $0 \leq \alpha < \beta \leq 1$, and let

$$K(x, g)(t) := U(t, 0)x + \int_0^t U(t, \tau)g(\tau)\, d\tau, \qquad 0 \leq t \leq T. \tag{2.12}$$

Then K is a continuous linear operator from $X_{\beta} \times C([0, T], X)$ into $C^{\gamma}([0, T], X_{\alpha})$ for every $\gamma \in [0, \beta - \alpha)$.

Proof The assertion follows immediately from (2.10) and (2.11). ●

Lastly we derive a useful generalization of the Gronwall inequality. For this purpose we let

$$m_{\alpha}(\xi) := \sum_{k=1}^{\infty} \frac{[\Gamma(1-\alpha)]^k}{\Gamma(k(1-\alpha))} \xi^{k-1}$$

for $\xi \in \mathbb{R}$ and $0 \leq \alpha < 1$. Observe that m_{α} is analytic on $\mathbb{R}$ and $m_{\alpha}(\xi) > 0$ for $\xi > 0$. Moreover, $m_0(\xi) = e^{\xi}$.

2.3 Lemma Let $0 \leq \alpha < 1$ and suppose that $g \in L_1(0, T)$ is nonnegative a.e. If $w \in L_1(0, T)$ satisfies the integral inequality

$$w(t) \leq g(t) + \kappa \int_0^t (t-\tau)^{-\alpha} w(\tau)\, d\tau, \tag{2.13}$$

for almost all $t \in (0, T)$ and some $\kappa > 0$, then

$$w(t) \leq g(t) + \kappa \int_0^t (t-\tau)^{-\alpha} m_{\alpha}(\kappa(t-\tau)^{1-\alpha}) g(\tau)\, d\tau$$

for a.a. $t \in (0, T)$.

Proof Let $q(\xi) := \kappa\xi^{-\alpha}$ for $\xi > 0$ and denote by Q the integral operator

$$Qu(t) := \int_0^t q(t-\tau)u(\tau)\, d\tau, \qquad 0 < t < T. \tag{2.14}$$

Then Q is a continuous endomorphism of $L_1(0, T)$. An easy computation shows that Q^k is an integral operator of the form (2.14) with kernel

$$q_k(\xi) := \kappa\xi^{-\alpha} \frac{[\Gamma(1-\alpha)]^k}{\Gamma(k(1-\alpha))} [\kappa\xi^{1-\alpha}]^{k-1}$$

for $k = 1, 2, 3, \ldots$. Hence

$$\|Q^k\|_{L_1} \leq \frac{[\kappa\Gamma(1-\alpha)T^{1-\alpha}]^k}{\Gamma(k(1-\alpha)+1)},$$

which shows that the spectral radius $\lim_{k\to\infty} \|Q^k\|_{L_1}^{1/k}$ is equal to zero. Consequently, the integral equation on $L_1(0, T)$

$$u - Qu = g \tag{2.15}$$

has a unique solution, which is given by the Neumann series. Hence

$$u(t) = (I - Q)^{-1}g(t) = g(t) + \kappa \int_0^t (t-\tau)^{-\alpha} m_\alpha(\kappa(t-\tau)^{1-\alpha}) g(\tau)\, d\tau \tag{2.16}$$

for $0 < t < T$.

It follows from (2.13) and (2.15) that

$$(u - w) - Q(u - w) = v,$$

where $v \in L_1(0, T)$ is nonnegative a.e. Hence $u - w = (I - Q)^{-1}v$, and (2.16) implies that $u - w \geq 0$ a.e. This proves the assertion. ●

The following corollary seems to be well known, although the author could not find a proof in the literature.

2.4 COROLLARY Suppose that $w \in L_1(0, T)$ satisfies

$$w(t) \leq c_0 t^{-\beta} + c_1 \int_0^t (t-\tau)^{-\alpha} w(\tau)\, d\tau$$

for almost all $t \in (0, T)$, where c_0 and c_1 are nonnegative constants, and $0 \leq \alpha, \beta < 1$. Then

$$w(t) \leq c_0 c(\alpha, c_1, T) t^{-\beta}$$

for almost all $t \in (0, T)$.

Proof Lemma 2.3 implies the estimate

$$w(t) \leq c_0 t^{-\beta} + c_0 c_2(\alpha, c_1, T) \int_0^t (t-\tau)^{-\alpha} \tau^{-\beta}\, d\tau.$$

Hence, by making the substitution $\tau \to \tau/t$ in the above integral, we find that

$$w(t) \leq c_0[1 + c_2(\alpha, c_1, T)t^{1-\alpha}]t^{-\beta} \leq c_0 c(\alpha, c_1, T)t^{-\beta}$$

for almost all $t \in (0, T)$. ●

3. Semilinear Evolution Equations

Throughout this section we presuppose hypotheses (A1)–(A3). In addition we make the following assumption:

(A4) There is given a function $g\colon [0, T] \times X_\alpha \to X$ for some $\alpha \in [0, 1)$ such that, for some $\nu \in (0, 1]$ and every $\rho \geq 0$,

$$\|g(t, x) - g(s, y)\| \leq c(\rho)(|s - t|^\nu + \|x - y\|_\alpha),$$

provided $(t, x), (s, y) \in [0, T] \times X_\alpha$ satisfy $\|x\|_\alpha, \|y\|_\alpha \leq \rho$.

We consider semilinear initial value problems (IVPs) of the form

$$\begin{aligned} u' + A(t)u &= g(t, u), \qquad 0 < t \leq T, \\ u(0) &= x \end{aligned} \tag{3.1}$$

with $x \in X_\alpha$. By a *solution* of (3.1) we mean a function of $u \in C([0, T], X_\alpha) \cap C^1((0, T], X)$ with $u(0) = x$ such that $u(t) \in D(A)$ and $u'(t) + A(t)u(t) = g(t, u(t))$ for $0 < t \leq T$.

3.1 Lemma The IVP (3.1) has at most one solution.

Proof Suppose that u_1 and u_2 are two solutions, and let $w := u_1 - u_2$. Then, by means of the representation formula (2.7),

$$w(t) = \int_0^t U(t, \tau)[g(\tau, u_1(\tau)) - g(\tau, u_2(\tau))]\, d\tau, \qquad 0 \leq t \leq T.$$

Hence (A4) and (2.8) imply

$$\|w(t)\|_\alpha \leq c(\alpha, \gamma) \int_0^t (t - \tau)^{-\gamma} \|w(\tau)\|_\alpha\, d\tau$$

for $0 \leq t \leq T$ and $\alpha < \gamma < 1$. Hence Corollary 2.4 implies $w = 0$. ●

In the following we denote by D the set of all $x \in X_\alpha$ for which the IVP (3.1) is solvable. Hence, by Lemma 3.1 we can define a map

$$S\colon D \to C([0, T], X_\alpha) \cap C^1((0, T], X),$$

which assigns to every $x \in D$ the unique solution $S(x)$ of (3.1). This map is called the *solution operator* for the IVP (3.1).

In the remainder of this section we study existence and continuity properties for S.

3.2 LEMMA Suppose that $x \in X_\beta$ for some $\beta \in (\alpha, 1]$. Then the IVP (3.1) is equivalent to the integral equation

$$u(t) = U(t, 0)x + \int_0^t U(t, \tau)g(\tau, u(\tau))\,d\tau, \qquad 0 \le t \le T, \tag{3.2}$$

in $C([0, T], X_\alpha)$.

Proof The assertion is an easy consequence of Corollary 2.2 and the above mentioned Sobolevskii–Tanabe results. ●

3.3 LEMMA Suppose that $x \in X_\beta$ for some $\beta \in (\alpha, 1]$. Then there exists a number $T_1 := T_1(\alpha, \beta, \|x\|_\beta) \in (0, T]$ such that the integral equation

$$u(t) = U(t, s)x + \int_s^t U(t, \tau)g(\tau, u(\tau))\,d\tau, \qquad s \le t \le T, \tag{3.3}$$

has a unique solution in $C([s, \min\{s + T_1, T\}], X_\alpha)$ for every $s \in [0, T)$.

Proof Denote the right-hand side of (3.3) by $G(u)(t)$, $s \le t \le T$. Then it follows from (2.10) and (2.11) that G maps $C([s, T], X_\alpha)$ into itself. Moreover, (2.10) implies

$$\|U(t, s)x - x\|_\alpha = \|[U(t, s) - U(s, s)]x\|_\alpha \le c(\alpha, \beta)(t - s)^\gamma \|x\|_\beta,$$

for $\gamma := (\beta - \alpha)/2$ and $s \le t \le T$.

Let $\rho := c(\alpha, \beta)T^\gamma \|x\|_\beta + 1$. Then $\|G(u)(t) - x\|_\alpha \le \rho$, provided

$$\left\| \int_s^t U(t, \tau)g(\tau, u(\tau))\,d\tau \right\|_\alpha \le 1.$$

It follows from (2.8) that the latter integral is estimated by

$$c_1(\alpha, \beta)(t - s)^\gamma \max_{s \le \tau \le T} \|g(\tau, u(\tau))\|.$$

Suppose now that $\|u(t) - x\|_\alpha \le \rho$ for $s \le t \le T$. Then by (A4), $\|g(\tau, u(\tau))\| \le c_2(\rho)$ for $s \le t \le T$. Hence, by taking $T_1 \in (0, T]$ such that

$$c_1(\alpha, \beta)c_2(\rho)T_1{}^\gamma \le \tfrac{1}{2},$$

it follows that G maps the closed ball in $C([s, \min\{s + T_1, T\}], X_\alpha)$ with center at x and radius ρ into itself. Moreover, on this ball G is a contraction with constant $\frac{1}{2}$. Hence the assertion follows from the contraction mapping principle. ●

After these preparations we can prove the existence of a solution of the IVP (3.1), provided an appropriate a priori estimate is known.

3.4 PROPOSITION Let $x \in X_\beta$ for some $\beta \in (\alpha, 1]$. Suppose that there exists a positive number ρ such that the possible solution of (3.1) satisfies the *a priori estimate* $\|u(t)\|_\alpha \le \rho$ for $0 \le t \le T$. Then the IVP (3.1) has a unique solution.

Proof The a priori estimate and (3.2) imply a uniform estimate for $\|u(t)\|_\gamma$ with $\alpha < \gamma < \beta$. By applying Lemma 3.3, with β replaced by γ, successively to the intervals $[0, T_1]$, $[T_1, 2T_1]$, ..., we obtain a unique solution of the integral equation (3.2) in $C([0, T], X_\alpha)$. Hence the assertion follows from Lemma 3.2. ●

It should be remarked that the above method of proof is of course well known (cf. Friedman [10], and Sobolevskii [24]).

The following lemma contains a simple but important continuity result for the solution operator S.

3.5 LEMMA Let $x, y \in D$ and suppose that $\|S(x)(t)\|_\alpha, \|S(y)(t)\|_\alpha \le \rho$ for $0 \le t \le T$. Then, for every $\gamma \in [0, 1]$, every $\beta \in [\alpha, 1)$, and every $\varepsilon \in (0, 1 - \beta)$,

$$\|S(x)(t) - S(y)(t)\|_\beta \le c(\rho, \alpha, \beta, \gamma, \varepsilon) t^{-\lambda} \|x - y\|_\gamma \tag{3.4}$$

for $0 \le t \le T$, where $\lambda := \max\{0, \beta - \gamma + \varepsilon\}$.

Proof Let $u := S(x)$, $v := S(y)$, and $w := u - v$. Then it follows from Lemma 3.2 that

$$\|w(t)\|_\beta \le \|U(t, 0)\|_{\gamma, \beta} \|x - y\|_\gamma + \int_0^t \|U(t, \tau)\|_{0, \beta} \|g(\tau, u_1(\tau)) - g(\tau, u_2(\tau))\| \, d\tau$$

for $0 \le t \le T$. Hence assumption (A4), Lemma 2.1(i), and the continuous imbedding of X_β in X_α imply

$$\|w(t)\|_\beta \le c(\beta, \gamma, \varepsilon) t^{-\lambda} \|x - y\|_\gamma + c_1(\alpha, \beta, \varepsilon, \rho) \int_0^t (t - \tau)^{-\beta - \varepsilon} \|w(\tau)\|_\beta \, d\tau,$$

for $0 \le t \le T$. The assertion follows now from Corollary 2.4. ●

Proposition 3.4 shows that the solvability of the IVP (3.1) can be guaranteed, provided we know an a priori estimate for the α-norm of the solution. The following lemma shows that such an a priori estimate can be obtained, provided the function g satisfies an appropriate growth condition and we known an a priori estimate *in some weaker norm*.

3.6 LEMMA Suppose that there exists a Banach space E with $X_\alpha \hookrightarrow E \hookrightarrow X$ and a constant $\lambda \in [1, 1/\alpha)$ such that

$$\|g(t, y)\| \le c(\rho)(1 + \|y\|_\alpha{}^\lambda) \tag{3.5}$$

for every $\rho \geq 0$ and every $(t, y) \in [0, T] \times X_\alpha$, satisfying $\|y\|_E \leq \rho$. Suppose that $x \in D$ and $\|S(x)(t)\|_E \leq \rho_0$ for $0 \leq t \leq T$. Then

$$\|S(x)(t)\|_\alpha \leq c(\alpha, \beta, \lambda, \varepsilon, \rho_0)(1 + t^{-\sigma}\|x\|_\beta) \tag{3.6}$$

for every $\varepsilon \in (0, 1 - \alpha\lambda)$, $\beta \in [0, 1]$, and $t \in [0, T]$, where

$$\sigma := \max\{0, \alpha\lambda - \beta + \varepsilon\}.$$

Proof Let $\gamma := \lambda\alpha$. Then by the "moment inequality"

$$\|y\|_\alpha \leq c(\alpha, \gamma)\|y\|_\gamma^{\alpha/\gamma}\|y\|_0^{1-\alpha/\gamma}$$

(cf. Sobolevskii [24, inequality (1.55)] or Friedman [10, Theorem II.14.1]) it follows that

$$\|y\|_\alpha{}^\lambda \leq c(\alpha, \lambda)\|y\|_0^{\lambda-1}\|y\|_\gamma \leq c_1(\alpha, \lambda)\|y\|_E^{\lambda-1}\|y\|_\gamma.$$

Hence it follows from (2.7), Lemma 2.1(i) and (3.5) that

$$\begin{aligned}\|S(x)(t)\|_\gamma &\leq c(\gamma, \beta, \sigma)t^{-\sigma}\|x\|_\beta \\ &\quad + c(\gamma, \varepsilon, \rho_0)\int_0^t (t-\tau)^{-\gamma-\varepsilon}(1 + \|S(x)(\tau)\|_\alpha{}^\lambda)\, d\tau \\ &\leq c(\beta, \gamma, \varepsilon, \rho_0)(1 + t^{-\sigma}\|x\|_\beta) \\ &\quad + c(\alpha, \gamma, \varepsilon, \rho_0)\int_0^t (t-\tau)^{-\gamma-\varepsilon}\|S(x)(\tau)\|_\gamma\, d\tau\end{aligned}$$

for $0 \leq t \leq T$. The assertion follows now from Corollary 2.4 and the fact that $X_\gamma \hookrightarrow X_\alpha$. ●

3.7 COROLLARY Let the hypotheses of Lemma 3.6 be satisfied and suppose that $x \in X_\beta$ for some $\beta \in (\alpha, 1]$. Then

$$\max_{0 \leq t \leq T}\|S(x)(t)\|_\alpha \leq c(\alpha, \beta, \lambda, \rho_0, \|x\|_\beta).$$

Proof It suffices to let $\varepsilon = (\beta - \lambda\alpha)/2$ in (3.6). ●

4. Semilinear Parabolic Equations

Let T be a fixed positive number. In this section we study parabolic initial boundary value problems (IBVPs) of the form

$$\begin{aligned} Lu := \frac{\partial u}{\partial t} + A(x, t, D)u &= f(x, t, u, \nabla u) && \text{in} \quad \Omega \times (0, T], \\ Bu &= 0 && \text{on} \quad \Gamma \times (0, T], \\ u(\cdot, 0) &= u_0 && \text{on} \quad \bar{\Omega}. \end{aligned} \tag{4.1}$$

We suppose that the differential operators L and B and the function f satisfy all the hypotheses of Section 1 on the interval $[0, T]$ with the sole exception of the periodicity assumption.

By a *classical solution* of the IBVP (4.1) we mean a function

$$u \in C(\bar{\Omega} \times [0, T]) \cap C^{2,1}(\Omega \times (0, T]) \cap C^{\delta,0}(\bar{\Omega} \times (0, T])$$

that satisfies (4.1) pointwise. A classical solution is called *regular* if

$$u \in C^{1,0}(\bar{\Omega} \times [0, T]) \cap C^{2,1}(\bar{\Omega} \times (0, T]).$$

Let $p > N$ be a real number and let

$$D(A) := \{u \in W_p{}^2(\Omega) \mid Bu = 0\},$$

where, of course, the boundary operator B is taken in the sense of traces. Hence $D(A)$ is a closed vector subspace of $W_p{}^2(\Omega)$ and a dense linear subspace of $X := L_p(\Omega)$. For every $u \in D(A)$ and $t \in (0, T]$, let

$$A(t)u := A(\cdot, t, D)u + ku,$$

where k is a sufficiently large positive number. Then it is well known (e.g., Sobolevskii [24, Section 3] or Friedman [10, Section (II.9)]) that the family $\{A(t) \mid 0 \le t \le T\}$ satisfies assumptions (A1) and (A2). Furthermore,

$$\|A(s)u - A(t)u\|_{L_p} \le c\,|s - t|^{\mu/2} \|u\|_{W_p{}^2}. \tag{4.2}$$

By the L_p-estimates for elliptic operators (cf. Agmon *et al.* [1]) there exist positive constants c_1 and c_2 such that

$$c_1 \|u\|_{W_p{}^2} \le \|A(t)u\|_{L_p} \le c_2 \|u\|_{W_p{}^2} \tag{4.3}$$

for $0 \le t \le T$ and $u \in D(A)$. Hence it follows from (4.2) and (4.3) that assumption (A3) is satisfied.

In order to verify assumption (A4) we need the following imbedding theorem.

4.1 Proposition Suppose that $(1/2) + (N/2p) < \alpha \le 1$. Then $X_\alpha \hookrightarrow C^{1+\lambda}(\bar{\Omega})$, where $0 \le \lambda < 2\alpha - 1 - N/p$.

Proof Fix a number $\lambda \in (0, 2\alpha - 1 - N/p)$, and let $\nu := (\lambda + 1 + N/p)/2$. Then $0 < \nu < \alpha \le 1$, and it follows from Friedman [10, Theorem (I.10.1)] that

$$\|u\|_{C^{1+\lambda}(\bar{\Omega})} \le c(\lambda, p)\|u\|_{W_p{}^2}^{\nu} \|u\|_{L_p}^{1-\nu}. \tag{4.4}$$

Consequently, (2.2)–(2.4), (4.3), and (4.4) imply that

$$\|A^{-\alpha}u\|_{C^{1+\lambda}(\bar{\Omega})} \leq \frac{1}{\Gamma(\alpha)} \int_0^\infty \tau^{\alpha-1} \|e^{-\tau A}u\|_{C^{1+\lambda}(\bar{\Omega})}\, d\tau$$

$$\leq c(\alpha, \lambda, p) \int_0^\infty \tau^{\alpha-1} \|Ae^{-\tau A}u\|_{L_p}^{\nu} \|e^{-\tau A}u\|_{L_p}^{1-\nu}\, d\tau$$

$$\leq c_1(\alpha, \lambda, p) \int_0^\infty \tau^{\alpha-\nu-1} e^{-\delta_0 \tau}\, d\tau\, \|u\|_{L_p}.$$

Hence $\|u\|_{C^{1+\lambda}(\bar{\Omega})} \leq c(\alpha, \lambda, p)\|u\|_\alpha$, and the assertion follows. ●

In the following we fix a number α with $(1/2) + (N/2p) < \alpha < 1$, and we denote by $j_\alpha\colon X_\alpha \to C^1(\bar{\Omega})$ the injection. Moreover, we let

$$g_1(t, u)(x) := f(x, t, u(x, t), \nabla u(x, t)) + ku(x, t)$$

for $(x, t) \in \bar{\Omega} \times [0, T]$. Then we define $g\colon [0, T] \times X_\alpha \to X$ by $g(t, u) := g_1(t, j_\alpha(u))$. It is obvious that g satisfies assumption (A4). Hence, with the above definitions, the IVP (3.1) is well defined.

Suppose that $u \in C(\bar{\Omega} \times [0, T])$ and let $v(t) := u(\cdot, t)$. Then $v \in C([0, T], C(\bar{\Omega}))$, and vice versa. Consequently, throughout the remainder of this paper we do not distinguish between these two functions, that is, we identify $u(t)$ with $u(\cdot, t)$.

In the following we make use of the Hölder spaces $C^{\sigma, \sigma/2}(\bar{\Omega} \times [0, T])$, $\sigma \in \mathbb{R}_+ \setminus \mathbb{N}$, where we refer to Ladyzenskaja *et al.* [19, p.7] for a precise definition of these spaces. Moreover, for any Banach space E of functions on Ω such that $E \hookrightarrow W_p^1(\Omega)$, we denote by E_B the closed subspace of those functions u which satisfy the boundary condition $Bu = 0$, that is, $E_B = \{u \in E \mid Bu = 0\}$. Using these notations, we obtain the following "equivalence result."

4.2 LEMMA Let $u_0 \in W^2_{p,B}(\Omega)$. Then every solution u of the IVP (3.1) belongs to $C^{1,0}(\bar{\Omega} \times [0, T]) \cap C^{2+\mu, 1+\mu/2}(\bar{\Omega} \times (0, T])$; hence u is a regular solution of the IBVP (4.1). Conversely, every regular solution of the IBVP (4.1) is a solution of the IVP (3.1).

Proof (i) Let u be a regular solution of the IBVP (4.1). Then u can be considered as a solution of the linear evolution equation

$$v' + A(t)v = h(t) + ku, \qquad 0 < t \leq T,$$
$$v(0) = u_0 \tag{4.5}$$

in $X = L_p(\Omega)$, where $h(x, t) := f(x, t, u(x, t), \nabla u(x, t))$ for $(x, t) \in \bar{\Omega} \times [0, T]$. Since the right-hand side belongs to $C([0, T], X)$, it follows from (2.7) and Corollary 2.2 that $u \in C([0, T], X_\alpha)$. Hence u is a solution of the integral equation (3.2) in $C([0, T], X_\alpha)$, and it follows from Lemma 3.2 that u is a solution of the IVP (3.1).

(ii) Let u be a solution of the IVP (3.1). Then by Lemma 3.2, u is a solution of the integral equation (3.2) in $C([0, T], X_\alpha)$. Hence Proposition 4.1 implies that $g(\cdot, u(\cdot)) \in C([0, T], X)$. Consequently, by Corollary 2.2, $u \in C^\gamma([0, T], X_\alpha)$ for $\gamma \in [0, 1 - \alpha)$. Hence, again by Proposition 4.1, $u \in C^\gamma([0, T], C^{1+\lambda}(\bar{\Omega}))$ for $0 \leq \lambda < 2\alpha - 1 - N/p$, which implies that $u \in C^{1+\sigma, (1+\sigma)/2}(\bar{\Omega} \times [0, T])$ for some $\sigma \in (0, \lambda)$. This gives

$$h \in C^{\sigma, \sigma/2}(\bar{\Omega} \times [0, T])$$

and, of course, $u \in C^{1, 0}(\bar{\Omega} \times [0, T])$.

Let $\varepsilon \in (0, T)$ be arbitrary and denote by $\phi \in C^1(\mathbb{R}_+, \mathbb{R}_+)$ a function such that $\phi(t) = 0$ for $0 \leq t \leq \varepsilon/4$ and $\phi(t) = 1$ for $\varepsilon/2 \leq t \leq T$.

Consider the IBVP

$$\begin{aligned} Lv &= \phi h + \phi' u =: m && \text{in} \quad \Omega \times (0, T], \\ Bv &= 0 && \text{on} \quad \Gamma \times (0, T], \\ v(\cdot, 0) &= 0 && \text{on} \quad \bar{\Omega} \end{aligned} \tag{4.6}$$

and observe that $m \in C^{\sigma, \sigma/2}(\bar{\Omega} \times [0, T])$. Then it follows from the Schauder theory for linear parabolic differential equations (cf. Ladyzenskaja *et al.* [19, Theorems IV.5.2 and IV.5.3.]) that (4.6) has a unique solution $v \in C^{2+\sigma, 1+\sigma/2}(\bar{\Omega} \times [0, T])$. By Part (i) of this proof it follows that v is the unique solution of the IVP

$$\begin{aligned} v' + A(t)v &= m + ku \qquad \text{in} \quad 0 < t \leq T, \\ v(0) &= 0. \end{aligned}$$

Since ϕu is also a solution of this IVP, it follows that $v = \phi u$, whence $u \in C^{2+\sigma, 1+\sigma/2}(\bar{\Omega} \times [\varepsilon/2, T])$.

This last result implies that $h \in C^{\mu, \mu/2}(\bar{\Omega} \times [\varepsilon/2, T])$. Choose a function $\psi \in C^1(\mathbb{R}_+, \mathbb{R}_+)$ with $\psi(t) = 0$ for $0 \leq t \leq \varepsilon/2$ and $\psi(t) = 1$ for $t \geq \varepsilon$, and consider the IBVP

$$\begin{aligned} Lv &= \psi h + \psi' u =: m_1 && \text{in} \quad \Omega \times (0, T], \\ Bv &= 0 && \text{on} \quad \Gamma \times (0, T], \\ v(\cdot, 0) &= 0 && \text{on} \quad \bar{\Omega}. \end{aligned}$$

Observe that $m_1 \in C^{\mu, \mu/2}(\bar{\Omega} \times [0, T])$. Hence an argument similar to the one above shows that $u \in C^{2+\mu, 1+\mu/2}(\bar{\Omega} \times [\varepsilon, T])$ and that u satisfies the differential equation (4.1) in $\Omega \times [\varepsilon, T]$. Since $\varepsilon \in (0, T)$ is arbitrary, the first assertion has been proved.

4.3 *Remark* It should be observed that it suffices to assume that $u_0 \in X_\beta$ for $\beta \in (\alpha, 1]$ in the above lemma. Moreover, it has been shown that every solution of the IVP (3.1) belongs to $C^{1+\sigma,(1+\sigma)/2}(\bar{\Omega} \times [0, T]) \cap C^{2+\mu, 1+\mu/2}(\bar{\Omega} \times (0, T])$ for some $\sigma \in (0, 1)$. ●

A function u is called a *subsolution* for the parabolic BVP

$$
\begin{aligned}
Lu &= f(x, t, u, \nabla u) && \text{in} \quad \Omega \times (0, T], \\
Bu &= 0 && \text{on} \quad \Gamma \times (0, T],
\end{aligned}
\tag{4.7}
$$

if $u \in C^{1,0}(\bar{\Omega} \times [0, T]) \cap C^{2,1}(\bar{\Omega} \times (0, T])$ and

$$
\begin{aligned}
Lu &\leq f(\cdot, \cdot, u, \nabla u) && \text{in} \quad \Omega \times (0, T], \\
Bu &\leq 0 && \text{on} \quad \Gamma \times (0, T].
\end{aligned}
\tag{4.8}
$$

A subsolution is called *strict* if there is at least one strict inequality sign in (4.8). Supersolutions and strict supersolutions are defined by reversing the above inequalities.

4.4 LEMMA Let v be a subsolution and let w be a supersolution for the BVP (4.7) such that $v(\cdot, 0) \leq w(\cdot, 0)$. Then $v \leq w$.

More precisely, if $v(\cdot, 0) < w(\cdot, 0)$ or $Bv(\cdot, 0) < Bw(\cdot, 0)$, then $v(x, t) < w(x, t)$ for every $(x, t) \in \Omega \times (0, T]$. Moreover, in the case of the first BVP (that is, if $\delta = 0$), $(\partial v/\partial \beta)(x, t) > (\partial w/\partial \beta)(x, t)$ for every $(x, t) \in \Gamma \times (0, T)$ with $v(x, t) = w(x, t)$, whereas in the case of the second BVP (that is, if $\delta = 1$), $v(x, t) < w(x, t)$ for every $(x, t) \in \Gamma \times (0, T)$.

Proof Recall that a generic point of $\bar{\Omega} \times [0, T] \times \mathbb{R}^{N+1}$ is denoted by (x, t, ξ, η), where $\eta = (\eta^1, \ldots, \eta^N) \in \mathbb{R}^N$. Let $\eta^0 := \xi$ and define functions $d_i \in C(\bar{\Omega} \times [0, T])$ by

$$d_i := d_i(v, w) := \int_0^1 \frac{\partial f}{\partial \eta^i}(\cdot, \cdot, (1-\tau)v + \tau w, (1-\tau)\nabla v + \tau \nabla w)\, d\tau$$

for $i = 0, 1, 2, \ldots, N$. Then we define the uniformly parabolic differential operator L_1 on $\bar{\Omega} \times [0, T]$ by

$$L_1 u := Lu - \sum_{i=1}^{N} d_i D_i u - d_0 u.$$

It should be observed that L_1 has continuous coefficients on $\bar{\Omega} \times [0, T]$.

It is now an immediate consequence of the intermediate value theorem that the function $u := w - v$ satisfies the inequalities

$$
\begin{aligned}
L_1 u &\geq 0 && \text{in} \quad \Omega \times (0, T], \\
Bu &\geq 0 && \text{on} \quad \Gamma \times (0, T], \\
u(\cdot, 0) &\geq 0 && \text{on} \quad \bar{\Omega}.
\end{aligned}
$$

Hence the assertion follows from the strong maximum principle for linear parabolic differential operators (cf. Protter and Weinberger [22, Section III.3] and Friedman [9, Chapter 2]). ●

After these preparations we are ready to prove that the IBVP (4.1) has a unique regular solution, provided $u_0 \in C_B{}^2(\bar{\Omega})$. For this purpose we observe that everything said in this section depends on the choice of p. So far we have only assumed that $p > N$. In the following proof we impose further restrictions on p, depending on the value of the exponent in the growth condition (1.1). To emphasize the dependence of $A(t)$ and X_β on the choice of p, we denote the latter space by $D(A_p{}^\beta)$, $0 \leq \beta \leq 1$.

4.5 THEOREM Let $\bar{v}$ be a subsolution and let $\hat{v}$ be a supersolution for the BVP (4.7) such that $\bar{v}(\cdot, 0) \leq \hat{v}(\cdot, 0)$. Then the IBVP (4.1) has a unique regular solution $S(u_0) \in C^{2+\mu, 1+\mu/2}(\bar{\Omega} \times (0, T])$ for every $u_0 \in C_B^2(\bar{\Omega})$, satisfying $\bar{v}(\cdot, 0) \leq u_0 \leq \hat{v}(\cdot, 0)$. Moreover, $\bar{v} \leq S(u_0) \leq \hat{v}$.

Proof The last part of the assertion is an immediate consequence of Lemma 4.4. Hence it follows that, a priori, $\|S(u_0)(t)\|_{C(\bar{\Omega})} \leq \rho$ for $0 \leq t \leq T$, where

$$\rho := \max\{\bar{v}(x, t), \hat{v}(x, t) \mid (x, t) \in \bar{\Omega} \times [0, T]\}. \tag{4.9}$$

Let $\varepsilon(\rho) \in (0, 1)$ be the corresponding number appearing in the growth condition (1.1), and fix a real number $p := p(\rho)$ such that $p > 2N/\varepsilon(\rho)$. Then we can find a number $\alpha := \alpha(\rho)$ such that

$$\frac{1}{2} + \frac{N}{2p} < \alpha < \frac{1}{2 - \varepsilon(\rho)} < 1. \tag{4.10}$$

Hence it follows from (1.1), Proposition 4.1, Lemma 4.2, and Corollary 3.7 that, a priori,

$$\|S(u_0)(t)\|_\alpha \leq c(\rho, \|u_0\|_{C^2(\bar{\Omega})}) \tag{4.11}$$

for $0 \leq t \leq T$. Now the assertion follows from Lemma 4.2 and Proposition 3.4. ●

5. Proof of the Main Results

Throughout this section we presuppose the assumptions of Section 1.

Let $\bar{v}$ be an ω-subsolution and let $\hat{v}$ be an ω-supersolution for the BVP (1.2), such that $\bar{v}_0 := \bar{v}(\cdot, 0) \leq \hat{v}(\cdot, 0) =: \hat{v}_0$. Then we can apply Theorem 4.5 with $T = \omega$. Hence it follows that the IBVP (4.1) has a unique regular solution $S(u)$ for every u with

$$u \in M_\nu(\bar{v}_0, \hat{v}_0) := \{u \in C_B^{2+\nu}(\bar{\Omega}) \mid \bar{v}_0 \le u \le \hat{v}_0\},$$

and every $\nu \in [0, 1)$. Moreover, $S(u)(t) \in C_B^{2+\mu}(\bar{\Omega})$ for $0 < t \le \omega$. Hence we can define a map

$$\Pi\colon M_\nu(\bar{v}_0, \hat{v}_0) \to C_B^{2+\nu}(\bar{\Omega}),$$

the *Poincaré operator* (or translation operator) for the ω-periodic BVP (1.2) by

$$\Pi(u) := S(u)(\omega).$$

It is an easy consequence of Lemma 4.2, Lemma 3.2, the properties of the evolution operator, and the periodicity assumptions that the fixed points of Π are in a one-to-one correspondence with the ω-periodic solutions of the BVP (1.2). Moreover, it follows from Lemma 4.3 that minimal and maximal ω-periodic solutions of (1.2) correspond to minimal and maximal fixed points of Π, respectively.

The following proposition is the basis for the proofs of the existence and multiplicity results. Recall that a map ϕ from a topological space Y into a topological space Z is called *compact* if ϕ is continuous and $\phi(Y)$ is relatively compact in Z.

5.1 PROPOSITION Let $\bar{v}$ be an ω-subsolution and let $\hat{v}$ be an ω-supersolution for the BVP (1.2) such that $\bar{v}_0 := \bar{v}(\cdot, 0) \le \hat{v}(\cdot, 0) =: \hat{v}_0$. Then the Poincaré operator is a compact self-map of $M_\nu(\bar{v}, \hat{v}_0)$ for every $\nu \in [0, \mu)$.

Proof Define ρ by (4.9) and fix real numbers

$$p > \max\{2N/\varepsilon(\rho), (N+2)/(1-\mu)\}$$

and α satisfying (4.10). Then (compare the proof of Theorem 4.5) the IBVP (4.1) (with $T = \omega$) has a unique regular solution $S(u)$ for every $u \in M_\gamma(\bar{v}_0, \hat{v}_0)$, $0 \le \gamma < 1$. Moreover, $S(u) \in C^{2+\mu, 1+\mu/2}(\bar{\Omega} \times (0, T])$,

$$\|S(u)(t)\|_\alpha \le c(\rho, \|u\|_{C^{2+\gamma}(\bar{\Omega})}) \tag{5.1}$$

and

$$\|S(u)(t)\|_{C(\bar{\Omega})} \le \rho \tag{5.2}$$

for $0 \le t \le T$. Hence we deduce from (5.2) and Lemma 3.6 that

$$\max_{t_0 \le t \le T} \|S(u)(t)\|_\alpha \le c(\rho), \tag{5.3}$$

where $t_0 \in (0, T)$ is an arbitrary fixed number.

Suppose that for some $\rho_0 > 0$ the functions $u, v \in M_\gamma(\bar{v}_0, \hat{v}_0)$ satisfy $\|u\|_E$,

$\|v\|_E \leq \rho_0$, where $E := C_B^{2+\gamma}(\bar{\Omega})$. Then it follows from (5.1) and Lemma 3.5 that

$$\max_{t_0 \leq t \leq T} \|S(u)(t) - S(v)(t)\|_\beta \geq c(\beta, \rho_0)\|u - v\|_E \tag{5.4}$$

for every $\beta \in (\alpha, 1)$. Hence by Proposition 4.1, the estimate (5.3) implies

$$\max_{t_0 \leq t \leq T} \|S(u)(t)\|_{C^{1+\lambda}(\bar{\Omega})} \leq c(\rho, \lambda), \tag{5.5}$$

and (5.4) gives

$$\max_{t_0 \leq t \leq T} \|S(u)(t) - S(v)(t)\|_{C^{1+\gamma}(\bar{\Omega})} \leq c(\lambda, \rho_0)\|u - v\|_E, \tag{5.6}$$

where $0 \leq \lambda < 2\alpha - 1 - N/p$.

We now fix numbers s, τ with $t_0 < s < \tau < T$, and we let $Q_{t,T} := \Omega \times (t, T)$ for $0 < t < T$. Moreover, we let $C^{(\sigma)}(\bar{Q}_{t,T}) := C^{\sigma,\sigma/2}(\bar{Q}_{t,T})$ for every $\sigma > 0$, $\sigma \notin \mathbb{N}$.

Then it follows from the local L_p-estimates for linear parabolic equations (cf. Ladyzenskaja *et al.* [19, inequality (IV.10.12)]), the uniform bounds (5.1) and (5.2), the continuous imbedding of X_α in $C^1(\bar{\Omega})$, the local Lipschitz continuity of f with respect to $(\xi, \eta) \in \dot{\mathbb{R}} \times \mathbb{R}^N$, and the estimates (5.5) and (5.6), that

$$\|S(u)\|_{W_p^{2,1}(Q_{s,t})} \leq c(\rho)$$

and

$$\|S(u) - S(v)\|_{W_p^{2,1}(Q_{s,T})} \leq c(\rho_0)\|u - v\|_E.$$

By a Sobolev-type imbedding theorem (cf. Ladyzenskaja *et al.* [19, Lemma II.3.3]), we find that

$$W_p^{2,1}(Q_{s,T}) \hookrightarrow C^{(1+\sigma)}(\bar{Q}_{s,T}) := F$$

for $\sigma := 1 - (N+2)/p$. Hence

$$\|S(u)\|_F < c(\rho) \tag{5.7}$$

and

$$\|S(u) - S(v)\|_F \leq c(\rho_0)\|u - v\|_E. \tag{5.8}$$

Finally, the regularity properties of f, the local Hölder estimate for linear parabolic equations (cf. Ladyzenskaja *et al.* [19, Theorem IV.10.1]), and the inequalities (5.7) and (5.8) imply

$$\|S(u)\|_{C^{(2+\kappa)}(\bar{Q}_{\tau,T})} \leq c(\rho) \tag{5.9}$$

and

$$\|S(u) - S(v)\|_{C^{(2+\kappa)}(\bar{Q}_{\tau, T})} \leq c(\rho_0)\|u - v\|_E \tag{5.10}$$

for $0 \leq \kappa \leq \mu$.

Now let $\nu \in [0, \mu)$ be arbitrary and let $\gamma = \nu$. Then the Poincaré operator Π is defined on $M_\nu(\bar{v}_0, \hat{v}_0)$ and (5.10) implies that Π maps $M_\nu(\bar{v}_0, \hat{v}_0)$ continuously (in fact, locally Lipschitz continuously) into $C_B^{2+\nu}(\bar{\Omega})$. Moreover, by (5.9), the image of Π is bounded in $C^{2+\mu}(\bar{\Omega})$. Since the latter space is compactly imbedded in $C^{2+\nu}(\bar{\Omega})$, it follows that the image of Π is relatively compact in $C^{2+\nu}(\bar{\Omega})$.

Lastly, it follows from Lemma 4.4, that $\bar{v} \leq S(u) \leq \hat{v}$ for every $u \in M_\nu(\bar{v}_0, \hat{v}_0)$. Hence $\bar{v}_0 \leq \bar{v}(\cdot, \omega) \leq \Pi(u) \leq \hat{v}(\cdot, \omega) \leq \hat{v}_0$, that is, Π maps $M_\nu(\bar{v}_0, \hat{v}_0)$ into itself. Now the assertion follows, since the latter set is closed in $C^{2+\nu}(\bar{\Omega})$. ●

5.2 *Remark* Let the hypotheses of Proposition 5.1 be satisfied. Then it follows from (5.10) that

$$\|S(u) - S(v)\|_{C^{(2+\mu)}(\bar{Q}_{t, T})} \leq c(\rho, t)\|u - v\|_{C^{2+\nu}(\bar{\Omega})}$$

for every $\rho \geq 0$, every $t \in (0, T]$, and every pair $u, v \in M_\nu(\bar{v}_0, \hat{v}_0)$ with $\|u\|_{C^{2+\nu}(\bar{\Omega})}, \|v\|_{C^{2+\nu}(\bar{\Omega})} \leq \rho$.

5.3 *Remark* It should be observed that the above proof implies that every *periodic* solutions of the BVP (1.2) belongs to $C^{2+\mu, 1+\mu/2}(\bar{\Omega} \times \mathbb{R})$.

5.4 Lemma Let $\bar{v}_1, \ldots, \bar{v}_k$ be ω-subsolutions and let $\hat{v}$ be a strict ω-supersolution, such that $\bar{v}_j(\cdot, 0) \leq \hat{v}(\cdot, 0)$ and $\bar{v}_j(\cdot, 0) \in C_B^{2+\nu}(\bar{\Omega})$ for $j = 1, \ldots, k$. Then the set

$$\bigcap_{j=1}^{k} M_\nu(\bar{v}_j(\cdot, 0), \hat{v}(\cdot, 0)) \tag{5.11}$$

has nonempty interior in $C_B^{2+\nu}(\bar{\Omega})$, $0 < \nu < 1$.

Proof It follows from Lemma 4.4, that $\hat{v}(x, 0) - \bar{v}_j(x, 0) \geq \hat{v}(x, \omega) - \bar{v}_j(x, \omega) > 0$ for $x \in \Omega$. Moreover, if $\delta = 0$ and $\hat{v}(x, 0) - \bar{v}_j(x, 0) = 0$ for some $x \in \Gamma$, then $(\partial(\hat{v} - \bar{v}_j)/\partial\beta)(x, 0) < 0$. If $\delta = 1$, then $\hat{v}(x, 0) - \bar{v}_j(x, 0) > 0$ for all $x \in \Gamma$. This implies that there exists a $\bar{v} \in C_B^{2+\nu}(\bar{\Omega})$ with $\bar{v} \geq \max\{\bar{v}_j(\cdot, 0)\}$ and $\hat{v}(x, 0) > \bar{v}(x)$ for $x \in \Omega$, and $(\partial(\hat{v} - \bar{v})/\partial\beta)(x, 0) < 0$ if $\hat{v}(x, 0) = \bar{v}(x)$ for some $x \in \Gamma$.

Let w be the unique solution of the elliptic BVP

$$-\Delta w + w = 1 \quad \text{in} \quad \Omega,$$

$$Bw = 0 \quad \text{on} \quad \Gamma.$$

Then $w \in C_B^{2+\nu}(\bar{\Omega})$ and by the strong maximum principle (cf. Protter and Weinberger [22]), it follows that $w(x) > 0$ for $x \in \Omega$. Moreover, if $\delta = 0$, then $(\partial w/\partial \beta)(x) < 0$ for all $x \in \Gamma$, and if $\delta = 1$, then $w(x) > 0$ for all $x \in \Gamma$. Hence it follows that there exists a positive number α such that $\bar{v} + \alpha w$ belongs to the set (5.11). Clearly, α can be chosen so small that $\bar{v} + \alpha w$ is an interior point in $C_B^{2+\nu}(\bar{\Omega})$ of (5.11). ●

5.5 *Remark* It should be observed that a similar argument shows that in the case of the first BVP an ω-subsolution $\bar{v}$ and an ω-supersolution $\hat{v}$ are always B-related if $\bar{v}(\cdot, 0) \le \hat{v}(\cdot, 0)$.

Proof of Theorem 1.2 Let $\nu \in [0, \mu)$ be fixed. By the remarks preceding Proposition 5.1, it suffices to show that the Poincaré operator Π has a minimal and a maximal fixed point in $M_\nu(\bar{v}_0, \hat{v}_0)$, where $\bar{v}_0 := \bar{v}(\cdot, 0)$ and $\hat{v} := \hat{v}(\cdot, 0)$.

Since $\bar{v}$ and $\hat{v}$ are B-related, $M_\nu(\bar{v}_0, \hat{v}_0)$ is nonempty. Since this set is obviously a closed convex subset of $C_B^{2+\nu}(\bar{\Omega})$, the existence of at least one fixed point of Π follows from Proposition 5.1 and Schauder's fixed point theorem.

Let F be the set of all fixed points of Π. In order to show that F possesses a minimal and a maximal element, we employ an idea of Akô [2].

To show that F has a maximal element, we can clearly assume that $\hat{v}$ is a strict ω-supersolution. Define $\sup F$ by $\sup F(x) := \sup\{u(x) \mid u \in F\}$ for $x \in \bar{\Omega}$. Then $\bar{v}_0 \le \sup F \le \hat{v}_0$.

Let (x_k) be a dense sequence in $\bar{\Omega}$. Then, for each $k \in \mathbb{N}$, we choose a sequence $(u_{k,j})$ in F with $u_{k,j}(x_k) \uparrow \sup F(x_k)$ for $j \to \infty$.

Let $v_0 := u_{0,0}$, and define inductively an increasing sequence (v_j) in F such that

$$\max\{v_{j-1}, u_{0,j}, \ldots, u_{j,j}\} \le v_j \le \sup F$$

for $j = 1, 2, \ldots$. For this we observe that by Lemma 5.4, each of the sets

$$M_\nu(v_{j-1}, \hat{v}) \cap \bigcap_{k=0}^{j} M_\nu(u_{k,j}, \hat{v}_0) \qquad (5.12)_j$$

is a nonempty closed convex subset of $M_\nu(\bar{v}_0, \hat{v}_0)$. Moreover, it follows from Lemma 4.4 that the Poincaré operator maps each of the sets $(5.12)_j$ into itself. Hence the existence of v_j follows from Proposition 5.1 and Schauder's fixed-point theorem.

Since the sequence (v_j) is contained in the range of Π, it is relatively compact. Hence, since it is increasing, and since Π is continuous, it follows that it converges in $C^{2+\nu}(\bar{\Omega})$ to some $v \in F$. By the construction of the sequence (v_j), it follows that $v(x_k) = \sup F(x_k)$ for all $k \in \mathbb{N}$. This implies that v is the maximal element in F. A similar argument shows that F possesses a minimal element. ●

Proof of Theorem 1.5 Let $\nu \in [0, \mu)$ be fixed and let $Y := M_\nu(\bar{v}_1(\cdot, 0), \hat{v}_2(\cdot, 0))$ and $Y_j := M_\nu(\bar{v}_j(\cdot, 0), \hat{v}_j(\cdot, 0))$, $j = 1, 2$. It follows from Theorem 1.2 (cf. the preceding proof) that Π has a maximal and a minimal fixed point in each of the sets Y, Y_1, and Y_2. Hence it suffices to show that Π has a fixed point in $Y \backslash (Y_1 \cup Y_2)$.

Let $\hat{u}_1$ be the maximal fixed point in Y_1. Then it follows from Lemma 5.4 that $M_\nu(\hat{u}_1, \hat{v}_1(\cdot, 0))$ has nonempty interior in $C_B^{2+\nu}(\bar{\Omega})$. Hence Y_1 has nonempty interior W_1 in Y.

Since $\hat{u}_1$ is a fixed point of Π, it follows from Lemma 4.4 that $\hat{u}_1(x) < \hat{v}_1(x, 0)$ for all $x \in \Omega$. Moreover, if $\hat{u}_1(x) = \hat{v}_1(x, 0)$ for some $x \in \Gamma$, then $(\partial(\hat{u}_1 - \hat{v}_1(\cdot, 0))/\partial\beta)(x) > 0$. This, together with the maximality of $\hat{u}_1$, implies that Π has no fixed points on $Y_1 \backslash W_1$. A similar argument shows that Y_2 has nonempty interior in Y, and that Π has no fixed point on the relative boundary of Y_2 in Y. Hence the assertion follows now from Amann [4, Lemma 14.1]. ●

5.6 *Remark* It should be noted that in the preceding proof we have implicitly used the fact that the Poincaré operator is strongly increasing, in the sense of ordered Banach spaces (e.g., Amann [3,4]). This is an immediate consequence of Lemma 4.4.

Proof of Theorem 1.1 Consider the linear IBVP

$$\begin{aligned} Lu &= w && \text{in } \Omega \times (0, T], \\ Bu &= 0 && \text{on } \Gamma \times (0, T], \\ u(\cdot, 0) &= u_0 && \text{on } \bar{\Omega}, \end{aligned} \tag{5.13}$$

where T is a fixed positive number, and w is a μ-Hölder continuous function on $\bar{\Omega} \times [0, T]$. By means of the transformation $v := e^{-kt}u$, we obtain the equivalent problem

$$\begin{aligned} (L + k)v &= e^{-kt}w && \text{in } \Omega \times (0, T], \\ Bv &= 0 && \text{on } \Gamma \times (0, T], \\ v(\cdot, 0) &= u_0 && \text{on } \bar{\Omega}. \end{aligned} \tag{5.14}$$

Hence it follows, that there exists a unique evolution operator $V(t, \tau)$, $0 \le \tau \le t \le T$, for problem (5.13) such that the solution u is given by

$$u(t) = V(t, 0)u_0 + \int_0^t V(t, \tau)w(\tau)\, d\tau, \qquad 0 \le t \le T.$$

In fact, $V(t, \tau) = e^{kt}U(t, \tau)e^{-k\tau}$, where U is the evolution operator for problem (5.14).

It follows that the Poincaré operator for (5.13) has the representation

$$\Pi(u_0) = V(\omega, 0)u_0 + \int_0^\omega V(\omega, \tau)w(\tau)\, d\tau.$$

Let $K := V(\omega, 0)$ and observe that K is a continuous linear operator from $L_p(\Omega)$ into $W_p{}^2(\Omega)$, $p > N$. Since $C_B^{2+\nu}(\bar{\Omega})$ is dense in $L_p(\Omega)$, it follows from the maximum principle that K is positive. Moreover, since $W_p{}^2(\Omega)$ is compactly imbedded in $C(\bar{\Omega})$, we can consider K as a positive compact endomorphism of $C(\bar{\Omega})$. Moreover, by the strong maximum principle, K is e-positive in the sense of Krasnosel'skii (cf. references [15,3,4]), where e is an arbitrary function in $C_B{}^1(\bar{\Omega})$ such that $e(x) > 0$ for $x \in \Omega$, $e(x) > 0$ for $x \in \Gamma$ if $\delta = 1$, and $(\partial e/\partial \beta)(x) < 0$ for $x \in \Gamma$ if $\delta = 0$.

It follows from the regularity theory that Π has a fixed point in $C_B^{2+\nu}(\bar{\Omega})$ if and only if the equation

$$u - Ku = \int_0^\omega V(\omega, \tau)w(\tau)\, d\tau \tag{5.15}$$

has a solution u in $C(\bar{\Omega})$. The maximum principle implies that the right-hand side of (5.15) is positive if $w > 0$. Hence reference [15, Theorem 2.16] and the Krein–Rutman theorem (cf. references [18,15,3,4]) imply that Eq. (5.15) has for every Hölder continuous $w > 0$ a unique solution $u_0 > 0$ if the spectral radius r of K is less than one, and no nonnegative solution if $r \geq 1$. Consequently, it remains to show that $r < 1$.

(i) Suppose first that $a_0 \geq 0$. If $\delta = 1$ and $b_0 = 0$, then suppose in addition that $a_0 > 0$. Let $u_0 \in C_B^{2+\nu}(\bar{\Omega})$ be arbitrary and let $u := V(\cdot, 0)u_0$. Then u is a regular solution of the linear IBVP

$$\begin{aligned} Lu &= 0 \quad &&\text{in} \quad \Omega \times (0, \infty), \\ Bu &= 0 \quad &&\text{on} \quad \Gamma \times (0, \infty), \\ u(\cdot, 0) &= u_0 \quad &&\text{on} \quad \bar{\Omega} \end{aligned}$$

Hence by the maximum principle, $|u(x, t)| \leq \|u_0\|_{C(\bar{\Omega})}$ for every $(x, t) \in \bar{\Omega} \times \mathbb{R}_+$. Consequently, $\|Ku_0\|_{C(\bar{\Omega})} \leq \|u_0\|_{C(\bar{\Omega})}$, and it follows that $r \leq \|K\| \leq 1$.

Suppose now that $r = 1$. Then, since K is e-positive, it follows that there exists a positive function $u_0 \in C_B^{2+\nu}(\bar{\Omega})$ with $Ku_0 = u_0$. Hence $V(\cdot, 0)u_0$ is a positive ω-periodic regular solution of the homogeneous BVP

$$\begin{aligned} Lu &= 0 \quad &&\text{in} \quad \Omega \times \mathbb{R}, \\ Bu &= 0 \quad &&\text{on} \quad \Gamma \times \mathbb{R}. \end{aligned}$$

But by the maximum principle, this BVP has only the trivial ω-periodic solution. This contradiction shows that $r < 1$.

(ii) Suppose now that the coefficients of $A(x, t, D)$ are independent of t and that the smallest eigenvalue λ_0 of the linear EVP

$$A(x, D)u = \lambda u \quad \text{in} \quad \Omega,$$
$$Bu = 0 \quad \text{on} \quad \Gamma$$

is positive. Then there exists a positive eigenfunction u_0 to the eigenvalue λ_0 (cf. Amann [3, Theorem (1.16)]). Let $u := V(\cdot, 0)u_0$. Then it follows that

$$L(u_0 - u) = \lambda_0 u_0 > 0 \quad \text{in} \quad \Omega \times (0, \infty),$$
$$B(u_0 - u) = 0 \quad \text{on} \quad \Gamma \times (0, \infty),$$
$$(u_0 - u)(\cdot, 0) = 0 \quad \text{on} \quad \bar{\Omega},$$

and $u_0 - u \in C^{2,1}(\bar{\Omega} \times \mathbb{R}_+)$. Hence by the maximum principle, $u_0 > u(\cdot, \omega) = Ku_0$. Now reference [15, Theorems (2.17) and (2.19)] (cf. also Amann [3, Theorem (1.3)]) implies that $r < 1$. This proves the Theorem. ●

5.7 *Remark* There is an essential distinction between the approach of Kolesov [14] and our treatment. Namely, Kolesov extends the Poincaré operator to the Banach space $C(\bar{\Omega})$ and studies fixed point equations in $C(\bar{\Omega})$. This space has the advantage that the order intervals are bounded. However, in order to carry through this extension of Π, one has essentially to assume that $\partial f/\partial \xi$ also satisfies the growth condition (1.1) (cf. Kolesov [14, inequality (15)]), which seems to be rather unnatural and restrictive. (In fact, in the case of the second BVP with $b_0 = 0$, it seems that one has also to assume that all of the functions $\partial f/\partial \eta^i$, $i = 1, \ldots, N$ satisfy condition (1.1).)

In our approach we use the space $C_B^{2+\nu}(\Omega)$ in which the sets $M_\nu(\bar{v}_0, \hat{v}_0)$ are neither bounded nor are they order intervals in general. This is the reason for the relatively difficult proofs. However, since we can deal with classical solutions, we can use the full strength of the strong maximum principle. This implies in particular that Π is strongly increasing, which is at the heart of the multiplicity results.

Addendum

The fundamental existence theorem (Theorem 4.5) remains true if $\varepsilon(\rho) = 0$, provided we can establish a uniform a priori estimate for the maximum norm of $|\nabla u(t)|$, given a uniform bound for the maximum norm of $u(t)$. In the meantime such an a priori estimate has been proved in the author's paper, Existence and multiplicity theorems for semi-linear elliptic boundary value problems. *Math. Z.* **150** (1976), 281–295.

The only other place where we have used the growth condition (1.1) is in the proof of Proposition 5.1, namely in establishing inequality (5.3) which, in turn, implies (5.5). Hence, Proposition 5.1 remains true without the growth restriction (1.1), provided an estimate of the form (5.5) can be established directly.

In the above-mentioned paper we have also established an a priori inequality of type (5.5) in the case that $\varepsilon(\rho) = 0$. Hence *all the results given in Sections 1, 4, and 5 of the present paper remain true, if the growth condition* (1.1) *is replaced by the weaker assumption* that

$$|f(x, t, \xi, \eta)| \leq c(\rho)(1 + |\eta|^2)$$

for every $\rho \geq 0$ *and every* $(x, t, \xi, \eta) \in \bar{\Omega} \times \mathbb{R} \times [-\rho, \rho] \times \mathbb{R}^N$.

References

1. S. Agmon, A. Douglis, and L. Nirenberg, Estimates near the boundary for solutions of elliptic partial differential equations satisfying general boundary conditions, I. *Comm. Pure Appl. Math.* XII (1959), 623–627.
2. K. Akô, On the Dirichlet problem for quasilinear elliptic differential equations of the second order. *J. Math. Soc. Japan* **13** (1961), 45–62.
3. H. Amann, Nonlinear operators in ordered Banach spaces and some applications to nonlinear boundary value problems. *In* "Nonlinear Operators and the Calculus of Variations," Lecture Notes in Mathematics, Vol. 543, pp. 1–55. Springer-Verlag, Berlin and New York, 1976.
4. H. Amann, Fixed point equations and nonlinear eigenvalue problems in ordered Banach spaces. *SIAM Rev.* **18** (1976), 620–709.
5. H. Amann, Nonlinear elliptic equations with nonlinear boundary conditions. *In* "New Developments in Differential Equations," *Proc. 2nd Scheveningen Conf. Differential Equations*, North-Holland Mathematics Studies. Vol. 21, pp. 43–64, 1976.
6. D. W. Bange, Periodic solutions of a quasilinear parabolic differential equation. *J. Differential Equations* **17** (1975), 61–72.
7. F. E. Browder, Periodic solutions of nonlinear equations of evolution in infinite dimensional spaces. *In* "Lectures in Differential Equations" (A. K. Aziz, ed.), Vol. 1. Van Nostrand-Reinhold, Princeton, New Jersey, 1969.
8. P. Fife, Solutions of parabolic boundary problems existing for all times. *Arch. Rational Mech. Anal.* **16** (1964), 155–186.
9. A Friedman, "Partial Differential Equations of Parabolic Type," Prentice-Hall, Englewood Cliffs, New Jersey, 1964.
10. A. Friedman, "Partial Differential Equations." Holt, New York, 1969.
11. R. Gaines and W. Walter, Periodic solutions to nonlinear parabolic differential equations. *Rocky Mountain J. Math.* **7** (1977), 297–312.
12. Ju. S. Kolesov, Certain tests for the existence of stable periodic solutions of quasilinear parabolic equations. *Soviet Math. Dokl.* **5** (1964), 1118–1120.
13. Ju. S. Kolesov, A test for the existence of periodic solutions to parabolic equations. *Soviet Math. Dokl.* **7** (1966), 1318–1320.
14. Ju. S. Kolesov, Periodic solutions of quasilinear parabolic equations of second order. *Trans. Moscow Math. Soc.* **21** (1970), 114–146.
15. M. A. Krasnosel'skii, "Positive Solutions of Operator Equations." Noordhoff, Groningen, 1964.
16. M. A. Krasnosel'skii, "Translation along Trajectories of Differential Equations," Amer. Math. Soc. Transl. of Math. Monographs, Vol. 19. Amer. Math. Soc., Providence, Rhode Island, 1968.

17. M. A. Krasnosel'skii, P. P. Zabreiko, E. I. Pustylnik, and P. E. Sobolevskii, "Integral Operators in Spaces of Summable Functions." Noordhoff, Leyden, 1976.
18. M. G. Krein and M. A. Rutman, Linear operators leaving invariant a cone in a Banach space. *Amer. Math. Soc. Transl. Ser. 1* **10** (1962), 199–325.
19. O. A. Ladyzenskaja, V. A. Solornikov, N. N. Ural'ceva, "Linear and Quasilinear Equations of Parabolic Type," Amer. Math. Soc. Trans. of Math. Monographs, Vol. 19. Amer. Math. Soc., Providence, Rhode Island.
20. M. Nakao, On boundedness, periodicity, and almost periodicity of solutions of some nonlinear parabolic equations. *J. Differential Equations* **19** (1975), 371–385.
21. A. Pazy, "Semi-Groups of Linear Operators and Applications to Partial Differential Equations," Univ. of Maryland Lecture Notes, No. 19. Univ. of Maryland, College Park, 1974.
22. M. H. Protter and H. F. Weinberger, "Maximum Principles in Differential Equations." Prentice-Hall, Englewood Cliffs, New Jersey, 1967.
23. T. I. Seidman, Periodic solutions of a non-linear parabolic equation. *J. Differential Equations* **19** (1975), 242–257.
24. P. E. Sobolevskii, Equations of parabolic type in a Banach space. *Amer. Math. Soc. Transl., Ser. 2* **49** (1966), 1–62.
25. H. Tanabe, On the equations of evolution in a Banach space. *Osaka J. Math.* **12** (1960), 363–376.
26. B. A. Ton, Periodic solutions of nonlinear evolution equations in Banach spaces. *Canad. J. Math.* **23** (1971), 189–196.

AMS (MOS) 1970 Subject Classification: 35K55.

Linear Maximal Monotone Operators and Singular Nonlinear Integral Equations of Hammerstein Type

H. Brézis

Université P. et M. Curie

F. E. Browder

University of Chicago

Introduction

Let X be a real Banach space that is paired to a second Banach space Y by a bilinear pairing $\langle w, u\rangle$. A mapping T of X into 2^Y is said to be monotone if for each pair w in $T(u)$, y in $T(x)$,

$$\langle y - w, x - u\rangle \geq 0.$$

T is said to be maximal monotone from X to 2^Y if it is maximal in the sense of inclusion of graphs among monotone maps from X to 2^Y.

In the application of the theory of monotone operators to the existence of solutions of nonlinear equations, the concept of maximal monotonicity plays a central role. It is therefore of importance to determine whether various concretely given monotone operators are maximal and to be able to generate maximal monotone operators satisfying given conditions. An important special case is that in which the operator T is linear, i.e., its graph $G(T)$ is a linear subset of $X \times Y$. For any linear operator L from X to 2^Y, we define its adjoint L^* from Y^* to 2^{X^*} by the condition that $x^* \in L^*(y^*)$ if and only if for all y in $L(x)$

$$\langle y^*, y \rangle = \langle x^*, x \rangle,$$

(cf. Arens [1]).

In this chapter, we prove and apply the following general theorem on linear maximal monotone operators.

THEOREM 1 Let X be a reflexive Banach space, L_0 and L_1 two linear monotone mappings from X to 2^{X^*} such that $L_0 \subseteq L_1{}^*$.

Then there exists a maximal monotone linear mapping L such that

$$L_0 \subseteq L \subseteq L_1{}^*.$$

One consequence of Theorem 1 is the following theorem established in 1968 by Brézis [2].

THEOREM 2 Let X be a reflexive Banach space, L a closed linear monotone operator from X to 2^{X^*}. Then L is maximal monotone if and only if L^* is monotone.

Proof of Theorem 2 from Theorem 1 Suppose first that L is maximal monotone and L^* is not monotone. Then there exists $w_0 \in L^*(u_0)$ with $\langle w_0, u_0 \rangle = -\delta < 0$. For all $[x, y]$ in $G(L)$,

$$\langle w_0, x \rangle = \langle y, u_0 \rangle.$$

For each such $[x, y]$ we compute

$$\langle -w_0 - y, u_0 - x \rangle = -\langle w_0, u_0 \rangle + \langle y, x \rangle + \langle w_0, x \rangle - \langle y, u_0 \rangle.$$

Since $\langle w_0, x \rangle - \langle y, u_0 \rangle = 0$ and $\langle y, x \rangle \geq 0$ by the monotonicity of L, we have

$$\langle -w_0 - y, u_0 - x \rangle \geq \delta > 0$$

for all $[x, y]$ in $G(L)$. Since L is maximal monotone, it follows that $[u_0, -w_0] \in G(L)$. Hence

$$0 = \langle -w_0 - (-w_0), u_0 - u_0 \rangle \geq \delta > 0,$$

which is a contradiction. Hence L^* is monotone if L is maximal monotone.

Suppose on the other hand that L^* is monotone. Since L is closed, $L = (L^*)^*$. By Theorem 1, there exists a maximal monotone mapping L' such that

$$L \subseteq L' \subseteq (L^*)^*.$$

Hence $L = L'$ is itself maximal monotone. Q.E.D.

A sketch of the results of the present paper is given in Brézis and Browder [4].

1. Proof of the Main Theorem

In the case in which X is a Hilbert space and L_0 and L_1 are single-valued and densely defined, the result of Theorem 1 was established by Phillips [9] using ideas due to Krein [8]. For later use, we give a simple direct proof for the multivalued Hilbert space case.

PROPOSITION 1 Let H be a Hilbert space, L_0 and L_1 monotone linear mappings from H to 2^H with $L_0 \subseteq L_1^*$. Then there exists a maximal monotone linear mapping L from H to 2^H such that

$$L_0 \subseteq L \subseteq L_1^*.$$

Proof For each linear operator L from H to 2^H, we introduce its Cayley transform

$$C(L) = 2(I + L)^{-1} - I,$$

which is also a linear operator from H to 2^H. Then L is monotone if and only if $C(L)$ is nonexpansive on its effective domain, i.e., if $v \in C(L)(u)$, then $C(L)(u) = \{v\}$ and $\|v\| \leq \|u\|$. Moreover, $L = C(C(L))$. It follows that $L \subseteq L^\#$ if and only if $C(L) \subseteq C(L^\#)$ and that $C(L^*) = (C(L))^*$. Finally, L is maximal monotone if and only if $C(L)$ is a nonexpansive linear map of H into H.

Let $U_0 = C(L_0)$, $U_1 = C(L_1)$. The assertion of Proposition 1 is equivalent to the assertion that if U_0 is a nonexpansive linear map of a closed subspace H_0 of H into H, and U_1 is a nonexpansive linear map of a closed subspace H_1 of H into H with $U_0 \subseteq U_1^*$, then there exists a nonexpansive linear map U of H into H such that $U_0 \subseteq U \subseteq U_1^*$. Let $\tilde{U}$ be U_1 considered as an element of $L(H_1, H)$, and let $\tilde{U}': H \to H_1$ be its adjoint map. Since $\|\tilde{U}\| \leq 1$, $\|\tilde{U}'\| \leq 1$. The condition that $U \subseteq U_1^*$ means that for all x_1 in H_1 and all x in H,

$$\langle U(x), x_1 \rangle = \langle x, U_1(x_1) \rangle = \langle x, \tilde{U}(x_1) \rangle = \langle (\tilde{U}')(x), x_1 \rangle,$$

i.e.,

$$U(x) = \tilde{U}'(x) + R(x),$$

where R is a linear map from H into $H_1{}^{\perp}$. Since $U_0 \subseteq U_1{}^*$, we see that there exists R_0: $H_0 \to H_1{}^{\perp}$ such that $U_0 = R_0 + \tilde{U}'|_{H_0}$.

The assertion of the proposition is then equivalent to the assertion of the following lemma:

LEMMA 1 Let H be a Hilbert space, H_0 and H_1 closed subspaces of H. Suppose U_0 is a nonexpansive linear map of H_0 into H and U_1 is another nonexpansive linear map of H_1 into H. Let $U_1{}'$ be the adjoint to U_1 mapping H into H_1. Suppose that for all x in H_0

$$U_0(x) - U_1{}'(x) \in H_1{}^{\perp}.$$

Then there exists a nonexpansive linear map U from H into H such that $U|_{H_0} = U_0$ and for all x in H,

$$U(x) - U_1{}'(x) \in H_1{}^{\perp}.$$

The proof of Lemma 1 depends upon the result of a second lemma.

LEMMA 2 Let H be a vector space and let $\varphi(x)$ be a nonnegative quadratic form on H. Let F be a Banach space with norm $\|\cdot\|$. Let H_0 be a subspace of H and let R_0: $H_0 \to F$ be a linear map such that

$$\|R_0 x\|^2 \leq \varphi(x) \qquad (x \in H_0).$$

Then there exists a linear map R: $H \to F$ such that $R_0 \subseteq R$ and $\|Rx\|^2 \leq \varphi(x)$ for all $x \in H$.

Proof Assume first that $\varphi(x) = 0$ implies $x = 0$ and let $\hat{H}$ be the Hilbert space obtained by completing H with respect to the norm $\sqrt{\varphi(x)}$. Let $\hat{\varphi}$ be the extension of φ to $\hat{H}$. Let $\bar{H}_0$ be the closure of H_0 in $\hat{H}$ and let $\bar{R}_0$: $\bar{H}_0 \to F$ be the continuous extension of R_0 to $\bar{H}_0$; it satisfies $\|\bar{R}_0 x\|^2 \leq \hat{\varphi}(x)$ for all $x \in \bar{H}_0$.

Let P be the orthogonal projection from $\hat{H}$ onto $\bar{H}_0$. The conclusion of the lemma follows by choosing $R = \bar{R}_0 \ P$ which maps H into F and satisfies $R_0 \subseteq R$ and $\|Rx\|^2 \leq \hat{\varphi}(x)$ for all $x \in \hat{H}$.

In the general case let $N = \{x \in H \mid \varphi(x) = 0\}$; N is a subspace of H. Let $H^\# = H/N$ and $\varphi^\#$ be the quadratic form induced by φ on $H^\#$, so that $\varphi^\#(x) > 0$ for all $x \in H^\#$, $x \neq 0$. Let $H_0{}^\# = H_0/N$. Since $R_0(x) = 0$ for all $x \in N \cap H_0$, R_0 induces a linear map $R_0{}^\#$: $H_0{}^\# \to F$ that satisfies $\|R_0{}^\# x\|^2 \leq \varphi^\#(x)$ for all $x \in H_0{}^\#$. By the previous result there exists a linear map $R^\#$: $H^\# \to F$ such that $R_0{}^\# \subseteq R^\#$ and $\|R^\# x\|^2 \leq \varphi^\#(x)$ for all

$x \in H^{\#}$. Finally let π be the canonical map of H onto $H^{\#} = H/N$; then $R = R \circ \pi$ yields the conclusion of the lemma. Q.E.D.

Proof of Lemma 1 from Lemma 2 Let $R_0 = U_0 - (U_1' |_{H_0})$. Since R_0 maps H_0 into H_1 and U_0 is nonexpansive, while U_1' has its image in H_1, we see that

$$\|R_0(x)\|^2 + \|U_1'(x)\|^2 \leq \|x\|^2$$

for all x in H_0. Let

$$\varphi(x) = \|x\|^2 - \|U_1'(x)\|^2$$

for all x in H. For x in H_0,

$$\|R_0(x)\|^2 \leq \varphi(x)$$

while φ is a nonnegative quadratic form since $\|U_1'\| = \|U_1\| \leq 1$. By Lemma 1, R_0 has an extension R mapping H into H_1 such that

$$\|R(x)\|^2 \leq \varphi(x)$$

for all x in H.

Let $U: H \to H$ be given by

$$U(x) = U_1'(x) + R(x) \qquad (x \in H).$$

Then U is an extension of U_1 and

$$\|U(x)\|^2 = \|U_1'(x)\|^2 + \|R(x)\|^2 \leq \|U_1'(x)\|^2 + \varphi(x) \leq \|x\|^2.$$

Hence $\|U\| \leq 1$. Q.E.D.

LEMMA 3 Let L be a linear monotone mapping from X to 2^Y with respect to the bilinear pairing $\langle w, u\rangle$. Let S be a linear subset of $X \times Y$ with $G(L) \subseteq S$.

Then L is maximal monotone in S if and only if L is maximal among linear monotone mappings of X to 2^Y with graphs contained in S.

Proof It suffices to show that if L is maximal among linear monotone maps with graphs in S, then L is maximal monotone in S. Suppose for some $[x_0, y_0]$ in S and all $[x, y]$ in $G(L)$

$$\langle y_0 - y, x_0 - x\rangle \geq 0.$$

Let

$$G(L_1) = G(L) + \mathbb{R}\{[x_0, y_0]\}.$$

Then L_1 is a linear map of X into 2^Y and $G(L_1) \subset S$. If L_1 is monotone, then $L = L_1$ and $[x_0, y_0] \in G(L)$. Hence it suffices to show that L_1 is monotone, i.e., for each $[u, w]$ in $G(L_1)$ that $\langle w, u\rangle \geq 0$. Each such $[u, w]$ may be written in the form

$$[u, w] = [x, y] + \rho[x_0, y_0]$$

with $[x, y]$ in $G(L)$. Thus

$$\langle w, u \rangle = \langle y + \rho y_0, x + \rho x_0 \rangle.$$

If $\rho = 0$, there is nothing to prove. If $\rho \neq 0$,

$$\langle w, u \rangle = \rho^2 \langle y_0 - (-\rho^{-1} y), x_0 - (-\rho^{-1} x) \rangle \geq 0$$

since the element $[-\rho^{-1}x, -\rho^{-1}y]$ also lies in $G(L)$. Hence L_1 is monotone. Q.E.D.

Proof of Theorem 1 Since $L_0 \subseteq L_1{}^*$, we may apply a Zorn's lemma argument to obtain a monotone linear operator L with $L_0 \subseteq L \subseteq L_1{}^*$ such that L is maximal among monotone linear operators with graphs contained in $G(L_1{}^*)$. If we apply Lemma 3, we see that L is maximal monotone in $G(L_1{}^*)$, i.e., if $[u_0, z_0]$ lies in $G(L_1{}^*)$ and

$$\langle z_0 - y, u_0 - z \rangle \geq 0$$

for all $[x, y]$ in $G(L)$, then $[u_0, z_0]$ lies in $G(L)$.

We shall now show that L_1 is maximal monotone from X to 2^{X^*}. We may assume that X is endowed with a locally uniformly convex norm such that the dual norm on X^* is also locally uniformly convex. Let J be the corresponding duality mapping, i.e., for all x in X,

$$\|J(x)\| = \|x\|,$$

$$\langle J(x), x \rangle = \|x\|^2.$$

Then it suffices to show that $(L + J)$ has all of X^* as its range [5].

Let w_0 be a fixed element of X^*. To show that $w_0 \in (L + J)(u_0)$ for some u_0, we shall form a convenient system of finite-dimensional approximants to the desired equation and take their limit.

Let M be a finite-dimensional subspace of X. Let φ_M be the injection map of M into X and $\varphi_M{}^*$ the corresponding projection mapping of X^* onto M^*.

For each x in M, let

$$L_M(x) = \varphi_M{}^*(L(x)), \qquad L_{1,M}(x) = \varphi_M{}^*(L_1(x)).$$

L_M and $L_{1,M}$ are linear mappings of M into 2^{M^*}. Each of these mappings is obviously monotone since if $w_M \in L_M(x)$, then $w_M = \varphi_M{}^*(w)$ with $w \in L(x)$ and

$$\langle w_M, x \rangle = \langle w, x \rangle \geq 0,$$

and similarly for $L_{1,M}$. Moreover, since $L \subseteq L_1{}^*$, it follows immediately that

$$L_M \subseteq (L_{1,M})^*.$$

Hence, we may apply Proposition 1 to the finite-dimensional space M, which is equivalent to a Hilbert space and obtain a maximal monotone linear operator K_M from M to 2^{M^*} such that

$$L_M \subseteq K_M \subseteq (L_{1,M})^*.$$

Since K_M is maximal monotone, $0 \in D(K_M)$ and the map $\varphi_M{}^*J$ of M into M^* is monotone, continuous, and coercive, the range of $K_M + \varphi_M{}^*J$ is all of M. Hence, there exists u_M in M such that

$$\varphi_M{}^*(w_0) \in K_M(u_M) + \varphi_M{}^*J(u_M).$$

Suppose that $[u, w]$ lies in $G(L)$ with u lying in M. Then $\varphi_M{}^*(w) \in L_M(u) \subseteq K_M(u)$ and by the monotonicity of K_M, we have

$$\langle \varphi_M{}^*(w_0) - \varphi_M{}^*J(u_M) - \varphi_M{}^*(w), u_M - u \rangle \geq 0,$$

i.e.,

$$\langle w_0 - J(u_M) - w, u_M - u \rangle \geq 0.$$

Similarly, suppose $[x, y]$ lies in $G(L_1)$ with x lying in M. Then $\varphi_M{}^*(y) \in L_{1,M}(x)$. Since $K_M \subseteq L_{1,M}^*$, it follows that

$$\langle \varphi_M{}^*(w_0) - \varphi_M{}^*J(u_M), x \rangle = \langle \varphi_M{}^*(y), u_M \rangle,$$

i.e.,

$$\langle w_0 - J(u_M), x \rangle = \langle y, u_M \rangle.$$

For every M, $[0, 0]$ lies in $G(L)$ with $0 \in M$. Hence

$$\langle w_0 - J(u_M), u_M \rangle \geq 0,$$

i.e.,

$$\|u_M\|^2 = \langle J(u_M), u_M \rangle \leq \langle w_0, u_M \rangle \leq \|w_0\| \cdot \|u_M\|$$

so that

$$\|u_M\| \leq \|w_0\|.$$

Let Λ be the family of finite-dimensional subspaces of X ordered by inclusion. For each M_0 in Λ, let

$$W_{M_0} = \bigcup_{M \supseteq M_0} [u_M, J(u_M)] \subset X \times X^*,$$

where the union ranges over all the solutions u_M of the systems of inequalities and equalities

(i) $_M$ $\langle w_0 - J(u_M) - w, u_M - u\rangle \geq 0$ for all $[u, w]$ in $G(L)$ with u in M;
(e) $_M$ $\langle w_0 - J(u_M), x\rangle = \langle y, u_M\rangle$ for all $[x, y]$ in $G(L_1)$ with x in M.

As M_0 ranges over Λ, the sets W_{M_0} have the finite intersection property and are all contained in a given weakly compact subset of $X \times X^*$. Hence their closures in the weak topology on $X \times X^*$ have a nonempty intersection, i.e., there exists $[u_0, y_0]$ with

$$[u_0, y_0] \in \text{weak closure}(W_{M_0})$$

for all M_0 in Λ.

Let $[x, y]$ be an element of $G(L_1)$, $[u, w]$ an element of $G(L)$. Choose M_0 containing both x and u. Then each M in Λ which contains M_0 must contain both x and u. Thus for each $[u_M, y_M]$ in W_{M_0} with $y_M = J(u_M)$, we have

$$\langle w_0 - y_M - w, u_M - u\rangle \geq 0, \qquad \text{and} \qquad \langle w_0 - y_M, x\rangle = \langle y, u_M\rangle.$$

Consider the real-valued function $g(v, z) = \langle w_0 - z, x\rangle - \langle y, v\rangle$. From the equality above, $g(u_M, y_M) = 0$ and g is continuous in the weak topology in $[v, z]$ on $X \times X^*$. Since g is null on W_{M_0}, it is null on the weak closure of W_{M_0}. Hence $g(u_0, y_0) = 0$, i.e.,

$$\langle w_0 - y_0, x\rangle = \langle y, u_0\rangle$$

for all $[x, y]$ in $G(L_1)$. Hence $[u_0, w_0 - y_0]$ lies in $G(L_1{}^*)$.

From the first inequality, we see that

$$\|u_M\|^2 = \langle J(u_M), u_M\rangle \leq \langle w_0 - w, u_M - u\rangle + \langle y_M, u\rangle.$$

Since $X \times X^*$ is a reflexive Banach space, there exists a sequence $\{[u_j, J(u_j)]\}$ in W_{M_0} converging weakly to $[u_0, y_0]$[6, p.81]. For each j and for the given $[u, w]$ in $G(L)$, we have

$$\begin{aligned} &\|u_j\|^2 - \|u_j\| \cdot \|u_0\| \\ &\leq \|u_j\|^2 - \langle J(u_j), u_0\rangle \leq \langle w_0 - w, u_j - u\rangle + \langle J(u_j), u - u_0\rangle. \end{aligned}$$

Hence

$$\begin{aligned} 0 &\leq \underline{\lim}\{\|u_j\|^2 - \|u_j\| \cdot \|u_0\|\} \leq \overline{\lim}\{\|u_j\|^2 - \|u_j\| \cdot \|u_0\|\} \\ &\leq \lim\{\langle w_0 - w, u_j - u\rangle + \langle J(u_j), u - u_0\rangle\} \\ &\leq \langle w_0 - y_0 - w, u_0 - u\rangle. \end{aligned}$$

Thus $[u_0, w_0 - y_0]$, which we have already shown to lie in $G(L_1{}^*)$, is monotonely related to $[u, w]$ in $G(L)$. This is true for each $[u, w]$ in $G(L)$. Hence since L is maximal monotone in $G(L_1{}^*)$,

$$(w_0 - y_0) \in L(u_0).$$

We now note that if M_0 were chosen to include u_0 and the above argument were repeated, it would follow that for a corresponding sequence $\{u_j\}$

$$\lim\{\|u_j\|^2 - \|u_j\| \cdot \|u_0\|\} = 0,$$

i.e.,

$$\lim\|u_j\| = \|u_0\|.$$

Since X and X^* are assumed locally uniformly convex, it follows that u_j converges strongly to u and $J(u_j)$ converges strongly to $J(u_0)$. Hence $y_0 = J(u_0)$, i.e., $w_0 = (w_0 - y_0) + y_0 \in (L + J)(u_0)$. Q.E.D.

2. Application to Hammerstein Equations

DEFINITION [3] Let X be a Banach space, F a mapping of X into X^*. Then F is said to be monotone angle-bounded if there exists a constant $c > 0$ such that for any triples $[u, v, x]$ in X,

$$\langle F(x) - F(u), v - x\rangle \leq c\langle F(u) - F(v), u - v\rangle.$$

THEOREM 3 Let (Ω, S, μ) be a measure space with a finite measure μ and let X be a reflexive Banach space such that

$$L^\infty(\mu) \subset X \subset L^1(\mu), \qquad L^\infty(\mu) \subset X^* \subset L^1(\mu)$$

with the pairing between w in X^* and u in X given by

$$\langle w, u\rangle = \int_\Omega wu \, d\mu.$$

Let K be a bounded linear mapping of $L^1(\mu)$ into $L^1(\mu)$ such that for each v in $L^\infty(\mu)$, we have

$$\langle Kv, v\rangle \geq 0.$$

Let F be a hemicontinuous monotone angle-bounded mapping of X into X^* with 0 lying in the interior of its range $R(F)$.

Then for each h in X, there exists u in X such that $u + KF(u) = h$ and $\langle Kv - KF(u), v - F(u)\rangle \geq 0$ for all $v \in L^\infty(\mu)$ with $Kv \in X$.

Proof We begin by constructing a linear maximal monotone mapping L from $D(L)$ in X^* to X which is a restriction of the given map K from $L^1(\mu)$ to

$L^1(\mu)$. Let $K'\colon L^\infty(\mu) \to L^\infty(\mu)$ be the mapping adjoint to K. We define L_1 from $D(L_1) = L^\infty(\mu) \subset X^*$ with values in X by setting

$$L_1(v) = K'(v) \qquad (v \in D(L_1) = L^\infty(\mu)).$$

If $[u, w]$ lies in $G(L_1{}^*)$, both u and w lie in $L^1(\mu)$ and for every v in $L^\infty(\mu)$

$$\langle u, L_1(v)\rangle = \langle w, v\rangle,$$

i.e.,

$$\langle w, v\rangle = \langle u, L_1 v\rangle = \langle u, K'(v)\rangle = \langle K(u), v\rangle.$$

Hence $w = K(u)$, i.e., $L_1{}^*$ is a restriction of K.

Let $K^\#$ be the mapping from X^* to X with domain $D(K^\#) = \{v \in L^\infty(\mu)$ and $Kv \in X\}$ and $K^\# v = Kv$ for v in $D(K^\#)$.

Let L be any maximal monotone linear operator such that $K^\# \subseteq L \subseteq L_1{}^*$. Such an L exists by Theorem 1 and is a restriction of K.

We shall obtain the desired solution u of the equation $u + KFu = h$ by solving instead the equation

$$h \in u + LFu.$$

If we take $v = u - h$ as a new variable, we note that the given equation for u is equivalent to

$$0 \in v + LF_h(v),$$

where $F_h(v) = F(v + h)$ and F_h satisfies the same conditions as F. Hence, we may assume that $h = 0$.

The equation

$$0 \in u + LF(u)$$

is equivalent to the equation

$$0 \in L^{-1}(u) + F(u).$$

Let J be the duality map of X into X^*. Since L^{-1} is maximal monotone, $L^{-1} + F + \varepsilon J$ is maximal monotone as well as coercive for each $\varepsilon > 0$. Hence there exists u_ε in X such that

$$0 = w_\varepsilon + F(u_\varepsilon) + \varepsilon J(u_\varepsilon) \in L^{-1}(u_\varepsilon) + F(u_\varepsilon) + \varepsilon J(u_\varepsilon).$$

Since $0 \in \operatorname{Int} F(X)$, there exists $\delta > 0$ such that for all y in X^* with $\|y\| \le \delta$,

$$y = F(v_y)$$

for some v_y in X. By the angle-boundedness of F

$$\langle F(u_\varepsilon) - F(v_y), u_\varepsilon - 0\rangle \ge -c\langle F(v_y) - F(0), v_y - 0\rangle = C_y,$$

C_y independent of ε. Thus

$$\langle -\varepsilon J(u_\varepsilon) - w_\varepsilon - y, u_\varepsilon \rangle \geq C_y$$

with $w_\varepsilon \in L^{-1}(u_\varepsilon)$, i.e.,

$$-(y, u_\varepsilon) \geq C_y + \varepsilon \|u_\varepsilon\|^2 + \langle w_\varepsilon, u_\varepsilon \rangle \geq C_y.$$

The uniform boundedness principle implies that $\|u_\varepsilon\| \leq M$.

Choose a sequence $u_{\varepsilon_j} = u_j$ in X^* with $\varepsilon_j \to 0$ and u_j converging weakly to an element u of X. Since

$$0 = \lim_{j \to +\infty} \{-\varepsilon_j J(u_{\varepsilon_j})\} \qquad \text{where} \quad -\varepsilon_j J(u_j) \in (L^{-1} + F)(u_j),$$

the maximal monotonicity of $(L^{-1} + F)$ implies that $0 \in (L^{-1} + F)(u)$. Q.E.D.

Using the result of Theorem 3, we can give a partial answer to a question raised by Browder [7] concerning the possibility of establishing existence theorems for singular Hammerstein nonlinear integral equations by methods based on the theory of monotone operators without the use of any kind of compactness assumption.

THEOREM 4 Let (Ω, S, μ) be a finite measure space. Consider the Hammerstein equation

$$u(x) + \int_\Omega k(x, y) f(y, u(y)) \mu(dy) = h(x) \qquad (x \in \Omega). \tag{1}$$

Suppose that

(1) $f(y, r)$ is nondecreasing in r and satisfies the Caratheodory condition.

(2) There exists $p > 1$ such that for suitable constants c_0 and c_1 and all x in Ω, r in $\mathbb{R}$,

$$c_0(1 + |r|^{p-1}) \leq |f(x, r)| \leq c_1(1 + |r|^{p-1}).$$

(3) The linear operator K given by

$$(Kv)(x) = \int_\Omega k(x, y) v(y) \mu(dy) \qquad (x \in \Omega)$$

is a bounded linear mapping of $L^1(\mu)$. For each v in $L^\infty(\mu)$, we have $\langle K(v), v \rangle \geq 0$.

Then for each h in $L^p(\mu)$, there exists u in $L^p(\mu)$ such that (1) holds.

Proof The operator F given by

$$(Fu)(y) = f(y, u(y)) \qquad (y \in \Omega)$$

is a mapping of $X = L^p(\mu)$ into $X^* = L^{p'}(\mu)$ which is continuous, monotone, and angle-bounded [3]. Moreover F maps X onto X^*. Hence the desired conclusion follows from Theorem 3. Q.E.D.

References

1. R. Arens, Operational calculus of linear relations. *Pacific J. Math.* **11** (1961), 9–23.
2. H. Brézis, On some degenerate nonlinear parabolic equations. "Nonlinear Functional Analysis" *Proc. Symp. Pure Math.* Vol. 18, P.1, pp. 28–38. American Mathematical Society Providence, Rhode Island, 1970.
3. H. Brézis and F. E. Browder, Nonlinear integral equations and systems of Hammerstein type. *Advances in Math.* **18** (1975), 115–147.
4. H. Brézis and F. E. Browder, Singular Hammerstein equations and maximal monotone operators. *Bull. Amer. Math. Soc.* **82** (1976), 623–625.
5. F. E. Browder, Nonlinear maximal monotone operators in Banach spaces. *Math. Ann.* **175** (1968), 89–113.
6. F. E. Browder, Nonlinear operators and nonlinear equations of evolution in Banach spaces. "Nonlinear Functional Analysis" *Proc. Symp. Pure Math.* Vol. 18, Part 2. American Mathematical Society, Providence, Rhode Island, 1975.
7. F. E. Browder, Strongly nonlinear integral equations of Hammerstein type. *Proc. Nat. Acad. Sci. U.S.A.* **72** (1975), 1937–1939.
8. M. A. Krein, The theory of self-adjoint extensions of semi-bounded hermitian transformations and its applications. *Math. USSR-Sb.* **20** (1947), 431–495; **21** (1947), 365–404.
9. R. S. Phillips, Dissipative operators and hyperbolic systems of partial differential equations. *Trans. Amer. Math. Soc.*, **90** (1959), 193–254.

AMS (MOS) 1970 Subject Classification: Primary 47H05, 47G05, Secondary 47H15

Nonlinear Problems across a Point of Resonance for Nonselfadjoint Systems

Lamberto Cesari

University of Michigan

1. Introduction

Let us consider nonlinear operational equations at resonance of the forms $Ex = Nx$ and $Ex + \alpha Ax = Nx$, where E: $\mathfrak{D}(E) \to Y$, $\mathfrak{D}(E) \subset X$ is a linear operator, N: $X \to Y$, A: $X \to Y$ are continuous operators, not necessarily linear, α a real parameter, X, Y Banach spaces. On the operator E we assume only that its kernel is not trivial and finite dimensional, or $1 \leq \dim \ker E < \infty$, and that E possesses a partial inverse which is linear, bounded, and compact (see Section 2 for notations and details). In any application, E may be a linear differential operator in some domain G in E^{ν}, $\nu \geq 1$, with linear homogeneous boundary conditions. In previous papers [7,8] we have shown that simple assumptions on N and A guarantee the existence of solutions to these abstract operational equations. Namely, conditions as

$$\|Nx\| \leq J_0 \quad \text{for all} \quad x \in X; \qquad N: X \to Y \quad \text{continuous,}$$

$$\langle QNx, x^* \rangle \leq 0 \; [\text{or} \geq 0]$$

$$\text{for all} \quad x \in X, \quad x^* \in X_0, \quad \|x^*\| \geq R_0, \quad \|x - x^*\| \leq K$$

(see Section 2 for notations and details), guarantee the existence of solutions to equation $Ex = Nx$ (existence at resonance). The slightly stronger assumptions

$$\|Nx\| \leq J_0 \quad \text{for all} \quad x \in X; \qquad N: X \to Y \text{ continuous,}$$

$$A: X \to Y \quad \text{continuous, mapping bounded sets into bounded sets,}$$

$$\langle QNx, x^* \rangle \leq -\varepsilon\|x^*\| \quad \text{or} \quad \geq \varepsilon\|x^*\|$$

$$\text{for all} \quad x \in X, \quad x^* \in X_0, \quad \|x^*\| \geq R_0, \quad \|x - x^*\| \leq K,$$

guarantee the existence of equibounded solutions to the equation $Ex + \alpha Ax = Nx$ at least for all $|\alpha|$ sufficiently small (existence across a point of resonance). Analogous statements hold in the case of N unbounded but of slow growth. All these abstract assumptions are satisfied under the usual specific hypotheses of theorems of the Landesman and Lazer type, which are thereby proved in a stronger form (see Cesari [5–8] for references and details).

In the present paper we consider specifically scalar equations of the forms

$$Ex = h(t) + f(t, x(t)), \qquad t \in G, \tag{1}$$

$$Ex + \alpha g(t, x(t)) = h(t) + f(t, x(t)), \qquad t \in G, \tag{2}$$

where G is a bounded domain in E^ν, $\nu \geq 1$, and E can well be thought of as a linear differential operator on G with linear homogeneous boundary conditions of ∂G, though E need not be specified, since on E we only require the abstract assumptions we have made in the previously mentioned references namely, that E have a nontrivial finite dimensional kernel and that E possess a bounded and compact partial inverse H. In (1) and (2), h denotes any element of $L_\infty(G)$, and $f(t, s)$, $g(t, s)$ are continuous real valued functions on $\text{cl}\, G \times R^1$. No other hypothesis is made on g besides continuity.

We show in this paper, by an extension of previous arguments, that the classical hypotheses on f

$$|f(t, s)| \leq J_0 \quad \text{on} \quad G \times R^1,$$

$$sf(t, s) \leq 0 \; [\geq 0] \quad \text{for all} \quad t \in G, \quad |s| \geq R_0$$

guarantee the existence of solutions to Eq. (1) at resonance. We also show that the slightly stronger hypotheses

$$|f(t, s)| \leq J_0 \qquad \text{on} \quad G \times R^1,$$

$$f(t, s) \leq -\eta\ [\geq \eta] \qquad \text{for all} \quad t \in G, s \geq R_0,$$

$$f(t, s) \geq \eta\ [\leq -\eta] \qquad \text{for all} \quad t \in G, s \leq -R_0, \qquad \eta > 0,$$

guarantee the existence of solutions to Eq. (2) for all $|\alpha|$ sufficiently small (existence of solutions across a point of resonance). The above properties of the real function f are transferred to properties of the Nemitsky operator $Nx = h(t) + f(t, x(t))$ in $L_\infty(G)$, namely, properties concerning the sign of $\langle Nx, x^* \rangle$ for $\|x^*\|_\infty \geq R_0$ and bounded $\|x - x^*\|_\infty$. Finally, in §3,4,5 we state and prove existence theorems across a point of resonance in a rather general situation, where the qualitative growth conditions above are replaced by quantitative relations on the surface of suitable balls in the spaces under consideration. The specific argument extends to a rather general situation a remark of Lazer and Leach for the problem of periodic solutions of second-order ordinary differential equations.

2. Notations

Let X, Y be Banach spaces over the reals with norms $\| \ \|_X$, $\| \ \|_Y$. We shall write only $\| \ \|$ when no explanation is needed. Let $P: X \to X$, $Q: Y \to Y$ be projection operators (i.e., linear, bounded, and idempotent) with ranges and null spaces

$$\mathscr{R}(P) = PX = X_0, \qquad \ker P = \mathscr{R}(I - P) = (I - P)X = X_1,$$

$$\mathscr{R}(Q) = QY = Y_0, \qquad \ker Q = \mathscr{R}(I - Q) = (I - Q)Y = Y_1.$$

Let E: $\mathfrak{D}(E) \to Y$ be a linear operator with domain $\mathfrak{D}(E) \subset X$, and let us assume here, for the sake of simplicity, that

$$\ker E = X_0, \qquad \mathscr{R}(E) = Y_1, \qquad 1 \leq \dim X_0 < \infty.$$

Then E, as an operator from $\mathfrak{D}(E) \cap X_1$ into Y_1, is one-one and onto, so that its partial inverse H: $Y_1 \to \mathfrak{D}(E) \cap X_1$ exists as a linear operator. We assume that H is a bounded compact linear operator, and that the usual axioms hold:

(k_1) $H(I - Q)E = I - P$,
(k_2) $EP = QE$,
(k_3) $EH(I - Q) = I - Q$.

We have depicted here, in an abstract way, a situation which is rather typical for a large class of differential operators E with boundary conditions and not necessarily self-adjoint. For a more general situation, see Cesari [3–8] and Hale [15].

The assumption above that dim ker $E \geq 1$ is the resonance hypothesis. Note that the theorem covers the case where E is of the form $E_0 + \lambda I$, with λ eigenvalue and finite dimensional eigenspace.

Let $N: X \to Y$ be a continuous operator, not necessarily linear, and let us consider the equation

$$Ex = Nx, \qquad x \in \mathfrak{D}(E) \subset X.$$

As we know from Cesari [4], this equation is equivalent to the system of auxiliary and bifurcation equations

$$x = Px + H(I - Q)Nx, \qquad Q(Ex - Nx) = 0.$$

Having assumed ker $E = X_0 = PX$, the bifurcation equation reduces to $QNx = 0$. Also, for $x^* = Px$, the auxiliary equation takes the form $x = x^* + H(I - Q)Nx$.

We shall now further assume that Y is a space of linear operators on X so that an operation $\langle y, x \rangle$ is defined from $X \times Y$ into the reals, is linear both in x and y, and has the following properties:

(π_1) $|\langle y, x \rangle| \leq K\|y\| \, \|x\|$ for all $x \in X$, $y \in Y$, and some constant $K > 0$.

(π_2) For $y \in Y_0$ we have $y = 0$ if and only if $\langle y, x^* \rangle = 0$ for all $x^* \in X_0$.

We have noted [8] that the existence of an operation $\langle y, x \rangle$ as shown previously is a rather general occurrence. It is always possible to choose the norms in X and in Y, or to choose the operation $\langle y, x \rangle$, in such a way that $K = 1$.

Let $L = \|H\|$, $k_0 = \|P\|$, $k' = \|I - P\|$, $\chi = \|Q\|$, $\chi' = \|I - Q\|$. Also, we shall choose a basis $w = (w_1, \ldots, w_m)$ in $X_0 = \ker E$, $m = \dim \ker E$. By $\langle y, w \rangle$ we shall denote below the m-vector $\langle y, w_i \rangle$, $i = 1, \ldots, m$. For every $x^* \in X_0$ we have $x^* = \sum_1^m c_i w_i$, or briefly $x^* = cw$, $c = (c_1, \ldots, c_m) \in R^m$, and there are constants $0 < \gamma' \leq \gamma < \infty$ such that $\gamma'|c| \leq \|cw\| \leq \gamma|c|$ for all $x^* = cw \in X_0$ and where $|\ \ |$ denotes the Euclidean norm in R^m. It would be always possible to choose the basis $w_1, \ldots, w_m$ in such a way as to make, say,

$\gamma = 1$. In any case we shall denote by γ_0 the constant $\gamma_0 = \min[1, \gamma]$. Finally, there is a constant $\mu > 0$ such that for every $y \in Y$ and $d = \langle y, w \rangle$, or $d = (d_1, \ldots, d_m) \in R^m$, $d_i = \langle y, w_i \rangle$, $i = 1, \ldots, m$, we have $|d| \leq \mu \|y\|$.

As mentioned, we shall also consider another continuous operator A: $X \to Y$, not necessarily linear, but bounded, that is, mapping bounded subsets of X into bounded subsets of Y, or equivalently satisfying a relation $\|Ax\| \leq \omega(\|x\|)$, for all $x \in X$ and some monotone nondecreasing function $\omega(\zeta) \geq 0$, $0 \leq \zeta < +\infty$ (ω not necessarily continuous, nor zero at the origin).

3. Some Abstract Existence Theorems

We state here a few of the existence theorems we have proved in reference [8].

A. *Existence at Resonance*

3.1 THEOREM (*existence at resonance*) (Cesari [8]) Let X, Y be Banach spaces; let E, H, F, P, Q be as in Section 2, satisfying properties (k_{123}) with $X_0 = \ker E$ nontrivial and finite dimensional, and H linear, bounded, and compact; let $\langle y, x \rangle$ be a linear operation satisfying (π_1) and (π_2); and let N: $X \to Y$ be a continuous operator, not necessarily linear. If (B_0) there is a constant $J_0 > 0$ such that $\|Nx\| \leq J_0$ for all $x \in X$; and if (N_0) there are constants $R_0 \geq 0$ and $K_0 > L\chi' J_0$ such that $\langle QNx, x^* \rangle \leq 0$ [or $\langle QNx, x^* \rangle \geq 0$] for all $x \in X$, $x^* \in X_0$ with $Px = x^*$, $\|x^*\| \geq R_0$ and $\|x - x^*\| \leq K_0$, then equation $Ex = Nx$ has at least a solution $x \in \mathfrak{D}(E) \subset X$.

For the case of limited growth of N, we need consider a suitable monotone function $\phi(\zeta) \geq 0$, $0 \leq \zeta < +\infty$, and assume that $\|Nx\| \leq \phi(\|x\|)$ for all $x \in X$. On $\phi(\zeta)$ we could simply require that $\phi(\zeta)/\zeta \to 0$ as $\zeta \to +\infty$. Actually, it is of some advantage to require less on ϕ.

Let $R_0 \geq 0$ denote the constant that will appear in the condition (N_ϕ) below, let $\sigma_1, \sigma_2, \sigma$ be arbitrary constants, $0 < \sigma_1 < \sigma_2 < \sigma < \min[1, \gamma^{-1}]$, and take positive numbers λ_0, λ_1 satisfying

$$\lambda_0 \geq \max[1, (\gamma')^{-1}], \qquad \lambda_1 \leq \min[(L\chi')^{-1}(1 - \gamma\sigma), (\mu\chi)^{-1}(\sigma - \sigma_2)].$$

The only requirement we need on the monotone function ϕ is that there be a number S satisfying

$$S \geq \sigma_1^{-1} \lambda_0 R_0, \qquad \phi(S)/S \leq \lambda_1. \tag{3}$$

Thus, if $\phi(\zeta)/\zeta \to 0$ as $\zeta \to +\infty$, then certainly such a constant S can be determined.

For instance, if $\|Nx\| \le J_0 + J_1 \|x\|^k$ for all $x \in X$ and some constants $J_0 \ge 0$, $J_1 > 0$, $0 < k < 1$, then $\phi(\zeta) = J_0 + J_1 \zeta^k$, $\phi(\zeta)/\zeta \to 0$ as $\zeta \to +\infty$, and the constant S can be found.

If $\|Nx\| \le J_0 + J_1 \|x\|^k$ for all $x \in X$, and constants $J_0 \ge 0$, $J_1 > 0$, $k > 1$, then $\phi(\zeta) = J_0 + J_1 \zeta^k$ and now $\phi(\zeta)/\zeta \to +\infty$ as $\zeta \to +\infty$. However, a constant S satisfying (3) can still be found provided J_1 is sufficiently small. Indeed it is enough to take, for instance,

$$S = \max[\sigma_1^{-1}\lambda_0 R_0, 2J_0\lambda_1^{-1}], \qquad J_1 \le 2^{-1}\lambda_1 S^{1-k},$$

so that $\phi(S)/S = J_0 S^{-1} + J_1 S^{k-1} \le \lambda_1/2 + \lambda_1/2 = \lambda_1$.

3.1* Theorem (*existence at resonance*) (Cesari [8]) Under the same general hypotheses as in Theorem 3.1, let $\phi(\zeta) \ge 0$, $\psi(\zeta) \ge 0$, $0 \le \zeta < +\infty$ be monotone nondecreasing functions. Let us assume that (B_ϕ) $\|Nx\| \le \phi(\|x\|)$ for all $x \in X$; and that (N_ϕ), $\langle QNx, x^*\rangle \le 0$ [or $\langle QNx, x^*\rangle \ge 0$] for all $x \in X$, $x^* \in X_0$ with $Px = x^*$, $\|x^*\| \ge R_0$, $\|x - x^*\| \le \psi(\|x\|)$. Let us assume further that there is a constant $S \ge \sigma_1^{-1}\lambda_0 R_0$ with $\phi(S)/S \le \lambda_1$ and

$$L\chi'\phi(S) \le \psi(k_0^{-1}\gamma'\sigma_1 S). \tag{4}$$

Then the equation $Ex = Nx$ has at least a solution $x \in \mathfrak{D}(E) \subset X$ with $\|x\| \le S$.

For instance, if we take $L\chi'\phi(\zeta) = \psi(k_0^{-1}\gamma'\sigma_1\zeta)$, then relation (N_ϕ) is required to hold for $\|x - x^*\| \le L\chi'\phi(\lambda_2\|x\|)$ with $\lambda_2 = \sigma_1^{-1}(\gamma')^{-1}k_0$, relation (4) is trivially satisfied for all S, and all we require on ϕ is that there be some S satisfying (3). For instance, in the case $\phi(\zeta) = J_0 + J_1\zeta^k$, $0 < k < 1$, this choice of ψ would yield $\psi(\zeta) = L\chi' J_0 + L\chi' J_1((\gamma')^{-1}k_0\sigma_1^{-1}\zeta)^k$.

B. Existence Theorems across a Point of Resonance

3.2 Theorem (*existence across a point of resonance*) (Cesari [8]) Let X, Y be Banach spaces; let E, H, N, P, Q be as in Section 2 satisfying (k_{123}), with $X_0 = \ker E$ nontrivial and finite dimensional, and H linear, bounded, and compact; let $\langle y, x\rangle$ be a linear operation satisfying (π_1) and (π_2); and let $N\colon X \to Y$, $A\colon X \to Y$ be continuous operators, not necessarily linear, A mapping bounded sets into bounded sets. If (B_0) there is a constant $J_0 > 0$ such that $\|Nx\| \le J_0$ for all $x \in X$; and if (N_ε) there are constants $R_0 \ge 0$, $\varepsilon > 0$, $K > L\chi' J_0$ such that $\langle QNx, x^*\rangle \le -\varepsilon\|x^*\|$ [or $\langle QNx, x^*\rangle \ge \varepsilon\|x^*\|$] for all $x \in X$, $x^* \in X_0$, $Px = x^*$, with $\|x^*\| \ge R_0$ and $\|x - x^*\| \le K$, then there are also constants $\alpha_0 > 0$, $C > 0$, such that, for every real α with $|\alpha| \le \alpha_0$, equations $Ex + \alpha Ax = Nx$ has at least a solution $x \in \mathfrak{D}(E) \subset X$ with $\|x\| \le C$.

For the case of slow growth of Nx an analogous theorem is as follows.

3.3 THEOREM (*existence of solutions across a point of resonance*) (Cesari [8]) Under the same general assumptions as in Theorem 3.2, if (B_k) there are constants $J_0 \geq 0$, $J_1 > 0$, $0 < k < 1$ such that $\|Nx\| \leq J_0 + J_1 \|x\|^k$ for all $x \in X$; and if (N_k), there are constants $R_0 \geq 0$, $\varepsilon > 0$, $K_0 > L\chi' J_0$, $K_1 > L\chi' J_1((\gamma')^{-1} k_0 \gamma_0)^k$ such that $\langle QNx, x^* \rangle \leq -\varepsilon \|x^*\|^{1+k}$ [or always $\langle QNx, x^* \rangle \geq \varepsilon \|x^*\|^{1+k}$] for all $x \in X$, $x^* \in X_0$ with $Px = x^*$, $\|x^*\| \geq R_0$, and $\|x - x^*\| \leq K_0 + K_1 \|x\|^k$, then there are constants $\alpha_0 > 0$, $C > 0$ such that for every real α, $|\alpha| \leq \alpha_0$, the equation $Ex + \alpha Ax = Nx$ has at least a solution $x \in \mathfrak{D}(E) \subset X$ with $\|x\| \leq C$.

Both statements 3.2 and 3.3 are actually particular cases of the unique statement 3.4 which contains also other cases of interest.

Let $R_0 \geq 0$ denote the constant which will appear in the assumption (N_ϕ) below. Let $\sigma_1, \sigma_2, \sigma$ be arbitrary constants, $0 < \sigma_1 < \sigma_2 < \sigma < \min[1, \gamma^{-1}]$, and let us consider two other constants

$$\lambda_0 \geq \max[1, (\gamma')^{-1} k_0], \qquad \lambda_1 < \min[(L\chi')^{-1}(1 - \gamma\sigma), (\mu\chi)^{-1}(\sigma - \sigma_2)].$$

3.4 THEOREM (*existence across a point of resonance*) (Cesari [8]) Under the same hypotheses of Section 2, let $\phi(\zeta)$, $\phi_1(\zeta)$, $\psi(\zeta) \geq 0$, $0 \leq \zeta < +\infty$, be monotone nondecreasing functions, both $\phi_1(\zeta)$, $\psi(\zeta)$ positive for $\zeta \geq R_0$. Let us assume that (B_ϕ) $\|Nx\| \leq \phi(\|x\|)$ for all $x \in X$; and that (N_ϕ) $\langle QNx, x^* \rangle \leq -\phi_1(\|x^*\|)$ [or $\langle QNx, x^* \rangle \geq \phi_1(\|x^*\|)$] for all $x \in X$, $x^* \in X_0$ with $Px = x^*$, $\|x^*\| \geq R_0$, $\|x - x^*\| \leq \psi(\|x\|)$. Let us assume further that there is a constant $S \geq \sigma_1^{-1} \lambda_0 R_0$ with $\phi(S)/S \leq \lambda_1$, and

$$L\chi' \phi(S) < \psi(k_0^{-1} \gamma' \sigma_1 S). \tag{4}$$

Then, there is a number $\alpha_0 > 0$ such that for every real $|\alpha| \leq \alpha_0$ the equation $Ex + \alpha Ax = Nx$ has at least a solution $x \in \mathfrak{D}(E) \subset X$ with $\|x\| \leq S$.

If we require that (N_ϕ) hold for $\|x - x^*\| \leq \psi(\|x^*\|)$, then we shall assume, instead of (4), that

$$L\chi' \phi(S) < \psi(\gamma' \sigma_1 S). \tag{4'}$$

For instance, if (B_ϕ) and (N_ϕ) of Theorem 3.4 hold with ϕ satisfying $\phi(\zeta)/\zeta \to 0$ as $\zeta \to -\infty$, with an arbitrary $\phi_1(\zeta)$ as stated, and any ψ satisfying

$$\psi(\zeta) \geq L\chi' \phi((\gamma')^{-1} k_0 \sigma_1^{-1} \zeta), \tag{5}$$

then all conditions of Theorem 3.4 hold. Indeed, inequality (5) implies that relation (4) holds for all S. Thus, it is enough to determine S in such a way that $S \geq \sigma_1^{-1} \lambda_0 R_0$ and $\phi(S)/S < \lambda_1$.

As a second example, let us assume that (B_0) and (N_ε) hold, that is, the conditions of Theorem 3.2. Then all conditions of Theorem 3.4 hold. Indeed, we have here $\phi(\zeta) = J_0 > 0$, $\phi_1(\zeta) = \varepsilon\zeta$, $\psi(\zeta) = K_0$ for some constants $\varepsilon > 0$ and $K_0 > L\chi' J_0$. Then relation (5) reduces here to $L\chi' J_0 < K_0$, which is satisfied by hypothesis, and we have only to determine S so that $S \geq \sigma_1^{-1}\lambda_0 R_0$, $J_0/S < \lambda_1$.

As a third example, let us assume that (B_k) and $(N_{\varepsilon k})$ hold, that is, the conditions of Theorem 3.3. Then all conditions of Theorem 3.4 hold. Indeed, we have here $\phi(\zeta) = J_0 + J_1\zeta^k$ for some $0 < k < 1$, $J_0 \geq 0$, $J_1 > 0$; $\phi_1(\zeta) = \varepsilon\zeta^{1+k}$, $\psi(\zeta) = K_0 + K_1\zeta^k$ for some constants $\varepsilon > 0$, $K_0 > L\chi' J_0$, $K_1 > L\chi' J_1((\gamma')^{-1}k_0)^k$. Then relation (5) reduces to $K_0 + K_1\zeta^k > L\chi'[J_0 + J_1((\gamma')^{-1}k_0\sigma_1^{-1}\zeta)^k]$, which is true because of the hypotheses on K_0, K_1, and thus (4) holds for every S. Thus we have only to determine $S \geq \sigma_1^{-1}\lambda_0 R_0$ satisfying $\phi(S)/S = J_0 S^{-1} + J_1 S^{k-1} < \lambda_1$, which is possible because $0 < k < 1$.

As a fourth example, let us assume that a relation (B_ϕ) holds with $\phi(\zeta) = J_0 + J_1\zeta^k$ for constants $k \geq 1$, $J_0 > 0$ fixed, and J_1 sufficiently small, and that (N_ϕ) holds with $\phi_1(\zeta) = \varepsilon\zeta$ and $\psi(\zeta) = K_0 > L\chi' J_0$. Then all conditions of Theorem 3.4 hold. Indeed, we can take S as in the second example so that $S > \sigma_1^{-1}\lambda_0 R_0$ and $J_0 S^{-1} < \lambda_1$. Then we can determine $J_1 > 0$ so small that we also have $\phi(S)/S = J_0 S^{-1} + J_1 S^{k-1} < \lambda_1$.

4. A Study on $\langle QNx, x^* \rangle$ in the Uniform Topology

Let G be a bounded domain in E^ν, $\nu \geq 1$, and let $|G|$ denote the measure of G. Let $X = Y = L_\infty(G)$, and let $\langle y, x \rangle$ denote the linear operation $\langle y, x \rangle = |G|^{-1} \int_G y(t)x(t)\,dt$, defined for all $x \in X$, $y \in Y$ and satisfying $|\langle y, x \rangle| \leq \|y\|_\infty \|x\|_\infty$.

Let $h(t)$, $t \in G$, be any element of $X = L_\infty(G)$, and let $f(t, s)$ be any continuous function on cl $G \times R^1$. Let Nx be the Nemitsky operator $Nx = h(t) + f(t, x(t))$ for $x \in X$. It is immediately seen that $N\colon X \to Y$.

A. *The Case of P and Q Mean Values*

Let $P\colon X \to X$, $Q\colon Y \to Y$ denote the projection operators defined by

$$Px = |G|^{-1} \int_G x(t)\,dt, \qquad Qy = |G|^{-1} \int_G y(t)\,dt. \tag{6}$$

In this situation $\|P\| = k_0 = 1$, $\|Q\| = \chi_0 = 1$, and $\|I - P\| = k' = 2$, $\|I - Q\| = \chi' = 2$.

We shall assume first that

(H_0) $\int_G h(t)\,dt = 0$;
(F_0) $sf(t, s) \leq 0$ for all $t \in G$ and $|s| \geq R_0$.

Let $K \geq 0$ be any constant, and take $R_1 \geq R_0 + K$.

4.1 LEMMA For P and Q defined by (6), and under the hypotheses (H_0), (F_0), then we have $\langle QNx, x^* \rangle \leq 0$ for all $x \in X$, $x^* \in X_0$ with $Px = x^*$, $\|x^*\|_\infty \geq R_1$, $\|x - x^*\|_\infty \leq K$. The same occurs if $\|x^*\|_\infty \geq R_0$ and $\|x - x^*\|_\infty \leq \|x^*\|_\infty - R_0$.

Indeed

$$x^* = Px = |G|^{-1} \int_G x(t)\,dt = c,$$

$$QNx = |G|^{-1} \int_G [h(t) + f(t, x(t))]\,dt = |G|^{-1} \int_G f(t, x(t))\,dt = d,$$

c, d real numbers, and hence $\langle QNx, x^* \rangle = cd$. For $\|x^*\| \geq R_1$, that is, $|c| \geq R_1$, we have either $c \geq R_1$ and then

$$x(t) = c - (c - x(t)) \geq c - \|x - x^*\|_\infty \geq R_1 - K \geq R_0,$$

or $c \leq -R_1$ and then

$$x(t) = c - (c - x(t)) \leq c + \|x - x^*\|_\infty \leq -R_1 + K \leq -R_0.$$

Correspondingly we have $f(t, x(t)) \leq 0$, $d \leq 0$ in the first case, and $cd \leq 0$, or $f(t, x(t)) \geq 0$, $d \geq 0$ in the second case, and again $cd \leq 0$.

Let η_1, η_2, ε be positive constants with $\eta_1 - \eta_2 \geq \varepsilon$. Let us assume now that

(H) $\int_G h(t)\,dt \leq \eta_2$;
(F_η) $f(t, s) \leq -\eta_1$ for $t \in G$, $s \geq R_0$, and $f(t, s) \geq \eta_1$ for $t \in G$, $s \leq -R_0$.

Let $K \geq 0$ be any constant, and take $R_1 \geq R_0 + K$.

4.2 LEMMA For P and Q defined by (6), under hypotheses (H), (F_η), and if $\eta_1 - \eta_2 \geq \varepsilon$, then $\langle QNx, x^* \rangle \leq -\varepsilon \|x^*\|_\infty$ for all $x \in X$, $x^* \in X_0$ with $Px = x^*$, $\|x^*\|_\infty \geq R_1$, $\|x - x^*\|_\infty \leq K$. The same occurs for $\|x^*\|_\infty \geq R_0$ and $\|x - x^*\|_\infty \leq \|x^*\|_\infty - R_0$.

Indeed, as before

$$x^* = Px = |G|^{-1} \int_G x(t)\,dt = c, \qquad \langle QNx, x^* \rangle = cd,$$

$$QNx = |G|^{-1} \int_G h(t)\,dt + |G|^{-1} \int_G f(t, x(t))\,dt = d,$$

and for $c \geq R_1 \geq R_0 + K$ we have $x(t) \geq R_0$, and

$$f(t, x(t)) \leq -\eta_1, \qquad d \leq -\eta_1 + \eta_2 \leq -\varepsilon, \qquad cd \leq -\varepsilon c = -\varepsilon\|x^*\|.$$

Analogously, for $c \leq -R_1 \leq -R_0 - K$ we have $x(t) \leq -R_0$, $f(t, x(t)) \geq \eta_1$, $d \geq \eta_1 - \eta_2 \geq \varepsilon$, $cd \leq -\varepsilon|c| = -\varepsilon\|x^*\|$.

B. *The Case of* P *and* Q *Orthogonal in* $L_2(G)$

Again let $X = Y = L_\infty(G)$, $X_0 = Y_0$, with $1 \leq m = \dim X_0 < \infty$. Let us think of X and Y as contained in $L_2(G)$, and let us take a basis $w = (w_1, \ldots, w_m)$ for X_0 and Y_0 made up of elements $w_i \in L_\infty(G)$ that are orthonormal in $L_2(G)$ (say, obtained by the Schmidt orthogonalization process). For $P = Q$ the same orthogonal projection, then

$$Px = \sum_1^m c_i w_i, \qquad c_i = \int_G x(t)w_i(t)\,dt, \qquad i = 1, \ldots, m, \quad x \in X,$$

$$Qy = \sum_1^m d_i w_i, \qquad d_i = \int_G y(t)w_i(t)\,dt, \qquad i = 1, \ldots, m, \quad y \in Y. \quad (7)$$

Let k_0, k_0', χ_0, χ_0', μ_0 be the norms of P, $I - P$, Q, $I - Q$ and the number μ of Section 2 in the uniform topologies of X and Y (and thus not necessarily equal to one). For $x^* \in X_0$, $x^*(t) = c_1 w_1 + \cdots + c_m w_m$ with $c = (c_1, \ldots, c_m) \neq 0$ we take $b = c/|c|$, or $b = (b_1, \ldots, b_m)$, $|b| = 1$, $b_i = c_i/|c|$, $i = 1, \ldots, m$, and $v(t) = b_1 w_1 + \cdots + b_m w_m$, $x^*(t) = |c|v(t)$, $t \in G$.

For any number $\sigma \geq 0$ let $D_b(\sigma)$ denote the set of all points $t \in G$ where $|v(t)| \leq \sigma$. Assumption (p^*) below is usually satisfied in applications with functions w_i, $i = 1, \ldots, m$, smooth in G:

(p^*) There is a constant $\rho > 0$ such that $|G|^{-1} \int_G |v(t)|\,dt \geq \rho$ for all functions $v(t) = b_1 w_1 + \cdots + b_m w_m$ with $|b| = 1$.

Let η_1, η_2, ε be positive constants with $\rho\eta_1 - \eta_2 > \varepsilon\gamma_0$, and take $Nx = h(t) + f(t, x(t))$, $x \in X = L_\infty(G)$, with $h \in L_\infty(G)$ satisfying

(H^*) $\left||G|^{-1} \int_G h(t)v(t)\,dt\right| \leq \eta_2$ for all $v(t) = b_1 w_1 + \cdots + b_m w_m$ with $|b| = 1$.

For the function $f(t, s)$, continuous on $\mathrm{cl}\, G \times R^1$, we consider the monotone nondecreasing function $\phi(\zeta)$, $0 \leq \zeta < +\infty$, defined by

$$\phi(\zeta) = \mathrm{Sup}\big[|f(t, s)|\ \big|\ t \in G,\ |s| \leq \zeta\big].$$

Furthermore, we assume that

(F^*) There is a constant $R_0 > 0$ such that $f(t, s) \leq -\eta_1$ for $t \in G$, $s \geq R_0$, and $f(t, s) \geq \eta_1$ for $t \in G$, $s \leq -R_0$.

Let K be a positive constant.

Statement 4.3 now extends the argument of Lazer and Leach to a rather general situation.

4.3 Lemma For $P = Q$ the orthogonal projection defined by (7), under hypotheses (p^*, (H^*), (F^*), and for $\rho\eta_1 - \eta_2 \geq \varepsilon\gamma_0$, then there is a constant $R_1 \geq R_0 + K$ such that $\langle QNx, x^*\rangle \leq -\varepsilon\|x^*\|_\infty$ for all $x \in X$, $x^* \in X_0$ with $Px = x^*$, $\|x^*\|_\infty \geq R_1$, $\|x - x^*\|_\infty \leq K$.

Proof For $Nx = h(t) + f(t, x(t))$,

$$QNx = \sum_1^m w_i(t) \int_G [h(\alpha) + f(\alpha, x(\alpha))]w_i(\alpha)\, d\alpha,$$

$$\langle QNx, x^*\rangle = |G|^{-1} \sum_1^m c_i \int_G [h(\alpha) + f(\alpha, x(\alpha))]w_i(\alpha)\, d\alpha$$

$$= |G|^{-1} \int_G [h(t) + f(t, x(t))]x^*(t)\, dt.$$

Let $\lambda > 0$ be a constant such that $\rho\eta_1 - \eta_2 - 2\lambda \geq \varepsilon\gamma_0$. Let R_1 be any number such that

$$R_1 \geq \max[\lambda^{-1}\gamma_0(R_0 + K)\phi(R_0 + 2K), \lambda^{-1}\gamma_0\eta_1(R_0 + K)].$$

Let us assume $\|x^*\|_\infty \geq R_1$ and $\|x - x^*\|_\infty \leq K$. Let $\sigma = (R_0 + K)|c|^{-1}$. We now have

$$x^* = |c|v, \qquad \gamma_0'|c| \leq \|x^*\|_\infty = \|cw\|_\infty \leq \gamma_0|c|,$$

$$\langle QNx, x^*\rangle = |c|\Bigg[|G|^{-1} \int_G h(t)v(t)\, dt + |G|^{-1} \int_{D_b(\sigma)} f(t, x(t))v(t)\, dt + \int_{G - D_b(\sigma)} f(t, x(t))v(t)\, dt\Bigg].$$

For $t \in D_b(\sigma)$ we have $|v(t)| \leq \sigma = (R_0 + K)|c|^{-1}$; hence $|x^*(t)| = |c|\,|v(t)| \leq R_0 + K$. Hence

$$\begin{aligned} |x(t)| &= |x^*(t) + (x(t) - x^*(t))| \\ &\leq |x^*(t)| + \|x - x^*\|_\infty \leq R_0 + K + K = R_0 + 2K, \end{aligned}$$

and $|f(t, x(t))| \leq \phi(R_0 + 2K)$. Hence

$$|G|^{-1} \int_{D_b(\sigma)} f(t, x(t))v(t)\, dt \leq |G|^{-1} |D_b(\sigma)| \phi(R_0 + 2K)\sigma$$
$$\leq \phi(R_0 + 2K)(R_0 + K)|c|^{-1}$$
$$\leq \phi(R_0 + 2K)(R_0 + K)\gamma_0 R_1^{-1} \leq \lambda.$$

Thus, for $t \in G - D_b(\sigma)$, either $x^*(t) \geq R_0 + K$, and

$$x(t) = x^*(t) - (x^*(t) - x(t)) \geq x^*(t) - \|x - x^*\|_\infty \geq R_0 + K - K = R_0,$$

and then $f(t, x(t)) \leq -\eta_1$; or $x^*(t) \leq -R_0 - K$, and

$$x(t) = x^*(t) - (x^*(t) - x(t))$$
$$\leq x^*(t) + \|x - x^*\|_\infty \leq -R_0 - K + K = -R_0,$$

and then $f(t, x(t)) \geq \eta_1$. In any case we have $f(t, x(t))v(t) \leq -\eta_1 |v(t)|$ for all $t \in G - D_b(\zeta)$. Thus

$$|G|^{-1} \int_{G - D_b(\sigma)} f(t, x(t))v(t)\, dt$$
$$\leq -|G|^{-1} \int_{G - D_b(\sigma)} \eta_1 |v(t)|\, dt$$
$$= -|G|^{-1} \int_G \eta_1 |v(t)|\, dt + |G|^{-1} \int_{D_b(\sigma)} \eta_1 |v(t)|\, dt$$
$$\leq -\eta_1 \rho + |G|^{-1} |D_b(\sigma)| \eta_1 \sigma$$
$$\leq -\eta_1 \rho + \eta_1 (R_0 + K)|c|^{-1} \leq -\eta_1 \rho + \eta_1 (R_0 + K)\gamma_0 R_1^{-1} \leq \lambda.$$

Finally,

$$\langle QNx, x^* \rangle \leq |c|[\eta_2 + \lambda + \lambda - \rho\eta_1] = -|c|[\rho\eta_1 - \eta_2 - 2\lambda]$$
$$\leq -\gamma_0^{-1}(\rho\eta_1 - \eta_2 - 2\lambda)\|x^*\|_\infty \leq -\varepsilon \|x^*\|_\infty.$$

This proves Lemma 4.3.

C. *Remarks on a General Type of Projection Operators*

Let $Y = L_\infty(G)$ and let Y_0 be a finite dimensional subspace of Y, say, $1 \leq \mu = \dim Y_0 < \infty$. Let $\bar{w} = (\bar{w}_1, \ldots, \bar{w}_\mu)$ denote a basis for Y_0. If $Q_j(t)$, $t \in G$, $j = 1, \ldots, q$, are q independent bounded measurable functions in G, and $[\bar{\gamma}_{ij}]$ is a given $\mu \times q$ real matrix, then the operation Q defined by

$$Qy = \sum_{i=1}^{\mu} \bar{w}_i(t) \left(\sum_{j=1}^{q} \bar{\gamma}_{ij} \int_G y(\alpha) Q_j(\alpha)\, d\alpha \right) \tag{8}$$

maps $Y = L_\infty(G)$ into Y_0, and obviously Q is linear and bounded in the uniform topology. We shall assume explicitly that the range of Q is exactly Y_0, or $Qy = Y_0$. Let $[\bar{\lambda}_{ij}]$ denote the $\mu \times q$ matrix defined by

$$\bar{\lambda}_{ij} = \int_G \bar{w}_i(t) Q_j(t)\, dt, \qquad i = 1, \ldots, \mu, \quad j = 1, \ldots, q.$$

It is easy to see that Q is idempotent, that is, $QQ = Q$, and Q is a projection operator if the condition is satisfied:

$$\sum_{i=1}^{\mu} \bar{\gamma}_{ij} \bar{\lambda}_{ih} = \delta_{jh}, \; j, h = 1, \ldots, \text{q}, \tag{Π}$$

that is, the q columns of $[\bar{\gamma}_{ij}]$ are orthonormal to the q columns of $[\lambda_{ij}]$.

The case of $\bar{w}_1, \ldots, \bar{w}_\mu$ orthonormal in $L_2(G)$, and Qy the usual projection operator $Qy = \sum_{i=1}^{\mu} \bar{w}_i(t) \int_G y(\alpha)\bar{w}_i(\alpha)\, d\alpha$, is a particular case of (8), namely, $q = \mu$, $Q_i = \bar{w}_i$, $\gamma_{ij} = \lambda_{ij} = \delta_{ij}$.

Let $X = L_\infty$, let X_0 be a finite dimensional subspace of X, say, $1 \le m = \dim X_0 < \infty$, and let $w = (w_1, \ldots, w_m)$ be a basis for X_0.

For $Nx = F(t) = h(t) + f(t, x(t))$, $x \in X$, and $x^* = Px$, thus $x^* = c_1 w_1 + \cdots + c_m w_m$ for some $c = (c_1, \ldots, c_m) \in R^m$, we have

$$\begin{aligned}
\langle QNx, x^* \rangle &= |G|^{-1} \int_G QF(t) x^*(t)\, dt \\
&= |G|^{-1} \int_G \left[\sum_{i=1}^{\mu} \bar{w}_i(t) \sum_{j=1}^{q} \bar{\gamma}_{ij} \int_G F(\alpha) Q_j(\alpha)\, d\alpha \right] x^*(t)\, dt \\
&= |G|^{-1} \int_G F(\alpha) \sum_{j=1}^{q} \left(\sum_{i=1}^{\mu} \bar{\gamma}_{ij} \int_G x^*(t) \bar{w}_i(t)\, dt \right) Q_j(\alpha)\, d\alpha.
\end{aligned}$$

Thus, we have

$$\begin{aligned}
\langle QNx, x^* \rangle &= |G|^{-1} \int_G F(t) X^*(t)\, dt, \\
X^*(t) &= C_1 Q_1(t) + \cdots + C_q Q_q(t), \\
C_j &= \sum_{i=1}^{\mu} \bar{\gamma}_{ij} \int_G x^*(t) \bar{w}_i(t)\, dt, \qquad j = 1, \ldots, q.
\end{aligned}$$

We have here a transformation $x^* \to X^*$, or $T^*\colon X_0 \to Y_0$, which is linear and bounded from the finite dimensional space X_0 into the finite dimensional space Y_0.

For instance, in the case of Q "orthogonal," that is, $\mu = q$, $\bar{w}_j = Q_j$, $\gamma_{ij} = \delta_{ij}$, $\bar{w}_1, \ldots, \bar{w}_\mu$ orthonormal in $L_2(G)$, then we have $X^*(t) = x^*(t)$ for all $t \in G$, that is, T^* is the identity map, and this occurs independently of P.

Note that in this situation

$$x^*(t) = \sum_1^m c_i w_i, \qquad x^*(t) = X^*(t) = \sum_1^\mu C_j \bar{w}_j, \qquad t \in G,$$

may be distinct developments of x^* in the two systems $(w_1, \ldots, w_m)$, an arbitrary basis for X_0, and $(\bar{w}_1, \ldots, \bar{w}_\mu)$, an orthonormal basis for Y_0 in $L_2(G)$. In this situation, if x^* has to describe X_0, then certainly we need $m \leq \mu$, $X_0 \subset Y_0$. Let $\gamma_0' \leq \gamma_0$, $\Gamma_0' \leq \Gamma_0$ be positive constants such that as usual

$$\gamma_0' |c| \leq \|x^*\|_\infty \leq \gamma_0 |c|, \qquad \Gamma_0' |C| \leq \|X^*\|_\infty \leq \Gamma_0 |C|, \tag{9}$$

where $c = (c_1, \ldots, c_m) \in R^m$, $C = (C_1, \ldots, C_q) \in R^q$. Furthermore we assume that there are positive constants $M' \leq M$ such that

$$M' \|x^*\|_\infty \leq \|X^*\|_\infty = \|T^*(x^*)\|_\infty \leq M \|x^*\|_\infty . \tag{10}$$

For $c \neq 0$ we take as usual $b = c/|c|$, or $b = (b_1, \ldots, b_m)$, $b_i = c_i/|c|$, $|b| = 1$, and analogously, for $C \neq 0$ we take $B = C/|C|$, or $B = (B_1, \ldots, B_q)$, $B_j = C_j/|C|$, $|B| = 1$. We then write

$$v(t) = b_1 w_1 + \cdots + b_m w_m, \qquad V(t) = B_1 Q_1 + \cdots + B_q Q_q;$$
$$x^*(t) = |c| v(t), \qquad X^*(t) = |C| V(t), \qquad t \in G.$$

By definition the functions $V(t)$ have a common bound, say, $|V(t)| \leq V_0$ for all $t \in G$, $V(t) = B_1 Q_1 + \cdots + B_m Q_q$ and all $|B| = 1$.

As before we shall need the assumption

(P^*) There is a constant $\rho > 0$ such that $|G|^{-1} \int_G |V(t)|\, dt \geq \rho$ for all functions $V(t) = B_1 Q_1(t) + \cdots + B_q Q_q(t)$ with $|B| = 1$.

Let $\eta_1, \eta_2, \varepsilon$ be positive constants with $\rho\eta_1 - \eta_2 > M_0'^{-1} \Gamma_0 \varepsilon$, and assume that the function $f(t, s)$, continuous on cl $G \times R^1$, satisfies the conditions

(B_0) There is a constant $J_0 > 0$ such that $|f(t, s)| \leq J_0$ for all $(t, s) \in G \times R^1$.

(F_η) There is a constant R_0 such that $f(t, s) \leq -\eta_1$ for $t \in G$, $s \geq R_0$, and $f(t, s) \geq \eta_1$ for $t \in G$, $s \leq -R_0$.

Concerning $h(t)$, $t \in G$, $h \in L_\infty(G)$, let us assume that the following relation holds:

(H^{**}) $\left| |G|^{-1} \int_G h(t) V(t)\, dt \right| \leq \eta_2$ for all functions $V(t) = B_1 Q_1 + \cdots + B_q Q_q$ with $|B| = 1$.

Let $\lambda > 0$ be a constant such that $\rho\eta_1 - \eta_2 - 2\lambda \geq M'^{-1} \Gamma_0 \varepsilon$.

We now need the basic assumption that the functions $v(t)$ and $V(t)$ have "about the same signs" on G. More precisely, we assume that

(p^{**}) There are constants $\sigma_1 > 0$, $\sigma_2 > 0$ such that if in the set D^*, where for any $|b| = 1$ the function $v(t)$ and the corresponding function $V(t)$ have different signs, or $v(t)V(t) < 0$, then $|v(t)| \leq \sigma_1$, $|V(t)| \leq \sigma_2$ on D^*, and $|D^*| \leq \lambda\sigma_2^{-1}J_0^{-1}|G|$. Furthermore, if for any $\sigma > 0$ we denote by $D(\sigma)$ the set of all points $t \in G$ where $v(t)$ and $V(t)$ have the same signs (or $v(t)V(t) \geq 0$) and $|v(t)| \leq \sigma$, then $|D(\sigma)| \to 0$ as $\sigma \to 0+$ (that is, given $\delta > 0$, there is $\bar{\sigma} > 0$ such that $|D(\sigma)| \leq \delta$ for $\sigma \leq \bar{\sigma}$ and all b with $|b| = 1$).

For instance, in the case of Q "orthogonal," as mentioned above, that is, $\mu = q$, $\bar{w}_j = Q_j$, $\gamma_{ij} = \delta_{ij}$, $\bar{w}_1, \ldots, \bar{w}_\mu$ orthonormal in $L_2(G)$, then $X^* = x^*$, $V = v$, and (p^{**}) reduces to the sole requirement that σ can be chosen so that $|D(\sigma)| \to 0$ as $\sigma \to 0+$, a very light requirement indeed, and all this is independent of the operation P in the space X.

Let K be any given constant, and let R_0 be the constant in (F_η). Let $\bar{\sigma} > 0$ be a number such that for $0 < \sigma < \bar{\sigma}$ we have $|D(\sigma)| \leq \lambda J_0^{-1}V_0^{-1}|G|$, and let $R_1 \geq \gamma_0^{-1}\bar{\sigma}^{-1}(R_0 + K)$. Since $0 < \eta_1 \leq J_0$, we also have $|D^*| \leq \lambda\sigma_2^{-1}\eta_1^{-1}|G|$ and, for $0 < \sigma < \bar{\sigma}$, also $|D(\sigma)| \leq \lambda\eta_1^{-1}V_0^{-1}|G|$. Statement 4.4 is now an extension of 4.3:

4.4 Lemma Let $Y = L_\infty(G)$, let Y_0 be finite dimensional, and let Q be the general type projection defined by (8) (and satisfying (Π)). Under hypotheses (H^{**}), (B_0), (F_η), (P^*), (p^{**}), R_1 as above, and $\rho\eta_1 - \eta_2 > M_0'^{-1}\Gamma_0\varepsilon$, then $\langle QNx, x^*\rangle \leq -\varepsilon\|x^*\|_\infty$ for all $x \in X$, $x^* \in X_0$ with $Px = x^*$, $\|x^*\|_\infty \geq R_1$, $\|x - x^*\|_\infty \leq K$.

Proof Let $\sigma = (R_0 + K)|c|^{-1}$, and $G' = G - D^* - D(\sigma)$. Then

$$\begin{aligned}\langle QNx, x^*\rangle &= |G|^{-1}\int_G [h(t) + f(t, x(t))]X^*(t)\,dt,\\ &= |C|\,|G|^{-1}\left[\int_G h(t)V(t)\,dt\right.\\ &\quad\left.+\left(\int_{D^*} + \int_{D(\sigma)} + \int_{G'}\right) f(t, x(t))V(t)\,dt\right].\end{aligned}$$

For $t \in D^*$ we have $|f| \leq J_0$, $|V| \leq \sigma_2$, and

$$\left|\,|G|^{-1}\int_{D^*} f(t, x(t))V(t)\,dt\right| \leq |G|^{-1}|D^*|J_0\sigma_2 \leq \lambda.$$

For $t \in D(\sigma)$ we have $|f| \leq J_0$, $|v(t)| \leq \sigma = (R_0 + K)/|c| \leq$

$(R_0 + K)\gamma_0/R_1 \leq \bar{\sigma}$; hence $D(\sigma) \leq \lambda J_0^{-1} V_0^{-1} |G|$, and

$$\left| |G|^{-1} \int_{D(\sigma)} f(t, x(t))V(t)\, dt \right| \leq |G|^{-1} |D(\sigma)| J_0 V_0 \leq \lambda.$$

For $t \in G' = G - D^* - D(\sigma)$ we have $|v(t)| > \sigma = (R_0 + K)/|c|$, $|x^*(t)| = |c|\,|v(t)| \geq R_0 + K$; hence, either $v(t) > 0$, $x^*(t) \geq R_0 + K$, and $x(t) \geq R_0$ by the same argument used in the proof of Theorem 3.3, or $v(t) < 0$, $x^*(t) \leq -R_0 - K$, and $x(t) \leq -R_0$. Correspondingly, $f(t, x(t)) \leq -\eta_1$, or $f(t, x(t)) \geq \eta_1$. Since $V(t)$ has the same sign of $v(t)$ in G', we see that $f(t, x(t))V(t) \leq -\eta_1 |V(t)|$ for $t \in G'$, and

$$\begin{aligned}
|G|^{-1} \int_{G'} f(t, x(t))V(t)\, dt &\leq -\eta_1 |G|^{-1} \int_{G'} |V(t)|\, dt \\
&= -\eta_1 |G|^{-1} \left[\int_G |V(t)|\, dt \right. \\
&\qquad \left. + \eta_1 \left[\left(\int_{D^*} + \int_{D(\sigma)} \right) |V(t)|\, dt \right] \right. \\
&\leq -\eta_1 \rho + \eta_1 (\sigma_2 |D^*| + V_0 |D(\sigma)|) \\
&\leq -\eta_1 \rho + \lambda + \lambda = -\eta_1 \rho + 2\lambda.
\end{aligned}$$

Finally,

$$\begin{aligned}
\langle QNx, x^* \rangle &\leq -|C| [\rho\eta_1 - \eta_2 - 2\lambda] \\
&\leq -M[\rho\eta_1 - \eta_2 - 2\lambda] \|x^*\| \leq -\varepsilon \|x^*\|.
\end{aligned}$$

This proves Lemma 4.4.

5. Existence Theorems for the Scalar Case by Uniform Topology

Let G be a bounded domain in R^ν, $\nu \geq 1$, and let $X = Y = L_\infty(G)$. Let $|G|$ denote the measure of G, and let $\langle y, x \rangle$ be defined by $\langle y, x \rangle = |G|^{-1} \int_G y(t)x(t)\, dt$ for any $x \in X$ and $y \in Y$, so that $|\langle y, x \rangle| \leq \|y\|_\infty \|x\|_\infty$. Let us consider an equation of the usual form $Ex = Nx$, where we assume that ker E is a space of constants, and we take here

$$Px = |G|^{-1} \int_G x(t)\, dt, \qquad Qy = |G|^{-1} \int_G y(t)\, dt. \tag{11}$$

We understand here that for any $h(t)$, $t \in G$, $h \in L_\infty(G) = Y$ with mean value zero, or $\int_G h(t)\, dt = 0$, there exists one and only one solution $x \in \mathfrak{D}(E) \subset X$ of the homogeneous linear problem $Ex = h$, with $\int_G x(t)\, dt = 0$; in other words, if we take $X_0 = PX$, $X_1 = (I - P)X$, $Y_0 = QY$, $Y_1 = (I - Q)Y$, then there is an operator $H: Y_1 \to X_1$ such that for $h \in Y_1$, $x = Hh$ belongs to $\mathfrak{D}(E)$, and $Ex = h$. Equivalently, for any $h \in Y$, $x = H(I - Q)h$ belongs to X_1 and satisfies $Ex = (I - Q)h$. We assume here that $H: Y_1 \to X_1$ is a linear, bounded, compact operator satisfying (k_{123}). Let $f(t, s)$: cl $G \times R^1 \to R^1$ be a continuous real valued function, and for any $h \in Y$, let N denote the Nemitsky operator $Nx = h(t) + f(t, x(t))$, $t \in G$, $x \in X$. Then, certainly $N: X \to Y$. We take as usual $L = \|H\|$. The constants concerning P and Q are here $k_0 = 1$, $k' = 2$, $\chi = 1$, $\chi' = 2$. Here $m = \dim X_0 = 1$, $\mu = \dim Y_0 = 1$, and we take for X_0 and Y_0 the basis $w = \{w_1\}$ with $w = 1$, and thus $x^* = cw = c$ for $x^* \in X_0$, and $y_0 = dw = d$ for $y_0 \in Y_0$. Finally, $\langle y, w\rangle = |G|^{-1} \int_G y(t)\, dt = Qy$ for $y \in Y$, and $|\langle y, w\rangle| \leq \|y\|_\infty$, that is, $\mu = 1$.

A. *Existence at Resonance with P, Q Mean Values*

THEOREM I (*existence at resonance*) Let $X = Y = L_\infty(G)$; let P, Q be defined by (11); let E, H be as in Section 2, satisfying (k_{123}), with $X_0 = PX = \ker E$ nontrivial and finite dimensional, and H linear, bounded, and compact; and let $f(t, s)$ be a real valued continuous function on cl $G \times R^1$. Let us assume now that (B_0) there is a constant $J_0 > 0$ such that $|f(t, s)| \leq J_0$ for all $(t, s) \in G \times R^1$; and that (F_0) there is a constant $R_0 > 0$ such that $sf(t, s) \leq 0$ for all $t \in G$, $|s| \geq R_0$ [or $sf(t, s) \geq 0$ for $t \in G$, $|s| \geq R_0$]. Let $h(t)$, $t \in G$, be any element of Y with $\int_G h(t)\, dt = 0$. Then, the equation $Ex = h(t) + f(t, x(t))$, $t \in G$, has at least a solution $x \in \mathfrak{D}(E) \subset L_\infty(G)$.

Proof Since $h \in L_\infty$, we have $|h(t)| \leq J$ and $|h(t) + f(t, s)| \leq J_0 + J$ for some J and all $(t, s) \in G \times R^1$, $\|Nx\|_\infty \leq J + J_0$. Since $x^* = c$, we have

$$\langle QNx, x^*\rangle = |G|^{-1} \int_G [h(t) + f(t, x(t))]c\, dt = |G|^{-1} \int_G f(t, x(t))c\, dt,$$

and the remarks of Section 4.A apply. Thus (B_0) of Theorem 3.1 holds, and for $K > L\chi' J_0$, by force of Lemma 4.1, also (N_0) of Theorem 3.1 holds. Theorem I is now a corollary of Theorem 3.1.

We take into consideration now the case of f of limited growth. Here h is bounded on G and $f(t, s)$ is continuous on cl $G \times R^1$. We may consider, therefore, the monotone function $\phi(\zeta) = \mathrm{Sup}[\,|h(t) + f(t, s)|\,|\,t \in G, |s| \leq \zeta]$. A simple requirement on ϕ is that $\phi(s)/s \to 0$ as $s \to +\infty$. As pointed out in Section 3, it is convenient to require less.

Here $\gamma = \gamma' = 1$, $k_0 = 1$, $k' = 2$, $\chi = 1$, $\chi' = 2$, $\mu = 1$. By particularizing what we stated in Section 3. A, we consider here the constants $\sigma_1 = \frac{1}{4}$, $\sigma_2 = \frac{1}{2}$, $\sigma = \frac{3}{4}$, and we take $\lambda_0 = 1$, and $\lambda_1 = \min[(8L)^{-1}, 4^{-1}]$.

THEOREM I* (*existence at resonance*) Under the same general assumptions as in Theorem I, let $h(t)$, $t \in G$, be an element of $L_\infty(G)$ with $\int_G h(t)\,dt = 0$. Let $f(t, s)$ be continuous on cl $G \times R^1$, and let us assume that (F_0) $sf(t, s) \leq 0$ [or $sf(t, s) \geq 0$] for all $t \in G$, $|s| \geq R_0$. Let us assume further that there is a number $S \geq 4R_0$ with $\phi(S)/S \leq \min[(8L)^{-1}, 4^{-1}]$. Then the equation $Ex = h(t) + f(t, x(t))$, $t \in G$, has at least a solution $x \in \mathfrak{D}(E) \subset L_\infty(G)$, with $\|x\|_\infty \leq S$.

If $\phi(\zeta)/\zeta \to 0$ as $\zeta \to +0$, then the constant S above certainly can be determined.

Proof We shall apply the existence theorem at resonance that is stated in the remark following Theorem 3.4 in Section 3. We have here $\| Nx \|_\infty \leq \phi(\| x \|_\infty)$, and we know from Lemma 4.1 that $\langle QNx, x^* \rangle \leq 0$ for all $x \in X$, $x^* \in X_0$ with $\|x^*\|_\infty \geq R_0$, $\|x - x^*\|_\infty \leq \|x^*\|_\infty - R_0$. In other words, we have here $\phi_1(\zeta) = 0$, $\psi(\zeta) = \zeta - R_0$ for $\zeta \geq R_0$. Thus the existence follows from the inequalities stated in the remark. With $\chi' = 2$, $\sigma_1 = \frac{1}{4}$, $k_0 = 1$, $\gamma = \gamma' = 1$, $\mu = 1$, $\chi = 1$, $\sigma = \frac{3}{4}$, $\sigma_2 = \frac{1}{2}$, $\psi(\zeta) = \zeta - R_0$, those inequalities reduce to $S \geq 4R_0$, $\phi(S)/S \leq \min[(8L)^{-1}, 4^{-1}]$, and $2L\phi(4S) \leq 4S - R_0$. We have proved Theorem I*.

B. Existence across a Point of Resonance with P, Q Mean Values

THEOREM II (*existence across a point of resonance*) Under the same general assumptions as in Theorem I, let us assume that (B_0) there is a constant $J_0 > 0$ such that $|f(t, s)| \leq J_0$ for all $(t, s) \in G \times R^1$; and that there are constants $R_0 \geq 0$, $\eta_1 > 0$ such that $f(t, s) \leq -\eta_1$ for $t \in G$, $s \geq R_0$, and $f(t, s) \geq \eta_1$ for $t \in G$, $s \leq -R_0$. Let $h(t)$, $t \in G$, be any element of $L_\infty(G)$ with $||G^{-1}| \int_G h(t)\,dt| \leq \eta_2 < \eta_1$. Let $g(t, s)$ be any continuous real valued function on cl $G \times R^1$, and α a real parameter. Then there are constants $\alpha_0 > 0$ and $C > 0$ such that for every real α with $|\alpha| \leq \alpha_0$ the equation $Ex + \alpha g(t, x(t)) = h(t) + f(t, x(t))$ has at least a solution $x \in \mathfrak{D}(E) \subset L_1(G)$ with $\|x\| \leq C$.

This statement holds in the alternative case that $f(t, s) \geq \eta_1$ for $t \in G$, $s \geq R_0$, and $f(t, s) \leq -\eta_1$ for $t \in G$, $s \leq -R_0$.

Since $g(t, s)$ is continuous on cl $G \times R^1$, there is some monotone function $\omega(\zeta) \geq 0$ such that $|g(t, s)| \leq \omega(|s|)$ for all $(t, s) \in G \times R^1$. The numbers α_0, C of Theorem I* depend only on J_0, R_0, η_1, η_2, $\omega(\zeta)$, but not on h, f, g.

Proof First, $g(t, s)$ is continuous on cl $G \times R^1$; hence the operator A, defined by $Ax = g(t, x(t))$, maps X into Y, is continuous, and maps bounded sets into bounded sets. Second, the operator N, defined by $Nx = h(t) + f(t, x(t))$, also maps X into Y and is continuous. By force of (F_η) and Lemma 4.2, the operator N satisfies (N_η) of Theorem 3.2. Theorem II is now a consequence of Theorem 3.2.

THEOREM II* (*existence across a point of resonance*) Under the same general assumptions as in Theorem I, let $f(t, s)$ be continuous on cl $G \times R^1$, and let us assume that (F_η) there are constants $R_0 \geq 0$, $\eta_1 > 0$ such that $f(t, s) \leq -\eta_1$ for $t \in G, s \geq R_0$, and $f(t, s) \geq \eta_1$ for $t \in G, s \leq -R_0$. Let $h(t)$, $t \in G$, be any element of $L_\infty(G)$ with $||G|^{-1} \int_G h(t)\, dt| \leq \eta_2 < \eta_1$. Let $g(t, s)$ be any continuous function on cl $G \times R^1$, and α a real parameter. Let us further assume that for $\phi(\zeta) = \text{Sup}[|h(t) + f(t,s)| t \in G, |s| \leq \zeta]$ there is some number $S \geq 4R_0$ with $\phi(S)/S < \min[(8L)^{-1}, 4^{-1}]$ and $2L\phi(4S) + R_0 < 4S$. Then, there is a constant $\alpha_0 > 0$ such that, for any real $|\alpha| < \alpha_0$, the equation $Ex + \alpha g(t, x(t)) = h(t) + f(t, x(t))$, $t \in G$, has at least a solution $x(t)$, $t \in G$, $x \in \mathfrak{D}(E) \subset L_\infty(G)$ with $\|x\|_\infty \leq S$.

The same statement holds even if $f(t, s) \geq \eta_1$ for $t \in G$, $s \leq R_0$, and $f(t, s) \leq -\eta_1$ for $t \in G$, $s \leq -R_0$.

If $\phi(\zeta)/\zeta \to 0$ as $\zeta \to +\infty$, then a constant S as above can certainly be determined (by taking $S \geq 4R_0$ with $\phi(S)/S < \min[(8L)^{-1}, 4^{-1}]$ and $2L\phi(4S)/4S + R_0/4S < 1$.

Proof Note that for $Nx = h(t) + f(t, x(t))$, $t \in G$, we have $\|Nx\|_\infty \leq \phi(\|x\|_\infty)$ for every $x \in X = L_\infty(G)$. Take $\phi_1(\zeta) = \varepsilon\zeta$ for $\varepsilon = \eta_1 - \eta_2$, $0 \leq \zeta < +\infty$, and $\psi(\zeta) = \zeta - R_0$ for $R_0 \leq \zeta < +\infty$. For $x \in L_\infty(G)$, $x^* = Px = c$ with $|c| \geq R_0$, and $|x(t) - c| \leq |c| - R_0$, we have, either $c \geq R_0$ and $x(t) = c + (x(t) - c) \geq c - (c - R_0) = R_0$ and then $f(t, x(t)) \leq -\eta_1$; or $c \leq -R_0$ and

$$x(t) = c + (x(t) - c) \leq -|c| + (|c| - R_0) = -R_0$$

and then $f(t, x(t)) \geq \eta_1$. In other words, for $\|x^*\|_\infty \geq R_0$ and $|x(t) - x^*| \leq \|x^*\| - R_0$ we have $f(t, x(t))x^* \leq -\eta_1 \|x^*\|$. Correspondingly we have then $\langle QNx, x^* \rangle = |G|^{-1}[\int_G h(t)x^*\, dt + \int_G f(t, x(t))x^*\, dt] \leq -(\eta_1 - \eta_2)\|x^*\|$. In other words, for $\|x^*\|_\infty \geq R_0$ and $\|x - x^*\|_\infty \leq \psi(\|x^*\|_\infty)$ we have $\langle QNx, x^* \rangle \leq -\phi_1(\|x^*\|_\infty)$ with $\psi = \zeta - R_0$, $\zeta \geq R_0$, and $\phi_1 = \varepsilon\zeta$, $\zeta \geq 0$. Here $\gamma = \gamma' = 1$, $k_0 = 1$, $k' = 2$, $\chi = 1$, $\chi' = 2$, $\mu = 1$. We take, for instance, $\sigma_1 = \frac{1}{4}$, $\sigma_2 = \frac{1}{2}$, $\sigma = \frac{3}{4}$, so that $1 - \gamma\sigma = \frac{1}{4}$, $\sigma - \sigma_2 = \frac{1}{4}$, and we take $\lambda_0 = \max[1, (\gamma')^{-1}k_0] = 1$. Here $\min[(L\chi')^{-1}(1 - \gamma\sigma), (\mu\chi)^{-1} \times (\sigma - \sigma_2)]$ reduces to $\min[(8L)^{-1}, 4^{-1}]$ and we take λ_1 so that $\phi(S)/S \leq \lambda_1 < \min[(8L)^{-1}, 4^{-1}]$. Finally, relation $L\chi'\phi(\sigma_1^{-1}S) < \psi(k_0^{-1}\gamma'\sigma_1^{-1}S)$ reduces to $2L\phi(4S) < \psi(4S)$, or $2L\phi(4S) < 4S - R_0$. Theorem II* is now a corollary of Theorem (3.4).

Applications

The following ordinary differential equations have 2π-periodic solutions:

(a) $x'' + \arctan x = \sin t$,
(b) $x'' + \arctan x + 2^{-1} \sin x = \sin t$,

(c) $x'' + (1 - \cos t) \arctan x + x \exp(-x^2) = \sin t$,
(d) $x'' + \arctan(x + \sin t) + 2^{-1} \sin x = \sin t$,
(e) $x'' + |x|^{1/2} \arctan x = \sin t$,
(f) $x'' + |x|^{1/2}(1 - \cos t) \arctan x = \sin t$,
(g) $x'' + |x|^{1/2} \arctan(x + \sin t) = \sin t$,
(h) $x'' + (\operatorname{sgn} x)5^{-2}(\exp(10^{-1}|x|) - 1) = 5^{-2} \sin t$.

On equations (a)–(d), Theorem I applies. On equations (e)–(g), Theorem I* applies with $\phi(\zeta) = \gamma_0 + \gamma_1 \zeta^{1/2}$ for some constants γ_0, γ_1; hence $\phi(\zeta)/\zeta \to 0$ as $\zeta \to +\infty$. On equation (h) Theorem I* applies with $\phi(\zeta) = 5^{-2}(\exp(10^{-1}\zeta) - 1) + 5^{-2}$, $0 \le \zeta < +\infty$, $L = \pi^2$, $R_0 = 0$, $S = 10$. Indeed, $\phi(10)/10 = 5^{-2}10^{-1}e < 0.0109 < 0.0125 < 1/8\pi^2 = \min[(8L)^{-1}, 4^{-1}]$. Concerning the estimate $L = \pi^2$, we may note that if $h(t)$, $0 \le t \le 2\pi$, is bounded with mean value zero, then the unique primitive $u = Uh$ of h with mean value zero is given by $u(t) = \int_0^{2\pi} K(t, s)h(s)\, ds$ with $K(t, s) = (2\pi)^{-1}s$ for $0 \le s < t$, $K(t, s) = -1 + (2\pi)^{-1}s$ for $t \le s \le 2\pi$. Thus $\|U\| = \max_t \int_0^2 |K(t, s)|\, ds = \pi$. Then, $H = U^2$, and $\|H\| = L = \pi^2$.

The following ordinary differential equations have 2π-periodic solutions for any $|\eta| < 2$ and for any $|\alpha|$ sufficiently small:

(a′) $x'' + \alpha x^2 \sin t + \arctan x = \sin t + \eta \sin^2 t$,
(b′) $x'' + \alpha \exp(x^3) \sin t + 2 \arctan x + \sin x = \sin t + \eta \sin^2 t$,
(c′) $x'' + \alpha x^3 \sin t + (2 - \cos t) \arctan x + \exp(-x^2) = \sin t + \eta \sin^2 t$,
(d′) $x'' + \alpha x^3 \sin t + 2 \arctan(x + \sin t) + \sin x = \sin t + \eta \sin^2 t$,
(e′) $x'' + \alpha x^3 \sin t + |x|^{1/2} \arctan x = \sin t + \eta \sin^2 t$,
(f′) $x'' + \alpha x^3 \sin t + |x|^{1/2} \arctan(x + \sin t) = \sin t + \eta \sin^2 t$,
(g′) $x'' + \alpha x^3 \sin t + (\operatorname{sgn} x)5^{-2}(\exp(10^{-1}|x|) - 1) = 5^{-3} \sin t$.

Since $(2\pi)^{-1} \int_0^{2\pi} \sin^2 t\, dt = 2^{-1}$, we can take $\eta_2 = \eta/2$. In all these equations we can take for η_1 any number < 1 and close to 1 as we want. To equations (a′)–(d′), Theorem II applies. To equations (e′) and f′), Theorem II* applies with $\phi(\zeta) = \zeta^{1/2}$. To equation (g′) Theorem II* also applies, with $\phi(\zeta) = 5^{-2}(\exp(10^{-1}\zeta) - 1) + 5^{-3}$, $0 \le \zeta < +\infty$, R^0 any positive number as small as we want, and $S = 2.5$. Indeed, $\phi(2.5)(2.5)^{-1} = 5^{-2}(\exp(0.25) - 1)(2.5)^{-1} = (5^{-2}(0.2840) + 5^{-3})(2.5)^{-1} < 0.0078 < 0.0125 < (8\pi^2)^{-1}$. Also,

$$2L\phi(4S) = 2\pi^2(5^{-2}(e - 1) + 5^{-3}) < 1.515 < 10 = 4S,$$

and all inequalities of Theorems II* are satisfied.

6. Extensions to the Case of P and Q Orthogonal

Let G be a bounded domain in R^ν, $\nu \ge 1$, and let $X = Y = L_\infty(G)$. Let $|G|$ denote the measure of G, and let $\langle y, x\rangle$ be defined by $\langle y, x\rangle =$

$|G|^{-1} \int_G y(t)x(t)\,dt$ for any $x \in X$ and $y \in Y$, so that $|\langle y, x\rangle| \le \|y\|_\infty \|x\|_\infty$. Let us consider an equation of the usual form $Ex = Nx$, where we assume that ker E is finite dimensional, $1 \le m = \dim \ker E < \infty$. As in Section 3.B, let $X_0 = Y_0 = \ker E$, and let $w = (w_1, \ldots, w_m)$ be a basis for X_0 and Y_0 made up of elements of $L_\infty(G)$, which are orthonormal in $L_2(G)$. For P and Q orthogonal projection then

$$\begin{aligned} Px &= \sum_1^m c_i w_i, \quad c_i = \int_G x(t)w_i(t)\,dt, \quad i = 1, \ldots, m, \quad x \in X, \\ Qy &= \sum_1^m d_i w_i, \quad d_i = \int_G y(t)w_i(t)\,dt, \quad i = 1, \ldots, m, \quad y \in Y. \end{aligned} \tag{12}$$

Let $k_0, k_0', \chi_0, \chi_0', \mu_0$ be the norms of $P, I - P, Q, I - Q$, and the number μ of Section 2 in the uniform topologies of X and Y (and thus not necessarily equal to one). Let $0 < \gamma_0' \le \gamma_0 < \infty$ be constants so that $\gamma_0'|c| \le \|x^*\|_\infty = \|cw\|_\infty \le \gamma_0|c|$ for any $c \in R^m$. For $x^* \in X_0$, $x^*(t) = c_1 w_1 + \cdots + c_m w_m$ with $c = (c_1, \ldots, c_m) \ne 0$ we take $b = c/|c|$, or $b = (b_1, \ldots, b_m)$, $|b| = 1$, $b_i = c_i/|c|$, $i = 1, \ldots, m$, and $v(t) = b_1 w_1 + \cdots + b_m w_m$, $x^*(t) = |c| v(t)$, $t \in G$. We shall need assumption (p) of Section 3B:

(p) There is a constant $\rho > 0$ such that $|G|^{-1} \int_G |v(t)|\,dt \ge \rho$ for all functions $v(t) = b_1 w_1 + \cdots + b_m w_m$, $|b| = 1$.

Let $\eta_1, \eta_2, \varepsilon$ be positive constants with $\rho\eta_1 - \eta_2 > \varepsilon\gamma_0$.

THEOREM III (*existence at resonance*) Let $X = Y = L_\infty(G)$; let $X_0 = Y_0 = \ker E$ be finite dimensional; let P, Q be the projection operations defined by relations (12); let E, H be as in Section 2, satisfying (k_{123}), and H linear, bounded, and compact; and let $f(t, s)$ be a real valued continuous functions on cl $G \times R^1$. Let us assume that (B_0) there is a constant $J_0 > 0$ such that $|f(t, s)| \le J_0$ for all $(t, s) \in G \times R^1$; and that (F_0) there is a constant $R_0 \ge 0$ such that $sf(t, s) \le 0$ for $t \in G$, $|s| \ge R_0$ [or $sf(t, s) \ge 0$ for $t \in G$, $|s| \ge R_0$]. Let $h(t)$, $t \in G$, be any element of $L_\infty(G)$ with $\int_G h(t)w_i(t)\,dt = 0$, $i = 1, \ldots, m$. Then equation $Ex = h(t) + f(t, x(t))$, $t \in G$, has at least a solution $x \in \mathfrak{D}(E) \subset L_\infty(G)$.

Proof Since $h \in L_\infty$, we have $|h(t)| \le J$ and $|h(t) + f(t, s)| \le J_0 + J$ for some J and all $(t, s) \in G \times R^1$, $\|Nx\| \le J + J_0$. Since $x^* = \sum_1^m c_i w_i$, we have

$$\begin{aligned} \langle QNx, x^*\rangle &= |G|^{-1} \int_G [h(t) + f(t, x(t))]x^*(t)\,dt \\ &= |G|^{-1} \int f(t, x(t))x^*(t)\,dt, \end{aligned}$$

and the remarks of Section 3.B apply. The theorem is now a corollary of Theorem 3.1.

For f of slow growth we consider the monotone nondecreasing function $\phi(\zeta) = \text{Sup}[\,|h(t) + f(t, s)|\,|\,t \in G, |s| \le \zeta]$. A simple requirement on ϕ is that $\phi(s)/s \to 0$ as $s \to +\infty$. As pointed out in Section 3 it is convenient to require less.

Let $R_0 \ge 0$ denote the constant that will appear in condition (N_ϕ) below; let $\sigma_1, \sigma_2, \sigma$ be arbitrary constants, $0 < \sigma_1 < \sigma_2 < \sigma < \min[1, \gamma_0^{-1}]$, and take

$$\lambda_0 = \max[1, (\gamma_0')^{-1}], \qquad \lambda_1 = \min[(L\chi_0')^{-1}(1 - \gamma_0\sigma), (\mu_0\chi_0)^{-1}(\sigma - \sigma_2)].$$

The only requirement we shall need on the monotone function ϕ is that there be a number S satisfying $S \ge \sigma_1^{-1}\lambda_0 R_0$, $\phi(S)/S \le \lambda_1$. Thus, if $\phi(\zeta)/\zeta \to 0$ as $\zeta \to +\infty$, then certainly such a constant S can be determined.

As pointed out before, the case $\phi(\zeta) = J_0 + J_1\zeta^k$, $0 < k < 1$, as well as the case $\phi(\zeta) = J_0 + J_1\zeta^k$, $k > 1$, with J_1 sufficiently small, satisfies these requirements.

THEOREM III* (*existence at resonance*) Under the same general assumptions as in Theorem III, let us assume that (N_0) there is a constant $R_0 > 0$ such that $sf(t, s) \le 0$ [or $sf(t, s) \ge 0$] for all $t \in G$, $|s| \ge R_0$, and that for the monotone function $\phi(\zeta) \ge 0$ bounding $|h(t) + f(t, s)|$ above we can determine a constant $S \ge \sigma_1^{-1}\lambda_0 R_0$ satisfying $\phi(S)/S < \lambda_1$, and $L\chi_0'\phi(\sigma_1^{-1}S) + R_0 < k_0^{-1}\gamma_0'\sigma_1^{-1}S$. Let $h(t)$, $t \in G$ be any element of $L_\infty(G)$ with $\int_G h(t)w_i(t)\,dt = 0$, $i = 1, \ldots, m$. Then equation $Ex = h(t) + f(t, x(t))$, $t \in G$, has at least a solution $x \in \mathfrak{D}(E) \subset L_\infty(G)$ with $\|x\|_\infty \le S$.

If $\phi(\zeta)/\zeta \to 0$ as $\zeta \to +\infty$, then the constant S above can certainly be determined.

Proof The remarks in the proof of Theorem III still apply. Theorem III is now a consequence of the considerations of Section 3.B and of the existence theorem at resonance in the remark after Theorem 3.4 with $\phi_1 = 0$, $\psi = \zeta - R_0$.

THEOREM IV (*existence across a point of resonance*) Under the same general assumptions as in Theorem III, let us assume that (B_0) there is a constant $J_0 > 0$ such that $\|Nx\| \le J_0$ for all $x \in X$; and that (F_η) there are constants $R_0 > 0$, $\eta_1 > 0$ such that $f(t, s) \le -\eta_1$ for $t \in G$, $s \ge R_0$, and $f(t, s) \ge \eta_1$ for $t \in G$, $s \le -R_0$. Let $h(t)$, $t \in G$, be any element of $L_\infty(G)$ with $|G|^{-1}\int_G h(t)v(t)\,dt \le \eta_2 < \rho\eta_1$ for every function $v(t) = b_1w_1 + \cdots + b_mw_m$ with $|b| = 1$. Let $g(t, s)$ be any continuous function on $\text{cl}\, G \times R^1$, and α a real parameter. Then there are constants $\alpha_0 > 0$, $M > 0$ such that for any real α with $|\alpha| \le \alpha_0$ the equation $Ex + \alpha g(t, x(t)) = h(t) + f(t, x(t))$, $t \in G$, has at least a solution $x(t)$, $t \in G$, $x \in \mathfrak{D}(E) \subset L_\infty(G)$ with $\|x\|_\infty \le M$.

Theorem IV is a corollary of Theorem 3.2.

THEOREM IV* (*existence across a point of resonance*) Under the hypotheses of the present section, let $f(t, s)$ be continuous on cl $G \times R^1$, and let us assume that (F_ϕ) there are constants $R_0 \geq 0$, $\eta_1 > 0$ such that $f(t, s) \leq -\eta_1$ for $t \in G, s \geq R_0$, and $f(t, s) \geq \eta_1$ for $t \in G, s \leq -R_0$. Let $h(t)$, $t \in G$, be any element of L_∞ with $||G|^{-1} \int_G h(t)v(t)\, dt| \leq \eta_2 < \rho\eta_1$ for every function $v(t) = b_1 w_1 + \cdots + b_m w_m$ with $|b| = 1$. Let $g(t, s)$ be any continuous function on cl $G \times R^1$, and α a real parameter. Let us further assume that for the monotone function $\phi(\zeta) \geq 0$ bounding $|h(t) + f(t, s)|$ above we can determine a constant $S \geq \sigma_1^{-1}\lambda_0 R_0$ with $\phi(S)/S < \min[(L\chi_0')(1 - \gamma_0\sigma), (\mu_0\chi_0')^{-1}(\sigma - \sigma_2)]$ and $L\chi_0'\phi(\sigma_1^{-1}S) + R_0 \leq k_0{}^{-1}\gamma'\sigma_1{}^{-1}S$. Then there is a constant $\alpha_0 > 0$ such that, for any real $|\alpha| \leq \alpha_0$, the equation $Ex + \alpha g(t, x(t)) = h(t) + f(t, x(t))$, $t \in G$, has at least a solution $x(t)$, $t \in G$, $x \in \mathfrak{D}(E) \subset L_\infty(G)$ with $\|x\|_\infty \leq S$. The same statement holds even if $f(t, s) \geq \eta_1$ for $t \in G$, $s \geq R_0$, and $f(t, s) \leq -\eta_1$ for $t \in G$, $s \leq -R_0$.

If $\phi(\zeta)/\zeta \to 0$ as $\zeta \to +\infty$, then a constant S as above can certainly be determined.

Theorem IV is a corollary of Theorem 3.4 and of the considerations of Section 4.B.

Application

Let us consider the ordinary differential equation

$$x'' + m^2 x + f(t, x(t)) = h(t),$$

where m is an integer and we seek 2π-periodic solutions. If we assume that f and h satisfy the conditions of Theorem III and we take $Ex = x'' + m^2 x$, then $X_0 = \ker E = \text{sp}(\cos mt, \ \sin mt)$, $v(t) = b_1 \cos mt + b_2 \sin mt$ for $b_1{}^2 + b_2{}^2 = 1$, or $v(t) = \sin(mt - \xi)$, ξ arbitrary. Hence $\int_0^2 |v(t)|\, dt = 4$ for all $|b| = 1$, that is, we can take $\rho = 4$. Thus relation $\rho\eta_1 > \eta_2$ reduces to $4\eta_1 > \eta_2$. If we make use of the notations $A = \int_0^{2\pi} h(t) \cos mt\, dt$, $B = \int_0^{2\pi} h(t) \sin mt\, dt$, $C = -\eta_1$, $D = \eta_1$, then $D - C = 2\eta_1$, and

$$\left| \int_0^{2\pi} h(t)(b_1 \cos mt + b_2 \sin mt)\, dt \right| = |b_1 A + b_2 B|$$

$$\leq (A^2 + B^2)^{1/2} \leq \eta_2.$$

Thus, the inequality $4\eta_1 > \eta_2$ reduces to the Lazer and Leach relation $2(D - C) > (A^2 + B^2)^{1/2}$.

References

1. L. Cesari, Functional analysis and periodic solutions of nonlinear differential equations. *Contrib. Differential Equations* **1** (1963), 149–187.

2. L. Cesari, Functional analysis and Galerkin's method. *Michigan Math. J.* **11** (1964), 385–414.
3. L. Cesari, Alternative methods in nonlinear analysis. *In Internat. Conf. Differential Equations*, (H. Antosiewicz, ed.), pp. 95–148. Academic Press, New York, 1975.
4. L. Cesari, Nonlinear oscillations in the frame of alternative methods. *In Internat. Conf. Dynamical Systems*, (L. Cesari, J. Hale, and J. LaSalle, eds.), pp. 29–50. Academic Press, New York, 1976.
5. L. Cesari, Functional analysis and nonlinear differential equations. *In* "Functional Analysis and Nonlinear Differential Equations" (L. Cesari, R. Kannan, and J. Schuur, eds.). Dekker, New York, 1976, 1–199.
6. L. Cesari, Existence theorems across a point of resonance. *Bull. Amer. Math. Soc.* **82** (1976), 903–906.
7. L. Cesari, An abstract existence theorem across a point of resonance. *Internat. Symp. Dynamical Systems*, (A. R. Bednarek and L. Cesari eds.), Academic Press, New York 1976, pp. 11–26.
8. L. Cesari, Nonlinear oscillations across a point of resonance. *J. Differential Equations* (to appear).
9. L. Cesari and R. Kannan, An abstract existence theorem at resonance. *Proc. Amer. Math. Soc.* **53** (1977), 221–225.
10. S. H. Chang, Existence of periodic solutions to second order nonlinear equations. *J. Math. Anal. Appl.* **52** (1975), 255–259.
11. D. G. DeFigueiredo, On the range of nonlinear operators with linear asymptotes which are not invertible. *Univ. Brasilia Trab. Mat.* **59** (1974),
12. D. G. DeFigueiredo, The Dirichlet problem for nonlinear elliptic equations: A Hilbert space approach. *In* "Partial Differential Equations and Related Topics" (A. Dold and B. Eckham, eds.), Lecture Notes in Mathematics, Vol. 446, pp. 144–165. Springer Verlag, Berlin and New York, 1975.
13. S. Fučik, Further remarks on a theorem of E. M. Landesman and A. C. Lazer. *Comment. Math. Univ. Carolinae* **15** (1974), 259–271.
14. S. Fučik, M. Kučera, and J. Nečas, Ranges of nonlinear asymptotic operators (to appear).
15. J. K. Hale, Applications of alternative problems. Brown University, Providence, Rhode Island, Lect. Notes, 1971.
16. P. Hess, On a theorem by Landesman and Lazer. *Indiana Univ. Math. J.* **23** (1974), 827–829.
17. R. Kannan and J. Locker, On a class of nonlinear boundary value problems. *J. Differential Equations*, **26** (1977), 1–8.
18. R. Kannan and P. J. McKenna, An existence theorem by alternative method for semilinear abstract equations. *Boll. Un. Mat. Ital.* (5) **14** (1977), 355–358.
19. E. M. Landesman and A. C. Lazer, Nonlinear perturbations of linear elliptic boundary value problems at resonance. *J. Math. Mech.* **19** (1970), 609–623.
20. A. C. Lazer and D. E. Leach, Bounded perturbations of forced harmonic oscillations at resonance. *Ann. Mat. Pura Appl.* **72** (1969), 49–68.
21. J. Mawhin, Equivalence theorems for nonlinear operator equations and coincidence degree theory for some mappings in locally convex topological vector spaces. *J. Differential Equations* **12** (1972), 610–636.
22. J. Mawhin, Topology and nonlinear boundary value problems. *In Internat. Conf. Dynamical Systems*, (L. Cesari, J. Hale, and J. LaSalle, eds.), pp. 51–82. Academic Press, New York, 1976.
23. J. McKenna, Nonselfadjoint nonlinear problems in the alternative method, Ph.D. Thesis. University of Michigan, Ann Arbor, 1976.

24. J. Nečas, On the range of nonlinear operators with linear asymptotes which are not invertible. *Comment. Math. Univ. Carolinae* **14** (1973), 63–72.
25. H. Shaw, A nonlinear elliptic boundary value problem at resonance. "Functional Analysis and Nonlinear Differential Equations" (L. Cesari, R. Kannan, and J. Schuur, eds.), Dekker, New York, 1976, 339–346.
26. H. Shaw, A nonlinear elliptic boundary value problem at resonance. *J. Differential Equations* **26** (1977), 335–346.
27. S. A. Williams, A connection between the Cesari and Leray–Schauder methods. *Michigan Math. J.* **15** (1968), 441–448.
28. S. A. Williams, A sharp sufficient condition for solutions of a nonlinear elliptic boundary value problem. *J. Differential Equations* **8** (1970), 580–586.

This research was partially supported by AFOSR Research Project 71-2122 at the University of Michigan.

AMS (MOS) 1970 Subject Classification: 47H15, 34B15, 34C15, 35G30, 35J40.

Branching of Periodic Solutions of Nonautonomous Systems

Jane Cronin

Rutgers—The State University

Introduction

The problem of branching of periodic solutions of a nonautonomous system is the following: given the system

$$x' = f(x, t, \varepsilon), \qquad (E - \varepsilon)$$

where f has period T in t, then if there exist periodic solutions of period T of $(E - 0)$, to determine if for small nonzero ε, Eq. $(E - \varepsilon)$ has a solution of period T near some periodic solution (of period T) of $(E - 0)$. This problem arises in many mechanical and electrical studies (see, e.g., Minorsky [7]), and there is a large literature of long standing on the qualitative and quantitative aspects of the problem. We describe here a study of certain qualitative aspects, i.e., existence and stability of periodic solutions. The first purpose of the description is to give an account of new results obtained by using topological degree. The usual procedure in applying degree theory is to show that if the degree of a certain mapping M is nonzero, then if ε is sufficiently small, Eq. $(E - \varepsilon)$ has a periodic solution. Here we show in addition that if the degree of M is not defined or is zero and certain smoothness conditions are imposed, then Eq. $(E - \varepsilon)$ has a periodic solution if ε is small and positive or $(E - \varepsilon)$ has a periodic solution if $|\varepsilon|$ is small and ε is negative. (This is an extension of an earlier result [4].) The same technique is applicable to other branching or bifurcation problems.

We also give a stability study. Such studies have been made for the two-dimensional case by Lefschetz [5] and Cronin [3]. (Analytic aspects of the problem have been studied by Malkin [6].) Here we give a considerably shorter and more transparent description which includes the n-dimensional case. (The practical value of the results is still limited to applications in the case where the space of periodic solutions of the linear variational equation is less than or equal to 2.) For brevity, we omit the proof of a lemma; for self-containment and coherence, we include statements of standard results.

1. Existence of Periodic Solutions

We consider the n-dimensional equation

$$u' = f(u, t, \varepsilon), \tag{1}$$

where the n-vector function f has continuous third derivatives at each point

$$(u, t, \varepsilon) \in R^n \times R \times I,$$

where I is an open interval on the real line with midpoint zero, and f has period $T(\varepsilon)$ in t where $T(\varepsilon)$ is a positive-valued differentiable function. We assume that for $\varepsilon = 0$, Eq. (1) has a solution $u(t)$ of period $T(0)$ and we study the following problem.

Problem 1 If $|\varepsilon|$ is sufficiently small, does Eq. (1) have a solution $u(t, \varepsilon)$ of period $T(\varepsilon)$ such that for each real t

$$\lim_{\varepsilon \to 0} u(t, \varepsilon) = u(t)?$$

First the problem is simplified by introducing the variable

$$s = \left[\frac{T(0)}{T(\varepsilon)}\right] t$$

and letting

$$g(u, s, \varepsilon) = \frac{T(\varepsilon)}{T(0)} f\left[u, \frac{T(\varepsilon)}{T(0)} s, \varepsilon\right].$$

Note that the function $g(u, s, \varepsilon)$ has period $T(0)$ in s. Then Eq. (1) becomes

$$\frac{du}{ds} = g(u, s, \varepsilon) \tag{2}$$

and Problem 1 becomes

PROBLEM 2 If $|s|$ is sufficiently small, does Eq. (2) have a solution $u(s, \varepsilon)$ of period $T(0)$ such that for all $s \in [0, T(0)]$,

$$\lim_{\varepsilon \to 0} u(s, \varepsilon) = u(s)?$$

In order to stay with conventional notation we rewrite Eq. (2) as

$$\frac{du}{dt} = g(u, t, \varepsilon), \tag{3}$$

and we investigate Problem 2 by investigating solutions of Eq. (3) of the form

$$u(t, \varepsilon) = u(t) + \varepsilon x(t, \varepsilon). \tag{4}$$

Substituting from (4) into (3) and using a Taylor's expansion with a remainder, we obtain

$$\frac{dx}{dt} = g_u[u(t), t, 0]x + \varepsilon \mathcal{G}[x, t, \varepsilon] + g_\varepsilon[u(t), t, 0], \tag{5}$$

where the $n \times n$ matrix $g_u[u(t), t, 0]$ has period $T = T(0)$ in t, the function $\mathcal{G}$ has continuous first derivatives in all variables and has period T in t, and the function $g_\varepsilon[u(t), t, 0]$ is a differentiable function of t which has period T. By using Floquet theory (see, e.g., Lefschetz [5]) we may change variables so that Eq. (5) becomes

$$\frac{dx}{dt} = Ax + \varepsilon F(x, t, \varepsilon) + G(t), \tag{6}$$

where A is a constant matrix in real Jordan canonical form (see Coddington and Levinson [1, p.358]), the functions F and G have continuous first derivatives in all variables, and F and G have period T in t. Thus Problem 2 can be rephrased as

PROBLEM 3 If $|\varepsilon|$ is sufficiently small, does Eq. (6) have solutions of period T?

To solve Problem 3, we follow a conventional procedure. By a standard existence theorem, if c is a fixed real n-vector and if $|\varepsilon|$ is sufficiently small, there exists a solution $x(t, \varepsilon, c)$ of (6) defined on an open interval which contains $[0, T]$ and $x(0, \varepsilon, c) = c$. By the variation of constants formula, solving (6) for $x(t, \varepsilon, c)$ is equivalent to solving the following integral equation for $x(t, \varepsilon, c)$:

$$x(t, \varepsilon, c) = e^{tA}c + e^{tA} \int_0^t e^{-sA}\{\varepsilon F[x(s, \varepsilon, c), s, \varepsilon] + G(s)\}\, ds. \tag{7}$$

A necessary and sufficient condition that $x(t, \varepsilon, c)$ have period T is that

$$x(T, \varepsilon, c) - c = 0. \tag{8}$$

Substituting from (7) into (8), we obtain

$$(e^{TA} - I)c + e^{TA} \int_0^T e^{-sA}\{\varepsilon F[x(s, \varepsilon, c), s, \varepsilon] + G(s)\}\, ds = 0. \tag{9}$$

Thus to solve Problem 3, it is sufficient to solve Eq. (9) for c as a function of ε. If $\varepsilon = 0$, Eq. (9) becomes

$$(e^{TA} - I)c = -e^{TA} \int_0^T e^{-sA} G(s)\, ds.$$

Thus if the matrix $e^{TA} - I$ is nonsingular, then (9) has the initial solution

$$\varepsilon = 0, \qquad c_0 = -(e^{TA} - I)^{-1} e^{TA} \int_0^T e^{-sA} G(s)\, ds. \tag{10}$$

Also since $e^{TA} - I$ is nonsingular, the implicit function theorem can be applied to solve (9) uniquely for c as a function of ε in a neighborhood of the initial solution (10). That matrix $e^{TA} - I$ is nonsingular is equivalent to the condition that the equation

$$x' = Ax$$

has no nontrivial solutions of period $2n\pi/T$. Thus we obtain the classical result

THEOREM 1 If the equation

$$y' = \left\{\frac{T(\varepsilon)}{T(0)} f_u[u(t), t, 0]\right\} y$$

(i.e., the linear variational system of (1) relative to the given periodic solution $u(t)$) has no nontrivial solutions of period $2n\pi/T$, then there exist $\eta_1 > 0$, $\eta_2 > 0$ such that for each ε with $|\varepsilon| < \eta_1$ there is a unique $c = c(\varepsilon)$ such that

$$|c(\varepsilon) - c_0| < \eta_2$$

and the solution

$$u(t, \varepsilon) = u(t) + \varepsilon x(t, \varepsilon, c)$$

has period $T(\varepsilon)$.

Next we consider the degenerate or resonant case, i.e., we suppose that $(e^{TA} - I)$ is a singular matrix. Let E_{n-r} denote the null space of $e^{TA} - I$ and let E_r denote the complement in R^n of E_{n-r}. Let P_{n-r}, P_r denote the projections of R^n onto E_{n-r} and E_r, respectively. We assume that E_{n-r} is properly contained in R^n, i.e., we assume E_r is an r-dimensional space where $0 < r < n$. Later we point out the simplification that occurs if $E_{n-r} = R^n$. There exists a real nonsingular matrix H such that

$$H(e^{TA} - I) = P_r.$$

(Since A is in real Jordan canonical form, it is easy to compute H explicitly and later we will do so for certain cases.) Applying H to Eq. (9), we obtain

$$P_r c + He^{TA} \int_0^T e^{-sA}\{\varepsilon F[x(s, \varepsilon, c), s, \varepsilon] + G(s)\}\, ds = 0. \tag{11}$$

Applying P_{n-r} to (11) yields

$$P_{n-r} He^{TA} \int_0^T e^{-sA}\{\varepsilon F[x(s, \varepsilon, c), s, \varepsilon] + G(s)\}\, ds = 0. \tag{12}$$

Setting $\varepsilon = 0$ in (12), we obtain

THEOREM 2 A necessary condition that Problem 3 can be solved (i.e., Eq. (11) solved for c in terms of ε if $|\varepsilon|$ is small) is that

$$P_{n-r} He^{TA} \int_0^T e^{-sA} G(s)\, ds = 0.$$

In the remainder of this discussion, we assume that this condition is satisfied.

Now we proceed to solve (11) for c in terms of ε. Applying P_r and P_{n-r} to (11), we conclude first that this is equivalent to solving the equations

$$P_r c + P_r He^{TA} \int_0^T e^{-sA}\{\varepsilon F[x(s, \varepsilon, c), s, \varepsilon] + G(s)\}\, ds = 0, \tag{13}$$

$$P_{n-r} He^{TA} \int_0^T e^{-sA} \varepsilon F[x(s, \varepsilon, c), s, \varepsilon]\, ds = 0 \tag{14}$$

for $P_r c$ and $P_{n-r} c$ in terms of ε or, dividing (14) by ε, solving the equations

$$P_r c + P_r He^{TA} \int_0^T e^{-sA} \varepsilon F[x(s, \varepsilon, c), s, \varepsilon]\, ds$$

$$= -P_r He^{TA} \int_0^T e^{-sA} G(s)\, ds, \qquad (15)$$

$$P_{n-r} He^{TA} \int_0^T e^{-sA} F[x(s, \varepsilon, c), s, \varepsilon]\, ds = 0. \qquad (16)$$

Let $\bar{M}_\varepsilon$ denote the mapping from R^n into R^n defined by

$$\begin{aligned} \bar{M}_\varepsilon(c) &= M_\varepsilon(P_r c, P_{n-r} c) \\ &= (P_r c + P_r He^{TA} \int_0^T e^{-sA} \varepsilon F[x(s, \varepsilon, c), s, \varepsilon]\, ds, \\ &\qquad P_{n-r} He^{TA} \int_0^T e^{-sA} F[x(s, \varepsilon, c), s, \varepsilon]\, ds). \end{aligned}$$

We investigate the existence of solutions of (15) and (16) by studying the topological degree of $\bar{M}_\varepsilon$ at the point

$$p = \left(P_r He^{TA} \int_0^T e^{-sA} G(s)\, ds, 0 \right).$$

Since $\bar{M}_\varepsilon$ is not given very explicitly, the only possibility for actually computing the topological degree lies in studying $\bar{M}_0$ which can be determined quite explicitly. The mapping $\bar{M}_0$ is

$$\bar{M}_0 : (P_r c, P_{n-r} c) \to \left(P_r c, P_{n-r} He^{TA} \int_0^T e^{-sA} F[x(s, 0, c), s, 0]\, ds \right).$$

From the definition of topological degree it follows that if B^n is a ball of radius ρ and center p in R^n, then

$$\deg[\bar{M}_0, B^n, p] = \deg[M, B^n \cap E_{n-r}, 0]$$

in which M is the mapping from E_{n-r} into E_{n-r} defined by

$$M: (c_{r+1}, \ldots, c_n) \to P_{n-r} He^{TA} \int_0^T e^{-sA} F[x(s, 0, c), s, 0]\, ds, \qquad (17)$$

where $(c_{r+1}, \ldots, c_n)$ denotes a point of E_{n-r} and

$$c = \left(P_r He^{TA} \int_0^T e^{-sA} G(s)\, ds, c_{r+1}, \ldots, c_n \right).$$

Since $x(s, 0, c) = e^{sA}c + e^{sA} \int_0^s e^{-\sigma A}G(\sigma)\, d\sigma$, then Eq. (17) can be written as

$$M\colon (c_{r+1}, \ldots, c_n) \to P_{n-r}He^{TA} \int_0^T e^{-sA}F\left[e^{sA}c + e^{sA} \int_0^s e^{-\sigma A}G(\sigma)\, d\sigma, s, 0\right] ds. \tag{18}$$

(Note that in the totally degenerate case, i.e., if $E_{n-r} = R^n$, then P_{n-r} is the identity mapping on R^n and $\bar{M}_0 = M$.) Now since

$$H(e^{TA} - I) = P_r,$$

then

$$P_{n-r}He^{TA} = P_{n-r}H.$$

Matrix A is in real Jordon canonical form, i.e., matrix A consists of a sequence of boxes along the main diagonal, each box associated with an eigenvalue of A. To determine $P_{n-r}H$, we consider first the case in which A has just one eigenvalue which occurs in one box. If the eigenvalue is different from $2n\pi i/T$ ($n = 0, \pm 1, \pm 2, \ldots$), it is easy to show that $P_{n-r}H = 0$. If the eigenvalue is 0, then

$$P_{n-r}H = \begin{bmatrix} 0 & \cdots & 0 & 1 \\ & 0 & & \end{bmatrix}.$$

That is, all elements in $P_{n-r}H$ are zero except the entry in the upper right-hand corner which is one. If the eigenvalue is $2n\pi i/T$, where $n = \pm 1, \pm 2, \ldots$, then

$$P_{n-r}H = \begin{bmatrix} 0 & \cdots & 0 & 0 & 1 \\ 0 & \cdots & 0 & 1 & 0 \\ & & 0 & & \end{bmatrix}.$$

That is, the first two rows are as indicated and all the other entries are zero. Now, instead of trying to describe explicitly the right-hand side of (18) in the general case, we just describe a typical case. Suppose

$$A = \begin{bmatrix} [\ ] & & & & & & & \\ & [\ ] & & & & & & \\ & & \ddots & & & & & \\ & & & 0 & 1 & & & \\ & & & & \ddots & \ddots & & \\ & & & & & \ddots & 1 & \\ & & & & & & 0 & \\ & & & & & & & 0 & -\dfrac{2n\pi}{T} \\ & & & & & & & \dfrac{2n\pi}{T} & 0 \end{bmatrix},$$

where the eigenvalues associated with all boxes except the last two are different from $2n\pi i/T$ ($n = 0, \pm 1, \pm 2, \ldots$). Then the mapping M is given by

$$M: (c_{n-2}, c_{n-1}, c_n) \to \int_0^T \begin{bmatrix} F_{n-2}(e^{sA}c, s, 0) \\ \begin{bmatrix} 0 & 1 \\ 1 & 0 \end{bmatrix} \begin{bmatrix} \cos \frac{2m\pi}{T} s & \sin - \frac{2m\pi}{T} s \\ -\sin \frac{2m\pi}{T} & \cos \frac{2m\pi}{T} s \end{bmatrix} \begin{bmatrix} F_{n-1}(e^{sA}c, s, 0) \\ F_n(e^{sA}c, s, 0) \end{bmatrix} \end{bmatrix} ds,$$

where

$$c = (P_r He^{TA} \int_0^T e^{-sA}G(s)\, ds, c_{n-2}, c_{n-1}, c_n)$$

and F_{n-2}, F_{n-1}, F_n are the last three components of function F.

Applying the standard properties of topological degree, we have

THEOREM 3 If

$$\deg[M, B^n \cap E_{n-r}, 0] \neq 0,$$

then if $|\varepsilon|$ is sufficiently small, Eq. (6) has a solution $x(t, \varepsilon)$ of period T and $x(0, \varepsilon) = c$, where $c \in B^n$.

The number $\deg[M, B^n \cap E_{n-r}, 0]$ is a measure of the number of periodic solutions in the following sense: if $G(t)$ is varied arbitrarily slightly, then $\deg[M, B^n \cap E_{n-r}, 0]$ is a lower bound for the number of distinct periodic solutions and c depends continuously on ε. This result is a straightforward application of Sard's theorem (see Cronin [2,3]).

Next we investigate the cases in which $\deg[M, B^n \cap E_{n-r}, 0]$ is not defined or is zero. For simplicity of notation, we assume that $r = 0$ so that $\bar{M}_0 = M$ and $B^n \cap E_{n-r} = B^n$. As pointed out earlier, the mapping $\bar{M}_\varepsilon$ cannot be computed explicitly except at $\varepsilon = 0$. So if $\deg[M, B^n, 0]$ is not defined or is zero, we do not have recourse to studying $\bar{M}_\varepsilon$ with $\varepsilon \neq 0$. The only possible direction seems to be somehow to utilize the condition that $\deg[M, B^n, 0]$ is undefined or zero. That is the purpose of the next two theorems.

THEOREM 4 Suppose

$$\deg[M, B^n, 0]$$

is not defined, i.e., suppose there exists a point $q \in \partial B^n$ such that $M(q) = 0$. Assume that the following condition holds: there is a neighborhood N of q such that

(i) $M[(\partial B^n) \cap N]$ is the underlying point set of a smooth $(n-1)$-dimensional surface Σ;

(ii) the vector $(\partial M_\varepsilon/\partial\varepsilon)(q)$ is nonzero and is not tangent to Σ at $M(q) = 0$;

(iii) If N_0 is a spherical neighborhood of q, then there exists a neighborhood N_1 of q such that $N_1 \subset N_0$ and $0 \notin M(\partial N_1)$.

Conclusion: There is a positive number r_0 such that *either* for each $\varepsilon \in (0, r_0)$, Eq. (6) has a solution $x(t, \varepsilon)$ of period T with the same properties as the solution described in Theorem 3 *or* for each $\varepsilon \in (-r_0, 0)$, Eq. (6) has such a solution $x(t, \varepsilon)$ of period T.

Proof This theorem is a consequence of the following lemma.

LEMMA 1 If $M = \bar{M}_0$ satisfies the hypotheses in the statement of Theorem 4, then there exists a neighborhood N_2 of q and a positive number μ_1 such that if $\varepsilon \in (0, \mu_1)$, then

$$\deg[M_\varepsilon, \overline{N_2 \cap B^n}, 0]$$

is defined and constant; if $\varepsilon \in (-\mu_1, 0)$, then

$$\deg[\bar{M}_\varepsilon, \overline{N_2 \cap B^n}, 0]$$

is defined and constant; and if $\varepsilon \in (0, \mu_1)$ and $\varepsilon \in (-\mu_1, 0)$, then

$$\deg[\bar{M}_\varepsilon, \overline{N_2 \cap B^n}, 0] \neq \deg[M_\varepsilon, \overline{N_2 \cap B^n}, 0].$$

The proof of Lemma 1 is obtained by straightforward consideration of the intersection number definition of topological degree. The set B^n need not be a ball. It is sufficient that $\partial B^n \cap N$ be a smooth surface where N is some neighborhood of q.

THEOREM 5 Suppose

$$\deg[M, B^n \cap E_{n-r}, 0]$$

is zero and suppose that the following conditions hold:

(i) there exists a point $q \in \mathrm{Int}(B^n)$ such that $M(q) = 0$;

(ii) there is a smooth $(n-1)$-dimensional surface S in B^n such that $q \in S, M(S) = \bar{M}_0(S)$ is a smooth $(n-1)$-dimensional surface, and $(\partial M_\varepsilon/\partial\varepsilon)(q)$ is a nonzero vector which is not tangent to $\bar{M}_0(S)$ at 0;

(iii) If N_0 is a spherical neighborhood of q, then there exists a neighborhood N_1 of q such that $N_1 \in N_0$ and $0 \in M(\partial N_1)$.

Then the conclusion of Theorem 4 holds.

Proof This theorem is a consequence of the following lemma.

LEMMA 2 If $M = M_0$ satisfies the hypotheses in the statement of Theorem 5, then the conclusion of Lemma 1 holds.

Lemma 2 follows from Lemma 1 in this way: if N is a spherical neighborhood of q of sufficiently small radius, then

$$N - S = V_1 \cup V_2,$$

where V_1, V_2 are connected open sets and $N \cap S$ is contained in $(\partial V_1) \cap (\partial V_2)$. Then Lemma 1 can be applied with V_1 or V_2 as the set $B^n \cap N_2$.

Remark In many applications, the surface S is a subset of the set of critical points of M.

2. Stability of Periodic Solutions

We seek criteria for the asymptotic stability of the periodic solutions obtained in Section 1. Suppose that (c_0, ε_0) is a solution of the periodicity condition equation (Eq. (9)), i.e.,

$$\tilde{F}(c_0, \varepsilon_0) \stackrel{\text{def}}{=} (e^{TA} - I)c_0 + e^{TA} \int_0^T e^{-SA}\{\varepsilon F[x(s, e_0, c_0), s, \varepsilon_0] + G(s)\}\, ds = 0. \quad (19)$$

Our object is to study the stability properties of the periodic solution $x(t, c_0, \varepsilon_0)$, and our first step is to compare $x(t, c_0, \varepsilon_0)$ with another solution, say $x(t, c_1, \varepsilon_0)$, of (6) when $\varepsilon = \varepsilon_0$. Substituting $x(t, c_0, \varepsilon_0)$ and $x(t, c_1, \varepsilon_0)$ in (6) and subtracting we obtain

$$\begin{aligned} x'(t, c_1, \varepsilon_0) &- x'(t, c_0, \varepsilon_0) \\ &= A[x(t, c_1, \varepsilon_0) - x(t, c_0, \varepsilon 0)] \\ &\quad + \varepsilon_0 B(t, \varepsilon_0)[x(t, c_1, \varepsilon_0) - x(t, c_0, \varepsilon_0)] \\ &\quad + \varepsilon_0 H[x(t, c_1, \varepsilon_0) - x(t, c_0, \varepsilon_0)t, \varepsilon_0], \end{aligned} \quad (20)$$

where

$$|H(\xi, t, \varepsilon_0)| = o(|\xi|) \quad (21)$$

for all $t \in [0, T]$. From (21) and the fact that the matrix $B(t, \varepsilon_0)$ has period T, it follows that in order to study the asymptotic stability of $x(t, c_0, \varepsilon_0)$, it is sufficient to study the characteristic multipliers of $A + \varepsilon_0 B(t, \varepsilon_0)$.

Let $U(t)$ be a fundamental matrix of the equation

$$w' = [A + \varepsilon_0 B(t, \varepsilon_0)]w$$

such that $U(0) = I$, the identity matrix. The characteristic multipliers to be studied are the eigenvalues of $U(T)$.

If we rewrite Eq. (20) as

$$u'(t) = Au(t) + \varepsilon_0 B(t, \varepsilon_0)u(t) + \varepsilon_0 H[u(t), t, \varepsilon_0], \tag{22}$$

then by the variation of constants formula, we obtain

$$u(T) = U(T)u(0) + U(T)\int_0^T [U(s)]^{-1}\varepsilon_0 H[u(s), s, \varepsilon_0]\, ds. \tag{23}$$

If $u(0) = c_1 - c_0 = K$, a constant n-vector, then (23) may be written

$$u(T) = U(T)K + U(T)\int_0^T [U(s)]^{-1}\varepsilon_0 H[u(s, K), s, \varepsilon_0]\, ds, \tag{24}$$

where $u(t, K)$ denotes the solution of (22) such that

$$u(0, K) = K.$$

By the condition (21) on H, it follows that the differential at 0 of the mapping

$$\tilde{M}: K \to U(T)K + U(T)\int_0^T [U(s)]^{-1}\varepsilon_0 H[u(s, K), s, \varepsilon_0]\, ds$$

is the matrix $U(T)$. But we have also

$$\begin{aligned} u(T) &= x(T, c_1, \varepsilon_0) - x(T, c_0, \varepsilon_0) \\ &= e^{TA}c_1 + e^{TA}\int_0^T e^{-sA}\{\varepsilon_0 F[x(s, c_1, \varepsilon_0), s, \varepsilon_0] + G(s)\}\, ds \\ &\quad - e^{TA}c_0 - e^{TA}\int_0^T e^{-sA}\{\varepsilon_0 F[x(s, c_0, \varepsilon_0), s, \varepsilon_0] + G(s)\}\, ds \\ &= e^{TA}(c_1 - c_0) + e^{TA}\int_0^T e^{-sA}\{e_0 B(s, e_0)u(s, c_1 - c_0) \\ &\quad + \varepsilon_0 H[u(s, c_1 - c_0), s, \varepsilon_0]\}\, ds. \end{aligned} \tag{25}$$

From Eq. (19), it follows that

$$\begin{aligned} \tilde{F}(c_1, \varepsilon_0) - \tilde{F}(c_0, \varepsilon_0) = (e^{TA} - I)(c_1 - c_0) \\ + e^{TA}\int_0^T e^{-sA}\{\varepsilon_0 B(s, \varepsilon_0)u(s, c_1 - c_0) \\ + \varepsilon_0 H[u(s, c_1 - c_0), s, \varepsilon_0]\}\, ds. \end{aligned} \tag{26}$$

Equations (24)–(26) show that the differential at 0 of $\tilde{M}$, which is the matrix $U(T)$, equals the differential at c_0 of $\tilde{F}(c, \varepsilon_0)$ plus the identity map. That is, if $D_0\tilde{M}$ denotes the differential of $\tilde{M}$ at 0 and $D_{c_0}\tilde{F}$ denotes the differential of

$\tilde{F}(\cdot, \varepsilon_0)$ at c_0, then

$$U(T) = D_0 \tilde{M} = D_{c_0}\tilde{F} + I.$$

Thus if $w_1, \ldots, w_n$ are the eigenvalues of $U(T)$, then $w_1 - 1, \ldots, w_n - 1$ are the eigenvalues of $D_{c_0}\tilde{F}$. In order to show that $x(t, c_0, \varepsilon_0)$ is asymptotically stable, it is sufficient to show that all the characteristic multipliers have absolute value less than 1, i.e., that $|w_i| < 1$ for $i = 1, \ldots, n$.

Now consider the special case $n = 2$, $r = 0$. Since $r = 0$, then $e^{TA} - I = 0$. If

$$\deg[M, B^2, 0] \neq 0$$

and if the function $F(t, x, 0)$ in Eq. (6) is varied arbitrarily slightly as a function of t, then by application of Sard's Theorem, it follows that the Jacobian of M is nonzero at each solution $\bar{c}$ of

$$M(\bar{c}) = 0,$$

and hence that $JD_c\tilde{F}(c, 0) \neq 0$ at $c = \bar{c}$. Also if

$$\deg[M, B^2, 0] > 0 \; [< 0],$$

there is at least one solution $\bar{c}$ such that

$$JD_c\tilde{F} > 0 \; [< 0].$$

If $JD_{c_0}\tilde{F} \neq 0$ for $\varepsilon_0 = 0$, then from (26) we have

$$\tilde{F}(c_1, \varepsilon_0) - \tilde{F}(c_0, \varepsilon_0) = \varepsilon_0\{L + \tilde{L}(\varepsilon_0)\}(c_1 - c_0) + \varepsilon_0 R(c_1 - c_0),$$

where L is a constant nonsingular matrix and $\tilde{L}(\varepsilon_0)$ is a 2×2 matrix such that

$$\lim_{\varepsilon_0 \to 0} L(\varepsilon_0) = 0 \qquad \text{and} \qquad \lim_{|c| \to 0} |R(c)|/|c| = 0.$$

Also

$$JD_{c_0}\tilde{F} = (\operatorname{sgn} \varepsilon_0) \det[L + L(\varepsilon_0)].$$

But

$$\det[L + L(\varepsilon_0)] = \det[D\bar{M}_{\varepsilon_0}].$$

If ε_0 is sufficiently small, $\det[L + L(\varepsilon_0)]$ has the same sign as $\det(L)$. If the eigenvalues of L are positive and ε_0 is positive, then $x(t, c_0, \varepsilon_0)$ is unstable. If the eigenvalues of L are positive and ε_0 is negative and $|\varepsilon_0|$ is sufficiently small, then $x(t, c_0, \varepsilon_0)$ is asymptotically stable. If the eigenvalues of L are negative and ε_0 is positive, then $x(t, c_0, \varepsilon_0)$ is asymptotically stable. If the eigenvalues of L are negative and ε_0 is negative, then $x(t, c_0, \varepsilon_0)$ is unstable.

If the eigenvalues of L have opposite signs, then $x(t, c_0, \varepsilon_0)$ is unstable. If the eigenvalues of $D_{c_0}F$ are not real, a more delicate analysis must be made, i.e., the trace of $\varepsilon_0 L$ must be studied (see Cronin [3]).

Analogous results may be obtained for the case in which n is arbitrary and $n - r = 2$. No new ideas are involved, but the statements are more complicated.

References

1. E. A. Coddington and N. Levinson, "Theory of Ordinary Differential Equations." McGraw-Hill, New York, 1955.
2. J. Cronin, Families of solutions of a perturbation problem. *Proc. Amer. Math. Soc.* **12** (1961), 84–91.
3. J. Cronin, "Fixed Points and Topological Degree in Nonlinear Analysis" Math. Surveys Vol. **11**. American Mathematical Society, Providence, Rhode Island, 1964.
4. J. Cronin, One-sided bifurcation points. *J. Differential Equations* **9** (1971), 1–12.
5. S. Lefschetz, "Differential Equations: Geometric Theory," 2nd ed. Wiley (Interscience), New York, 1962.
6. I. G. Malkin, "Theory of Stability of Motion," transl. U.S. Atomic Energy Commission. AEC-tr-3352, Moscow, 1952.
7. N. Minorsky, "Nonlinear Oscillations." Van Nostrand-Reinhold, Princeton, New Jersey, 1962.

The work in this paper was supported by the U.S. Army Research Office (Durham) (Grant No. DAH-C04-75-G-0148).

AMS (MOS) 1970 Subject Classification: 34C15, 34C25, 34D20, 54C25.

Restricted Generic Bifurcation

Jack K. Hale

Brown University

Introduction

In the past few years, there has been considerable attention devoted to the existence of bifurcation for one-parameter families of mappings. Concurrent with this development has been the extensive theory of the universal unfolding of mappings or generic bifurcation for families of mappings which depend on a sufficiently large number of parameters. The purpose of this paper is to discuss methods for determining the nature of bifurcation when the family of mappings has $k \geq 1$ parameters but k is generally smaller than the number of parameters necessary to describe the universal unfolding.

1. Motivation and Statement of the Problems

Suppose X, Z are Banach spaces, $\Omega \subset X$ is an open set containing the zero element θ, $C^r(\Omega, Z)$, $r \geq 1$, is the space of functions from Ω to Z which have all Frechet derivatives of order $\leq r$ continuous and bounded. The topology for $C^r(\Omega, Z)$ is the usual C^r-topology.

DEFINITION 1.1 Let τ be a subset of $C^r(\Omega, Z)$ and suppose $S0 = 0$ for all $S \in \tau$. If $T \in \tau$ is given, then the point $(T, 0) \in C^r(\Omega, Z) \times X$ is said to be a *bifurcation point with respect to* τ if for any open neighborhood U of $(T, 0)$ there is an $(S, x) \in U$, $S \in \tau$, $x \neq 0$, such that $Sx = 0$.

Loosely speaking, $(T, 0)$ is a bifurcation point if in any neighborhood of $x = 0$ there is nonuniqueness of the solution 0 of the equation $Sx = 0$ for some member S of τ which is close to T. Rather than talk only about bifurcation, we consider the more general problem of the nature of the zeros of a function which is close to T. We will always assume $(T0) = 0$, but will allow members of the family τ to be nonzero at zero.

The following three problems are of primary importance.

PROBLEM I For a given family $\tau \subset C^r(\Omega, Z)$, show there exists a bifurcation at $(T, 0)$ with respect to τ.

PROBLEM II For a given neighborhood U of $(T, 0)$ and a given family $\tau \in C^r(\Omega, Z)$, characterize the number of zeros x of $S \in \tau$, $(S, x) \in U$.

PROBLEM III For a given neighborhood U of $(T, 0)$, characterize the number of zeros x of $(S, x) \in U$.

A large quantity of the literature on bifurcation theory is concerned with the existence of bifurcation. Also, the typical family of mappings τ is a one-parameter family of the form $\tau = \{I - \lambda B, \lambda \in R\}$ where B is a fixed operator. The existence of a bifurcation is basically a problem in fixed-point theory and one can use all of the existing fixed-point theorems, degree theory, monotone operator theory, and the Luisternik–Schnirelmann theory.

The literature on this subject is extensive. One can obtain a fairly good idea of the problems and methods by consulting references [4,6,15]. References [6,15] also contain very detailed computational procedures for the bifurcating solutions for one-parameter families.

If $T0 = 0$ and the derivative $T'(0)$ has a bounded inverse, then the implicit function theorem implies there are neighborhoods V of T and W of $0 \in X$ such that every $S \in V$ has a unique solution in W. Therefore, there cannot be a bifurcation under these hypotheses and, in fact, Problem III is solved for $U = V \times W$.

Let us, therefore, assume that $T'(0)$ does not have a bounded inverse. Problem III, the characterizations of the number of zeros x of S for (S, x) in a neighborhood of $(T, 0)$ is the basic problem in the theory of singularities of mappings. To understand well the basic difference between Problems II and III, it is beneficial to discuss in detail a few elementary aspects of Problem III.

If $P: X \to X$ is a continuous projection, we let X_p denote the range of P. Let $\mathfrak{N}(T'(0))$ be the null space of $T'(0)$, $\mathscr{R}(T'(0))$ be the range of $T'(0)$, and suppose $\mathfrak{N}(T'(0)) = X_U$, $\mathscr{R}(T'(0)) = Z_E$, where U and E are continuous projections. In a neighborhood of $(T, 0)$, the method of Liapunov–Schmidt allows one to reduce the discussion of the zeros of S to the discussion of a set of equations involving only an element of $\mathfrak{N}(T'(0))$ as the unknown variable. In fact, the equation $Sx = 0$ is equivalent to the pair of equations

$$ES(y + z) = 0, \qquad (I - E)S(y + z) = 0,$$

where $y \in X_U$, $z \in X_{I-U}$. The implicit function theorem implies there is a $z^*(y, S)$ such that $z^*(0, T) = 0$ and $ES(y + z^*(y, S)) = 0$. Therefore, one need only discuss the zeros of the *bifurcation equation*

$$(I - E)S(y + z^*(y, S)) = 0. \tag{1.1}$$

Of course, one never knows $z^*(y, S)$ exactly, but it can be obtained as accurately as desired by successive approximations.

For our general discussion, we will therefore be concerned only with Eq. (1.1) and the particular case where $\dim \mathfrak{N}(T'(0)) = p = \operatorname{codim} \mathscr{R}(T'(0))$ and, in appropriate basis vectors, Eq. (1.1) represents p equations in p unknowns. Therefore, Eq. (1.1) can be written as

$$f(u, S) = 0, \qquad f: R^p \times V \to R^p, \quad V \subset C^r(\Omega, Z). \tag{1.2}$$

A characterization of the number of zeros of (1.2) as a function of the mapping S will give a complete solution of Problem III. Such a characterization will be impossible unless some restrictions are imposed upon $f(u, T)$, the value of f at the original fixed map T.

If all functions are assumed sufficiently smooth, then

$$f(u, T) = H_q(u) + H_{q+1}(u) + \cdots,$$

where $H_q(u)$ represents homogeneous polynomials of degree q in the components of the vector u. For S close to T, the function $f(u, S)$ has the form

$$f(u, S) = \sum_k J_k(u, S),$$

where each $J_k(u, S)$ is homogeneous of degree k in the components of u with

$$J_k(u, T) = 0, \quad k < q; \qquad J_k(u, T) = H_q(u), \quad q \le k.$$

The theory of universal unfoldings of the singularity zero of the function f is concerned with the possibility of determining, by suitable transformations in the space of mappings as well as the u-space, a polynomial function

$$\bar{f}(\bar{u}, S) = \bar{J}_0(\bar{u}, S) + \bar{J}_1(\bar{u}, S) + \cdots + \bar{J}_q(\bar{u}, S) \tag{1.3}$$

such that the number of the zeros of $f(u, S)$ in (1.2) coincides with the number of zeros of $\bar{f}(\bar{u}, S)$ in (1.3). The function in (1.3) contains only a finite number of parameters $\lambda_1, \ldots, \lambda_r$ which, of course, are functions of S. The surfaces in the λ-space across which the number of zeros of $\bar{f}$ change are referred to as the *bifurcation surfaces* or *catastrophe sets*. The theory of universal unfoldings in this general setting was begun by René Thom in the 1950s and is a very active area of investigation at the present time. For gradient mappings, the theory is extensively developed.

For references on the theory as well as applications to physics, engineering, biology, and the social sciences, see Thom and Zeeman [12].

Similar ideas were encountered in elastic stability by Koiter, Buliansky, and Thompson and Hunt (see Thompson and Hunt [13] for references), and in algebraic geometry (see Walker [9]).

One can understand the ideas by looking at specific examples and applying the Weierstrass preparation theorem for analytic functions. Suppose $p = 1$; that is, the bifurcation equation (1.2) is a scalar equation in the scalar variable u. Let us also assume that all functions are analytic and

$$f(u, T) = cu^2 + \text{h.o.t.}, \qquad c \neq 0, \tag{1.4}$$

where h.o.t. designates higher order terms. The Weierstrass preparation theorem implies there is a polynomial $p_2(u, S) = u^2 + \mu_1 u + \mu_2$ and an analytic function $q(u, S)$ such that $q(0, S) \neq 0$, and

$$f(u, S) = p_2(u, S)q(u, S).$$

If we let $u = \bar{u} - \mu_1/2$, $\lambda = \mu_2 - {\mu_1}^2/4$, and

$$\bar{p}_2(u, S) = \bar{u}^2 + \lambda, \tag{1.5}$$

then the number of zeros of (1.4) coincides with the number of zeros of $\bar{p}_2 = 0$. The catastrophe set for this example is the set in λ space consisting of the point $\lambda = 0$. If $\lambda > 0$, there are no zeros, if $\lambda < 0$, there are two simple zeros, and if $\lambda = 0$, there is one double zero. Of course, λ is a function of S and this gives a complete description of the zeros of S. There are always two pictures of interest in these problems, the catastrophe set and the solution $\bar{u}$ as a function of the parameters (see Fig. 1).

As a second illustration, suppose $p = 1$ and

$$f(u, T) = cu^3 + \text{h.o.t.}, \qquad c \neq 0.$$

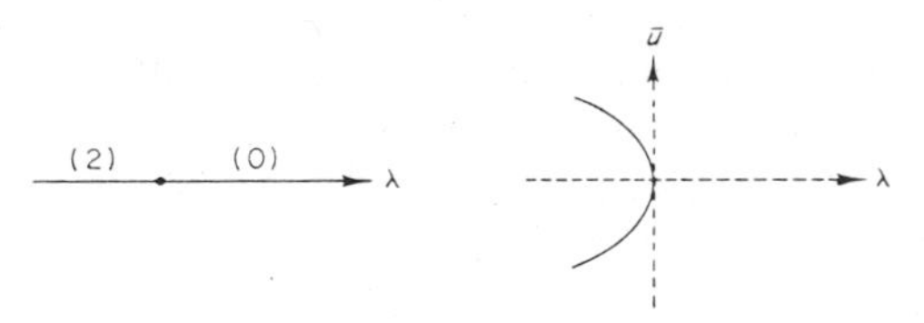

Figure 1

The Weierstrass preparation theorem implies

$$f(u, S) = p_3(u, S)q(u, S), \qquad q(0, S) \neq 0,$$

where $p_3(u, S) = u^3 + \mu_1 u^2 + \mu_2 u + \mu_3$ and q is analytic. Letting $u = \bar{u} - \mu_1/3$, one obtains a cubic polynomial

$$\bar{p}_3(\bar{u}, \lambda_1, \lambda_2) = \bar{u}^3 + \lambda_1 \bar{u} + \lambda_2$$

with the number of zeros of $\bar{p}_3(\bar{u}, \lambda_1(S), \lambda_2(S))$ coinciding with the number of zeros of $f(u, S)$. The catastrophe set is determined by the discriminant of $\bar{p}_3$ and the pictures of this set as well as the solution versus the parameters are shown in Fig. 2. To the left of the cusp catastrophe set there is one zero, to the right there are three, and on the cusp there are two. The light line in the solution diagram represents the locus of the double zero and the heavy lines represent the solutions for fixed $\lambda_2 > 0$.

The case in which $p = 1, f(u, T) = cu^k +$ h.o.t., $c \neq 0$, can be treated by the Weierstrass preparation theorem in a similar manner and the corresponding polynomial $\bar{p}_k$ involves $(k \dot{-} 1)$ parameters. The catastrophe sets are surfaces in R^{k-1} and naturally complicated (see Golubitsky and Guillemin [5]).

More interesting problems arise when $p = 2$; that is, f: $R^2 \times V \to R^2$. Let us also assume that our mappings represent the gradient of some potential function $V(u, S)$ and

$$f(u, T) = C(u) + \text{h.o.t.},$$

where C: $R^2 \to R^2$ is homogeneous and cubic in the components of the two vector $u = (u_1, u_2)$. The general theory of singularities of maps (see Wasserman [14]) implies (with few restrictions on C) that there is a potential

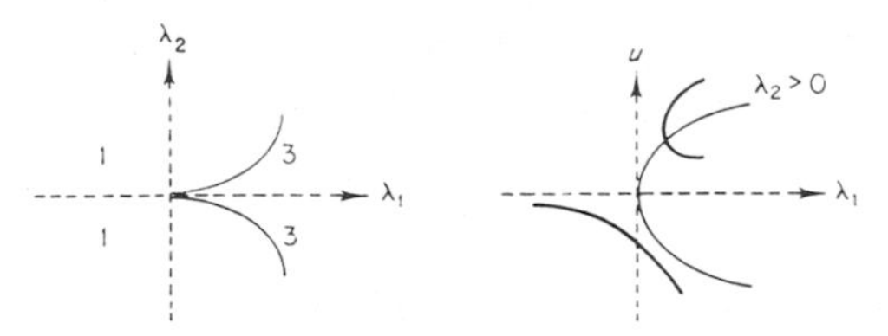

Figure 2

function $W(u, \lambda)$, depending on eight parameters $\lambda = (\lambda_1, \ldots, \lambda_8)$, such that λ is a function of S and the number of extreme points of $W(u, \lambda(S))$ coincides with the number of zeros of $f(u, S)$. The bifurcation surfaces or catastrophe sets are, therefore, surfaces in R^8.

As the preceding examples show, the catastrophe sets for problems for which the degeneracy is high (high degree polynomials for $p = 1$ or even low degree polynomials for $p > 1$) are in very high dimensional spaces. Extensive research on the nature of these surfaces has been and is continuing to be conducted.

It is not clear that one will ever obtain a description of the catastrophe surfaces which is sufficiently detailed to answer the questions posed by physical problems. In fact, an application contains some set of relevant parameters $\mu \in R^k$ for some value of k which may be small. One of the main challenges in applications is to explain phenomena qualitatively and reasonably accurately with as few parameters as possible. This implies that one must proceed in small steps. First one considers the most important parameters, say one or two, determines the effect of these parameters on the system, and then introduces more parameters to explain experimental evidence as it is obtained. Therefore, in the illustration above with $p = 2$ and two cubics, one does not begin the problem with eight independent parameters (independent as far as influencing the number of zeros), but with a number k much smaller than eight. Furthermore, the form of the k-parameter family is specified a priori. This means the interesting part of the general catastrophe set is its intersection with this particular k-parameter family. The theory presently available in catastrophe theory does not seem to provide too much assistance in the determination of this intersection. At the present time, each such problem involving k parameters with k small has been solved by ignoring the general catastrophe theory and concentrating directly on the special family.

From the preceding discussion, it should be clear that I believe Problem II is the most important one mentioned. More specifically, suppose $S\colon X \times R^k \to X$ is a C^r-family of mappings with $S(x, \lambda)$, $x \in X$, $\lambda \in R^k$, satisfying $S(0, 0) = 0$ and the derivative $S'(0, 0)$ of S with respect to x not having a bounded inverse. Characterize the number of zeros of $S(x, \lambda)$ for (x, λ) in some neighborhood of $(0, 0)$ and find the bifurcation or catastrophe surfaces in R^k across which the number of zeros change. We shall refer to this problem as the problem of *restricted generic bifurcation* or *restricted catastrophe theory*.

This restricted generic theory is in its infancy and has been concerned primarily with families of mappings whose form is motivated by specific applications, especially to plates and shells [1,2,7,8]. The remaining pages give a summary of some of the methods and results in this theory.

2. Restricted Generic Bifurcation, $p = 1$

In the previous section, it was indicated how the Weierstrass preparation theorem could be used to give a complete description of the number of zeros of analytic mappings S "near" a given map T when dim $\mathfrak{N}(T'(0)) = 1$. In this section, we also discuss the case when dim $\mathfrak{N}(T'(0)) = 1$, but consider only families τ of mappings in $C^r(\Omega, Z)$ with $T \in \tau$. The family τ will be taken to depend upon a finite number of parameters. The particular manner in which the family depends upon these parameters will be specified explicitly and is motivated by applications of the results to the buckling behavior of plates and shells. The choice of this particular application is not important since careful study of the ideas will indicate that similar procedures can be adopted to discuss other families of mappings.

The equations describing the steady-state solutions of plates and shells are partial differential equations of order 4. In an appropriate Banach space X which incorporates the boundary conditionss, the solutions of these partial differential equations are obtained as the solution of an operator equation of the form (see References [1],[2],[7],[8])

$$(I - \lambda A + \alpha^2 M)w + \alpha Q(w) + C(w) + \nu K = 0, \tag{2.1}$$

where K is a given element of X, A, M are bounded linear operators on X, Q is a quadratic form on X, and C is a cubic form on X. The symbols, λ, α, ν represent real numbers with α, ν small and λ close to some value to be specified later. In the plate problem, λ is a lateral loading and α, ν represent imperfections. In the shell problem, λ is lateral loading, ν is the imperfection, and α represents the curvature in the shell. Of course, the operators may depend upon other parameters which describe other physical parameters; for example, a parameter s which describes the relative sizes of the edges of a rectangular plate, a parameter t which describes a distortion of the rectangular plate to a parallelogram, etc.

The family of mappings to be considered is, therefore,

$$\tau = \{I - \lambda A + \alpha^2 M + \alpha Q + C + \nu K,\ \lambda, \alpha, \nu \in R\}. \tag{2.2}$$

The mapping T will be chosen as

$$T = I - \lambda_0 A + C, \tag{2.3}$$

where λ_0^{-1} is an eigenvalue of A. This choice is reasonable because we are interested in α, ν small, and we know there cannot be a bifurcation unless $T'(0)$ is singular.

In this section, we assume λ_0^{-1} is a simple eigenvalue of A. Applying the Liapunov–Schmidt procedure, the bifurcation equation is scalar and has the form

$$f(u, \alpha, \mu, \nu) \overset{\text{def}}{=} cu^3 + \alpha du^2 + (\mu e + \alpha^2 f)u + \nu k + \text{h.o.t.} = 0, \tag{2.4}$$

where c, d, e, f, k are constants, $\mu = \lambda - \lambda_0$, and h.o.t. designates terms of higher order of the form

$$\text{h.o.t.} \overset{\text{def}}{=} O(|u|^4 + \alpha^2 u^2 + |\alpha|^3 |u| + |\alpha| u^3 + \mu^2 |u| + |\mu| u^2 + \nu^2 + |\nu| |u|).$$

As mentioned in the previous section, if $c \neq 0$, the Weierstrass preparation theorem implies (in a neighborhood of $u = \alpha = \mu = \nu = 0$) there is a polynomial p of degree 3 and constants $\bar{\alpha}, \bar{\mu}, \bar{\nu}$ analytic in α, μ, ν such that

$$p(u) \overset{\text{def}}{=} u^3 + \bar{\alpha} u^2 + \bar{\mu} u + \bar{\nu} = 0 \tag{2.5}$$

implies u is a solution of (2.4) and conversely. Furthermore, the coefficients $\bar{\alpha}, \bar{\mu}, \bar{\nu}$ in p are computable in terms of α, μ, ν.

We now want to discuss the same bifurcation equation by proceeding in the spirit of Problem II as discussed in the previous section. More precisely, we will consider first a one-parameter family of mappings, then a two-parameter family.

This approach will also serve as motivation for the results in the next section where the eigenvalue λ_0^{-1} is no longer simple. Due to lack of space, we also only discuss special two-parameter families of mappings from the family (2.2).

Case 1 ($\alpha = \nu = 0, \tau = \{I - \lambda A + C, \lambda \in R\}$). The bifurcation equation is

$$f(u, \mu) = cu^3 + \mu e u + \text{h.o.t.} = 0, \tag{2.6}$$

where c, e are constants, $\mu = \lambda - \lambda_0$, and

$$\text{h.o.t.} = O(|u|^4 + \mu^2 |u| + |\mu| u^2).$$

If $c \neq 0$, one can easily obtain a priori bounds on the solutions of (2.6) in a neighborhood of $u = \mu = 0$:

$$|u| \leq \text{const}(|\mu|^{1/2}). \tag{2.7}$$

Inequality (2.7) suggests the scaling $u = |\mu|^{1/2} v$ for u in terms of μ. The equation for v for $\mu \neq 0$ is

$$g(v, \mu) \overset{\text{def}}{=} cv^3 \pm ev + O(|\mu|^{1/2}) = 0. \tag{2.8}$$

The + designates $\mu > 0$ and $-$ designates $\mu < 0$. If $e \neq 0$, the zeros of $g(v, 0)$ are simple. For each solution the implicit function theorem implies the existence of a unique solution for μ sufficiently small. If $v_j^*(\mu)$, $j = 1, 2, 3$, are these solutions, then $u_j^*(\mu) = \pm |\mu|^{1/2} v_j^*(\mu)$ are the solutions of the original equations. We may assume $u_1^*(\mu) = 0$ for all μ from the form of τ. The bifurcation diagram and solution diagram are shown in Fig. 3 for $ec < 0$.

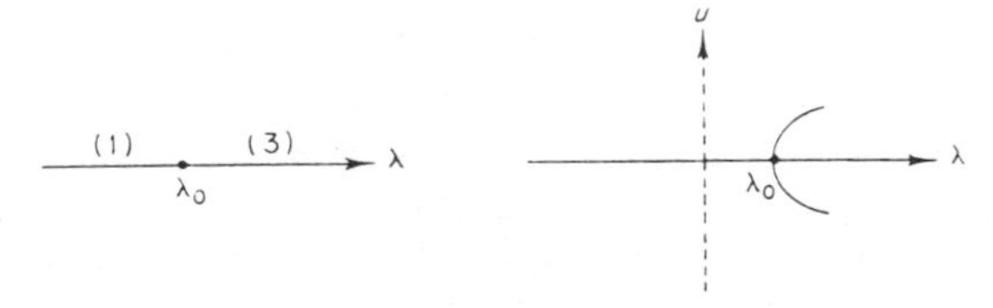

Figure 3

Summarizing the method, we see that the hypotheses were the following: (1) $c \neq 0$ to obtain a priori bounds, and (2) $e \neq 0$ to obtain simple zeros of the scaled equation.

Case 2 ($\alpha = 0$, $\tau = \{I - \lambda A + C + \nu K, \lambda, \nu \in R\}$). The bifurcation equation is

$$f(u, \mu, \nu) \overset{\text{def}}{=} cu^3 + \mu eu + \nu k + \text{h.o.t.} = 0, \tag{2.9}$$

where c, e, k are constants, $\mu = \lambda - \lambda_0$, and

$$\text{h.o.t.} = O(|u|^4 + \mu^2 |u| + |\mu| u^2 + \nu^2 + |\nu| |u|).$$

If $c \neq 0$, then the solutions of (2.9) have an a priori bound

$$|u| \leq \text{const}(|\mu|^{1/2} + |\nu|^{1/3}). \tag{2.10}$$

The bound (2.10) justifies two types of scalings:

$$u = |\mu|^{1/2} v, \qquad \nu = |\mu|^{3/2} n \tag{2.11}$$

and

$$u = \nu^{1/3} v, \qquad \mu = \nu^{2/3} m. \tag{2.12}$$

With the scaling (2.11), Eq. (2.9) becomes

$$cv^3 \pm ev + nk - g(v, n, \mu) = 0, \tag{2.13}$$

where $g = O(|\mu|^{1/2})$. Again, the + designates $\mu > 0$ and $-$ designates $\mu < 0$.

With the scaling (2.12), Eq. (2.9) becomes

$$cv^3 + mev + k = O(|\nu|^{1/3}). \tag{2.14}$$

We wish to exclude the possibility of having infinitely many solutions $v_j(\nu)$, $m_j(\nu)$ of (2.14) with $m_j(\nu) \to 0$ as $n \to \infty$. This will imply there cannot be infinitely many bifurcation curves $\mu = v^{2/3}m$ near the μ-axis in the (μ, v)-space. (One could equally well begin the process by excluding such curves from the μ-axis.) The hypothesis $k \neq 0$ will imply this fact since, in this case, the equation

$$cv^3 + k = 0$$

always has simple zeros (in fact, only one zero and it is simple). Therefore, the implicit function theorem implies there is a unique zero for m and v small. Consequently, no bifurcation can occur near the μ-axis.

If $c \neq 0$, $k \neq 0$, then no bifurcation occurs near the μ-axis. Thus there is a constant $\gamma > 0$ such that all bifurcation curves must lie in the region $|v| \leq |\mu|^{3/2}\gamma$. This justifies the consideration of only the scaling (2.11) for the determination of the bifurcation curves and we need only consider Eq. (2.13) for v as a function of μ, n.

On a bifurcation curve, we must have a double root of Eq. (2.13), that is, v satisfies (2.13) as well as the equation

$$3cv^2 \pm e = O(|\mu|^{1/2}). \tag{2.15}$$

The Jacobian of these two functions with respect to v and n for $\mu = 0$ is $-6ckv$. If we assume that $e \neq 0$, then (2.15) cannot have the solution $v = 0$ for $\mu = 0$. Therefore, the Jacobian is not zero at a solution of (2.13) and (2.15). Consequently, for any solution v_0, n_0 of Eqs. (2.13) and (2.15) for $\mu = 0$, there is a unique solution $v^*(\mu)$, $n^*(\mu)$, $v^*(0) = v_0$, $n^*(0) = n_0$, which satisfies (2.13) and (2.15) for μ sufficiently small. The curve $v = |\mu|^{1/2}n^*(\mu)$, $\mu = \lambda - \lambda_0$ is a possible candidate for a bifurcation.

One must check to see if the number of solutions changes by two as one crosses this curve. From (2.9) we see that along this curve, u can only be zero if $\mu = 0$. Also, $\partial^2 f/\partial u^2 = 6cu^*(\mu) + O(|\mu|)$. Since $u^*(\mu) = O(|\mu|^{1/2})$, it follows that $\partial^2 f/\partial u^2 \neq 0$ along this curve if $\mu \neq 0$ is sufficiently small. Also, $\partial f/\partial v = k + O(|\mu|) \neq 0$ along this curve. Consequently, f has a double zero along this curve, and as v changes from one side of the curve $v = |\mu|^{1/2}n^*(\mu)$, the number of zeros of f changes by two. The bifurcation diagram and solution diagram are shown in Fig. 4.

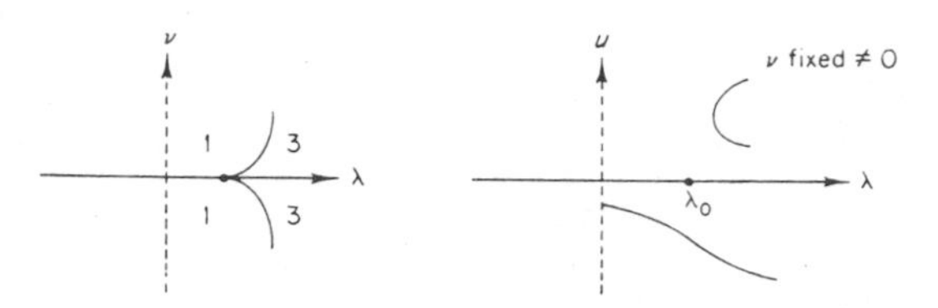

Figure 4

In summary, the preceding hypotheses and their implications are

(1) $c \neq 0$ implies a priori bounds on solutions as a function of λ, μ;
(2) $c \neq 0$, $k \neq 0$ imply no bifurcation curves near the μ-axis;
(3) $c \neq 0$, $k \neq 0$, $e \neq 0$ imply each multiple solution of the scaled equations (2.13) is double and yields the bifurcation curves in (λ, μ)-space.

Case 3 ($v = 0$, $\tau = \{I - \lambda A + \alpha^2 M + \alpha Q + C, \alpha, \lambda \in R\}$) The bifurcation equation is

$$f(u, \alpha, \mu) = cu^3 + \alpha du^2 + (\mu e + \alpha^2 f + \alpha^3 h + \mu^2 k)u + \text{h.o.t.} = 0, \tag{2.16}$$

where c, d, e, f, h, k are constants, $\mu = \lambda - \lambda_0$, and

$$\text{h.o.t.} = O(|u|^4 + \alpha^2 u^2 + |\alpha|\,|u|^3 + |\mu| u^2).$$

If $e \neq 0$, then the parameters

$$\alpha, \qquad \mu e + \alpha^2 f + \alpha^3 h + \mu^2 k \stackrel{\text{def}}{=} \beta$$

will play the same role as the parameters α, μ. Therefore, we consider the bifurcation equations

$$f(u, \alpha, \beta) = cu^3 + \alpha du^2 + \beta u + \text{h.o.t.} = 0. \tag{2.17}$$

If $c \neq 0$, the solutions of (2.17) have the a priori bound

$$|u| \leq \text{const}(|\alpha| + |\beta|^{1/2}). \tag{2.18}$$

As before, this bound (2.18) justifies the two scalings

$$u = |\beta|^{1/2} v, \qquad \alpha = |\beta|^{1/2} a \tag{2.19}$$

and

$$u = \alpha v, \qquad \beta = \alpha^2 b. \tag{2.20}$$

The corresponding scaled equations are

$$cv^3 + adv^2 \pm v = O(|\beta|^{1/2}) \tag{2.21}$$

and

$$cv^3 + dv^2 + bv = O(|\alpha|). \tag{2.22}$$

The assumptions $c \neq 0$ and the coefficients of v being $\neq 0$ exclude bifurcations near the α-axis in the (α, β)-plane. In fact, the zeros of the equation $cv^3 \pm v = 0$ are simple and so for (a, β) sufficiently small, there are unique solutions of (2.21), which means that along the curves $\alpha = |\beta|^{1/2} a$ for a small there are no bifurcations.

Therefore, if $c \neq 0$, it remains only to discuss Eq. (2.22) corresponding to the scaling (2.20) and $|b|$ bounded above by some constant. A multiple solution of (2.22) also must satisfy the equation

$$3cv^2 + 2dv + b = O(|\alpha|). \tag{2.23}$$

The Jacobian of (2.22) with respect to (v, b) is

$$\Delta_1 \stackrel{\text{def}}{=} -3cv^2 + b.$$

One verifies that $\Delta_1 \neq 0$ for all $(v, b) \neq (0, 0)$ which satisfy Eqs. (2.21) and (2.22) for $\alpha = 0$. It follows that for any $(v_0, b_0) \neq (0, 0)$ which satisfies (2.22) and (2.23) for $\alpha = 0$ there exists an $\alpha_0 > 0$, depending on (v_0, b_0), and $v^*(\alpha)$, $b^*(\alpha)$, defined for $|\alpha| < \alpha_0$, $v^*(0) = v_0$, $b^*(0) = b_0$ such that Eq. (2.22) has at least a double zero on the curve, $\beta = \alpha^2 b^*(\alpha)$ for $|\alpha| < \alpha_0$.

To show there is exactly a double zero on this curve, observe that $\partial^2 f/\partial u^2 = 6cu + O(\alpha^2)$ on this curve. If $b_0 \neq 0$, then $v_0 \neq 0$ and for $\alpha \neq 0$, but small, the zero $u^* = \alpha v^*(\alpha)$ is double. Also, $\partial f/\partial \beta = u + O(\alpha^2) \neq 0$ under the same hypothesis and all points on this curve are bifurcation points.

To discuss the other points on the curve, we must discuss the solutions of (2.22) and (2.23) near $v = 0$, $b = 0$. Near $v = 0$, the dominant terms become dv^2 if $d \neq 0$. Without the hypothesis $d \neq 0$, there could be infinitely many solutions of (2.22) and (2.23) near $v = 0$, $b = 0$. Observe first that the term $O(|\alpha|)$ is actually $O(|\alpha|\,|v|^2)$ as $|\alpha|$, $|v| \to 0$. If $d \neq 0$, then one can obtain a priori bounds on the solutions v of (2.22) near zero as

$$|v| \leq \text{const}(|b|).$$

If $v = b\bar{v}$, then Eq. (2.22) becomes

$$\pm d\bar{v}^2 + \bar{v} = O(|b|\,|v|) \tag{2.24}$$

as $|b| \to 0$.

If we assume $d \neq 0$, then Eq. (2.24) for $|b|$ sufficiently small has unique solutions near $v = 0$, $v = \pm d$. Consequently, no bifurcation occurs near $(v, b) = (0, 0)$.

Let us now summarize the preceding discussion. For the bifurcation function (2.17), the hypothesis $c \neq 0$ implies that one can obtain the a priori bound (2.18) on the solutions. The term βu in this function implies there can be no bifurcation near the α-axis in the (α, β)-plane. With the scaling (2.20), the zeros of the scaled equation (2.22) had to be discussed separately for v large and v small since the dominant term is v^3 for large v and dv^2 for small v if $d \neq 0$. The hypothesis $d \neq 0$ implies that there are simple zeros of (2.22) near $b = 0$ and, therefore, no possible bifurcation. The bifurcation curves were obtained from the double zeros (v, b) of (2.22) with the first approximations (v_0, b_0) obtained as solutions of these equations for $\alpha = 0$.

The approximate bifurcation curves and solutions on these curves are

$$u = \alpha v_0, \qquad \beta = \alpha^2 b_0$$

(see Fig. 5).

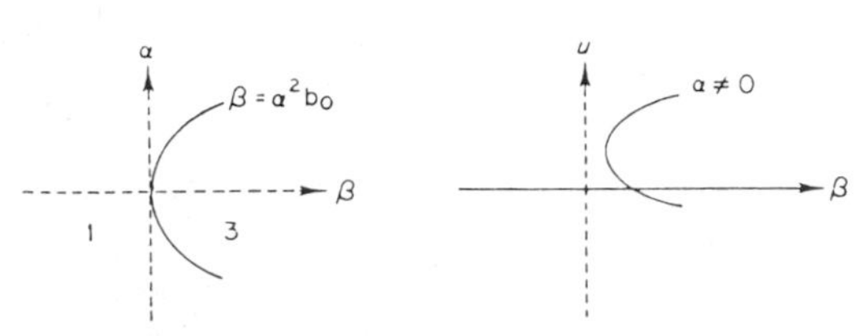

Figure 5

3. Restricted Generic Bifurcation, $p = 2$

In this section, we consider the same family of mappings (2.2) except under the hypotheses that λ_0^{-1} is an eigenvalue of the operator A of multiplicity 2. We also assume that

$$\mathfrak{N}(I - \lambda_0 A) \oplus \mathscr{R}(I - \lambda_0 A) = X.$$

An application of the Liapunov–Schmidt procedure yields the two bifurcation equations

$$c(u) + \alpha g(u) + \mu l^1(u) + s l^2(u) + t l^3(u) + v k + \text{h.o.t.} = 0, \tag{3.1}$$

where $u \in R^2$, c: $R^2 \to R^2$ is homogeneous cubic, q: $R^2 \to R^2$ is homogeneous quadratic, l^j: $R^2 \to R^2$ is linear, and $k \in R^2$ is constant. The parameters $\mu = \lambda - \lambda_0$, s, t, v, α are the ones described before.

Our objective in this section is to describe some results that have been obtained for some special cases of (3.1), using methods motivated by the discussion in Section 2.

Case 1 $(\alpha = s = t = 0)$ Consider the bifurcation equations

$$c(u) + \mu l(u) + v k + \text{h.o.t.} = 0, \tag{3.2}$$

where

$$\text{h.o.t.} = O(|u|^4 + |\mu|^2|u| + |\mu|\,|u|^2 + |v|\,|u| + |v|^2).$$

To discuss the complete bifurcation picture near $\mu = v = 0$, we follow Chow, Hale and Mallet-Paret [3] and make the hypotheses that generalize the ones in Section 2 for the scalar case.

For the scalar case, we required $c(u) = cu^3$ to have $c \neq 0$. The generalization of this hypothesis for the two-dimensional case is

(H_1) $c(u) = 0 \Rightarrow u = 0.$

Hypothesis (H_1) implies that any solution u of (3.2) for μ, ν small has an a priori bound

$$|u| \leq \mathrm{const}(|\mu|^{1/2} + |\nu|^{1/3}). \tag{3.3}$$

One can therefore make either the scaling

$$u = |\mu|^{1/2}v, \qquad \nu = |\mu|^{3/2}n \tag{3.4}$$

or the scaling

$$u = \nu^{1/3}v, \qquad \mu = \nu^{2/3}m. \tag{3.5}$$

With the scaling (3.4), Eqs. (3.2) become

$$h(v, n, \mu) \stackrel{\mathrm{def}}{=} c(v) \pm l(v) + nk = O(|\mu|^{1/3}) \tag{3.6}$$

and with the scaling (3.5), Eqs. (3.2) become

$$c(v) + ml(v) + k = O(|\nu|^{1/3}). \tag{3.7}$$

As for the scalar case, we wish to exclude the possibility of having infinitely many solutions $v_j(\nu)$, $m_j(\nu)$ of (3.7) with $m_j(\nu) \to 0$ as $j \to \infty$; that is, we wish to exclude the possibility of bifurcation curves being arbitrarily close to the μ-axis in the (μ, ν)-space. The generalization of the corresponding hypothesis for the scalar case is

(H_2) all zeros of $c(v) + k$ are simple.

This hypothesis and the implicit function theorem imply that there can be no double zero of (3.7) for m, ν small and, therefore, no bifurcation near the μ-axis.

It remains only to discuss the solutions of (3.6). The bifurcation curves in the (μ, ν)-space will be determined by the multiple solutions of (3.6). Let

$$\Delta(v, n, \mu) = \det[\partial h(v, n, \mu)/\partial v]. \tag{3.8}$$

The multiple solutions of (2.7) must satisfy

$$h(v, n, \mu) = 0, \qquad \Delta(v, n, \mu) = 0. \tag{3.9}$$

Our hypothesis for the scalar case was that all zeros (v_0, n_0) of (3.9) for $\mu = 0$ are simple, that is,

(H_3) $\Delta_1(v, \nu) \stackrel{\mathrm{def}}{=} \det[\partial(h, \Delta)/\partial(v, n)] \neq 0$ if $h(v, n, 0) = 0$, $\Delta(v, n, 0) = 0$.

If we make the same hypothesis for this more general case, then there is a

$\mu_0 > 0$ such that for any solution (v_0, n_0) of the equations

$$h(v_0, n_0, 0) = 0, \qquad \Delta(v_0, n_0, 0) = 0, \tag{3.10}$$

there is a unique solution $v^*(\mu)$, $n^*(\mu)$ of (3.9) for $|\mu| \leq \mu_0$, $v^*(0) = v_0$, $n^*(0) = n_0$. Also, one can show after more tedious calculations than those for the scalar case that the curve $v = |\mu|^{3/2} n^*(\mu)$ is a bifurcation curve and the number of zeros of (2.2) changes by exactly two as one crosses this curve.

The pictures of bifurcation curves and solution curves are the same as in Fig. 5 except for the coordinate $u \in R^2$.

Case 2 $(s = t = v = 0)$ Consider the bifurcation equations

$$g(u, \alpha, \mu) \stackrel{\text{def}}{=} c(u) + \alpha q(u) + \mu l(u) + \text{h.o.t.} = 0, \tag{3.11}$$

where

$$\text{h.o.t.} = O(|u|^4 + \alpha^2 |u|^2 + |\alpha|\,|u|^3 + |\alpha|^3 |u| + |\mu|\,|u|^2 + |\mu|^2 |u|).$$

For this case, we simply write down the hypotheses and leave it to the reader to make the connection with the scalar case.

(H_1') $c(u) = 0 \;\Rightarrow\; u = 0$.
(H_2') All zeros of $C(u) \pm l(u)$ are simple.
(H_3') If $G(u, b) \stackrel{\text{def}}{=} c(u) + q(u) + bl(u) = 0$,
$\Delta(u, b) \stackrel{\text{def}}{=} \det[\partial G(u, b)/\partial u] = 0$,

then

$$\Delta_1(u, b) \stackrel{\text{def}}{=} \det[\partial(G, \Delta)/\partial(u, b)] \neq 0 \qquad \text{unless} \quad u = 0, \quad b = 0.$$

(H_4') $q(u) = 0 \;\Rightarrow\; u = 0$.
(H_5') All zeros of $q(u) + l(u) = 0$ are simple.

Under these hypotheses, it is shown by Mallet-Paret [11] that each solution (u_0, b_0) of the equations

$$G(u_0, b_0) = 0, \qquad \Delta(u_0, b_0) = 0 \tag{3.12}$$

generates a bifurcation curve γ in (α, μ)-space given approximately by $\mu = \alpha^2 b_0$ and the solution u is given approximately by $u = \pm|\alpha| v_0$. Also, as the curve γ is crossed, the number of solutions changes by exactly two.

The case in which $\alpha = t = 0$ is discussed by List [10].

References

1. L. Bauer, E. Reiss, and H. Keller, Multiple eigenvalues lead to secondary bifurcation. *SIAM Rev.* **17** (1975), 101–122.
2. S. Chow, J. K. Hale, and J. Mallet-Paret, Application of generic bifurcation, I. *Arch. Rational Mech. Anal.* **59** (1975), 159–188.

3. S. Chow, J. K. Hale, and J. Mallet-Paret, Application of generic bifurcation, II. *Arch. Rational Mech. Anal.* **62** (1976), 209–235.
4. S. Fučík, J. Nečas, J. Souček, and V. Souček, "Spectral Analysis of Nonlinear Operators," Lecture Notes in Mathematics, Vol. 346. Springer-Verlag, Berlin and New York, 1973.
5. M. Golubitsky and V. Guillemin, "Stable Mappings and Their Singularities." Springer-Verlag, Berlin and New York, 1973.
6. J. B. Keller and S. Antman, "Bifurcation Theory and Nonlinear Eigenvalue Problems." Benjamin, New York, 1969.
7. G. H. Knightly, Some mathematical problems in plate and shell theory. "Nonlinear Functional Analysis and Differential Equations," pp. 245–268. Dekker, New York, 1976.
8. G. H. Knightly and D. Sather, Nonlinear buckled state of rectangular plates. *Arch. Rational Mech. Anal.* **54** (1974), 356–372.
9. R. J. Walker, "Algebraic Curves." Princeton Univ. Prèss, Princeton, New Jersey, 1950.
10. S. List, Generic bifurcation with application to the von Kármán equations. Ph.D. Thesis, Brown University, Providence, Rhode Island, 1976.
11. J. Mallet-Paret, Buckling of cylindrical plates with small curvature. *Quart. Appl. Math.* (to appear).
12. R. Thom and E. C. Zeeman, Bibliography on catastrophe theory. "Dynamical Systems—Warwick 1974." Lecture Notes in Mathematics, Vol. 468, pp. 390–401. Springer-Verlag, Berlin and New York, 1974.
13. J. M. T. Thompson and G. W. Hunt, "A General Theory of Elastic Stability." Wiley (Interscience), New York, 1973.
14. G. Wasserman, "Stability of Unfolding." Lecture Notes in Mathematics, Vol. 393. Springer-Verlag, Berlin and New York, 1974.
15. M. M. Vainberg and V. A. Trenogin, "Theory of Branching of Solutions of Nonlinear Equations." Noordhoff, The Netherlands, Leyden, 1974.

This research was supported in part by the Air Force Office of Scientific Research under AF-AFOSR 71-2078C, in part by the United States Army under AROD AAG 29-76-6-0052, and in part by the National Science Foundation under GP 28931X3.

AMS (MOS) 1970. Subject Classification: 47H15, 58C25.

On a Second-Order Nonlinear Elliptic Boundary Value Problem

Peter Hess

University of Zurich

To Professor Erich H. Rothe on the occasion of his 80th birthday.

1. Introduction and Statement of the Result

Let Ω denote a bounded domain in $\mathbb{R}^N$ $(N \geq 1)$ with smooth boundary Γ, and let $\mathscr{A}$ be a quasilinear elliptic differential operator of second order in divergence form,

$$(\mathscr{A}u)(x) = -\sum_{i=1}^{N} \frac{\partial}{\partial x_i} A_i(x, u(x), \nabla u(x)), \qquad x \in \Omega.$$

Let further $p\colon \Omega \times \mathbb{R} \times \mathbb{R}^N \to \mathbb{R}$ be a given function. To p we associate the Nemytskii operator P, defined by

$$(Pu)(x) = p(x, u(x), \nabla u(x)), \qquad x \in \Omega.$$

Finally let f and g be functions given on Ω and Γ, respectively. We are concerned with the solvability of the nonlinear Dirichlet problem

$$(D) \quad \mathscr{A}u + Pu = f \qquad \text{in } \Omega,$$
$$u = g \qquad \text{on } \Gamma,$$

provided we know the existence of a lower solution Φ and an upper solution ψ of (D) with $\Phi \leq \psi$ in Ω.

The following conditions of Leray–Lions type are imposed on the coefficient functions A_i $(i = 1, \ldots, N)$ of $\mathscr{A}$:

(A1) Each $A_i\colon \Omega \times \mathbb{R} \times \mathbb{R}^N \to \mathbb{R}$ satisfies the Caratheodory conditions (i.e., $A_i(x, t, \xi)$ is measurable in $x \in \Omega$ for all fixed $(t, \xi) \in \mathbb{R} \times \mathbb{R}^N$ and continuous in (t, ξ) for a.a. fixed x). There exist constants q: $1 < q < \infty$, $C_0 \geq 0$, and a function $k_0 \in L^{q'}(\Omega)$ $(q' = q/q - 1)$ such that

$$|A_i(x, t, \xi)| \leq k_0(x) + c_0(|t|^{q-1} + |\xi|^{q-1}), \qquad i = 1, \ldots, N,$$

for a.a. $x \in \Omega$, $\forall (t, \xi) \in \mathbb{R} \times \mathbb{R}^N$.

(A2) $\sum_{i=1}^N (A_i(x, t, \xi) - A_i(x, t, \xi^*))(\xi_i - \xi_i^*) > 0$ if $\xi \neq \xi^*$, for a.a. $x \in \Omega$, $\forall t \in \mathbb{R}$.

(A3) $\sum_{i=1}^N A_i(x, t, \xi)\xi_i \geq \alpha|\xi|^q$ $(\alpha > 0)$ for a.a. $x \in \Omega$, $\forall (t, \xi) \in \mathbb{R} \times \mathbb{R}^N$.

As a consequence of (A1), the semilinear form a,

$$a(u, v) = \sum_{i=1}^N \int_\Omega A_i(\cdot, u, \nabla u) \frac{\partial v}{\partial x_i}\, dx,$$

is defined on $W^{1,q}(\Omega) \times W^{1,q}(\Omega)$. Suppose further p satisfies the Caratheodory conditions, and

$$f \in W^{-1,q'}(\Omega), \qquad g \in W^{1-1/q,q}(\Gamma).$$

Definition 1 A function u is said to be a weak solution of problem (D) provided

$$u \in W^{1,q}(\Omega), \qquad u/\Gamma = g \quad \text{in} \quad W^{1-1/q,q}(\Gamma),$$
$$Pu \in L^1(\Omega),$$

and

$$a(u, v) + \int_\Omega Puv\, dx = (f, v) \qquad \forall v \in W_0^{1,q}(\Omega) \cap L^\infty(\Omega).$$

A natural extension of the classical concept of upper solution is given in

Definition 2 A function ψ is said to be a weak upper solution of problem (D) if

$$\psi \in W^{1,q}(\Omega), \qquad \psi/\Gamma \geq g \quad \text{in} \quad W^{1-1/q,q}(\Gamma),$$

$$P\psi \in L^1(\Omega),$$

and

$$a(\psi, v) + \int_\Omega P\psi v \, dx \geq (f, v) \qquad \forall v \in W_0^{1,q}(\Omega) \cap L^\infty(\Omega)$$

with $v \geq 0$ a.e. in Ω.

Similarly, a weak lower solution Φ is characterized by the reverse inequality signs in the above definition.

THEOREM Suppose Φ and ψ are weak lower and upper solutions of problem (D), respectively, with $\Phi, \psi \in L^\infty(\Omega)$, and such that $\Phi \leq \psi$ a.e. in Ω. Suppose further that there exist constants $c_1 \geq 0$, $\varepsilon > 0$ and a function $k_1 \in L^1(\Omega)$ such that

$$|p(x, t, \xi)| < k_1(x) + c_1 |\xi|^{q-\varepsilon} \tag{1}$$

for a.a. $x \in \Omega$, $\forall \xi \in \mathbb{R}^N$, $\forall t$: $\Phi(x) \leq t \leq \psi(x)$.

Then problem (D) admits a weak solution u with $\Phi \leq u \leq \psi$ a.e. in Ω.

This theorem generalizes all the classical results by Cohen, Keller, Shampine, Laetsch, Simpson, Sattinger, Amann, and others, which are proved by constructing a monotone iteration scheme (e.g., see Amann [1]). It also extends the main result of Deuel and Hess [3] in the case that $\Phi, \psi \in L^\infty(\Omega)$, as well as Theorem 2.1 of Puel [6] (where $\mathcal{A}$ is assumed to be linear).

In the proof of the theorem we roughly follow the method developed by Deuel and Hess [3]. The idea consists of first modifying the problem outside the interesting range of functions v: $\Phi \leq v \leq \psi$ a.e. in Ω. To the modified problem there is then associated a certain (coercive) variational inequality. Finally it is shown that any solution u of this variational inequality is a solution of problem (D) with $\Phi \leq u \leq \psi$ a.e. in Ω.

The suggestion to investigate this auxiliary variational inequality was given to us by H. Brézis.

For simplicity we restrict attention to the Dirichlet problem here. Similar results—with identical proofs—however hold also for other boundary conditions, such as mixed or nonlinear Neumann-type ones.

2. Proof of the Theorem

The proof is achieved in a number of steps.

(i) Let $\hat{g} \in W^{1,q}(\Omega)$ denote an extension of the function g to Ω, i.e., $\hat{g}/\Gamma = g$ in $W^{1-1/q,q}(\Gamma)$. We may choose $\hat{g}$ in such a way that $\Phi \leq \hat{g} \leq \psi$ a.e.

in Ω. Performing the change of variable $u \mapsto u - \hat{g}$, we reduce the problem to the case $\hat{g} = 0$ (possibly perturbing the right-hand side of (A3) by unessential lower order terms in $|\xi|$). *We thus assume in future that* $g = 0$ *and* $\Phi \le 0 \le \psi$ a.e. *in* Ω, and search for a solution in $V \equiv W_0^{1,q}(\Omega)$.

(ii) For $i = 1, \ldots, N$ let

$$\tilde{A}_i(x, t, \xi) = \begin{cases} A_i(x, \Phi(x), \xi), & t < \Phi(x), \\ A_i(x, t, \xi), & \Phi(x) \le t \le \psi(x), \\ A_i(x, \psi(x), \xi), & \psi(x) < t \end{cases}$$

for a.a. $x \in \Omega$, $\forall (t, \xi) \in \mathbb{R} \times \mathbb{R}^N$. The functions $\tilde{A}_i$ still satisfy conditions (A1)–(A3). Let

$$\tilde{a}(v, u, w) = \sum_{i=1}^{N} \int_\Omega \tilde{A}_i(\cdot, v, \nabla u) \frac{\partial w}{\partial x_i}\, dx \qquad \forall u, w \in W^{1,q}(\Omega), \quad \forall v \in L^q(\Omega).$$

(iii) Let the function $\gamma: \Omega \times \mathbb{R} \to \mathbb{R}$ be given by

$$\gamma(x, t) = \begin{cases} -(-t + \Phi(x))^{q-1}, & t < \Phi(x), \\ 0, & \Phi(x) \le t \le \psi(x), \\ (t - \psi(x))^{q-1}, & \psi(x) < t, \end{cases}$$

for a.a. $x \in \Omega$, $\forall t \in \mathbb{R}$. It is readily verified that γ satisfies the Caratheodory conditions. Moreover $\gamma(\cdot, v) \in L^{q'}(\Omega)$ $\forall v \in L^q(\Omega)$.

(iv) For $u \in W^{1,q}(\Omega)$ let Tu be the truncated function

$$(Tu)(x) = \begin{cases} \Phi(x), & u(x) < \Phi(x), \\ u(x), & \Phi(x) \le u(x) \le \psi(x), \\ \psi(x), & \psi(x) < u(x) \end{cases}$$

a.a. $x \in \Omega$. It is known that T is a bounded and continuous mapping from $W^{1,q}(\Omega)$ into itself. As a consequence of (1), $P \circ Tu \in L^1(\Omega)$ $\forall u \in W^{1,q}(\Omega)$.

(v) Finally, let $K \subset V$ denote the closed convex set

$$K = \{u \in V : \Phi - 1 \le u \le \psi + 1 \text{ a.e. in } \Omega\}$$

(note that K is different from the convex set considered by Puel [6]!).

LEMMA There exists $u \in K$ such that

$$\tilde{a}(u, u, w - u) + \int_\Omega \gamma(u)(w - u)\, dx + \int_\Omega (P \circ Tu)(w - u)\, dx$$
$$\ge (f, w - u) \qquad \forall w \in K. \tag{2}$$

We postpone the proof of the lemma for a moment and suppose we have $u \in K$ satisfying (2).

(vi) We claim that then necessarily $\Phi \le u \le \psi$ a.e. in Ω. In fact, introducing $w = \min(u, \psi)$ in (2) we get

$$\tilde{a}(u, u, (u-\psi)^+) + \int_\Omega \gamma(u)(u-\psi)^+ \, dx + \int_\Omega (P \circ Tu)(u-\psi)^+ \, dx$$
$$\leq (f, (u-\psi)^+)$$

(v^+ means the positive part of any function v on Ω). Since ψ is a weak upper solution,

$$\tilde{a}(\psi, \psi, (u-\psi)^+) + \int_\Omega P\psi(u-\psi)^+ \, dx \geq (f, (u-\psi)^+).$$

We infer that

$$\tilde{a}(u, u, (u-\psi)^+) - \tilde{a}(\psi, \psi, (u-\psi)^+)$$
$$+ \int_\Omega \gamma(u)(u-\psi)^+ \, dx + \int_\Omega (P \circ Tu - P\psi)(u-\psi)^+ \, dx \leq 0. \qquad (3)$$

As an immediate consequence of the definition of T,

$$\int_\Omega (P \circ Tu - P\psi)(u-\psi)^+ \, dx = 0.$$

Further, with the notation $\Omega_+ = \{x \in \Omega : u(x) > \psi(x)\}$,

$$\tilde{a}(u, u, (u-\psi)^+) - \tilde{a}(\psi, \psi, (u-\psi)^+)$$
$$= \int_\Omega \sum_{i=1}^N (\tilde{A}_i(\cdot, u, \nabla u) - A_i(\cdot, \psi, \nabla\psi)) \frac{\partial}{\partial x_i} (u-\psi)^+ \, dx$$
$$= \int_{\Omega_+} \sum_{i=1}^N (A_i(\cdot, \psi, \nabla u) - A_i(\cdot, \psi, \nabla\psi)) \frac{\partial}{\partial x_i} (u-\psi) \, dx$$
$$\geq 0 \qquad \text{by hypothesis (A2).}$$

We conclude from (3) that

$$\|(u-\psi)^+\|_{0,q}^q = \int_\Omega \gamma(u)(u-\psi)^+ \, dx \leq 0$$

(where $\| \ \|_{k,q}$ denotes the norm in $W^{k,q}(\Omega)$). Hence

$$u \leq \psi \qquad \text{a.e. in} \quad \Omega.$$

Similarly, one proves that $\Phi \leq u$ a.e. in Ω.

(vii) Since now $\tilde{A}_i(\cdot, u, \nabla u) = A_i(\cdot, u, \nabla u)$ $(i = 1, \ldots, N)$, $Tu = u$ and

$\gamma(\cdot, u) = 0$; relation (2) is reduced to

$$a(u, w - u) + \int_\Omega Pu(w - u)\, dx \geq (f, w - u) \qquad \forall w \in K. \tag{4}$$

Let $v \in V \cap L^\infty(\Omega)$ be arbitrarily given. Introducing $w = u \pm \delta v$ ($\delta > 0$ sufficiently small) in (4) we obtain

$$a(u, v) + \int_\Omega Puv\, dx = (f, v).$$

Thus u is a weak solution of problem (D) with $\Phi \leq u \leq \psi$ a.e. in Ω.

It remains to give the proof of the lemma. It is based on a fixed-point argument and employs a refined version of a method we found in Puel [6, Proof of Theorem 2.1].

Proof of the Lemma

(viii) We first show that given $f \in W^{-1,q'}(\Omega)$, $z \in L^1(\Omega)$, and $v \in L^q(\Omega)$, there exists a unique $u \in K$:

$$\tilde{a}(v, u, w - u) + \int_\Omega \gamma(u)(w - u)\, dx + \int_\Omega z(w - u)\, dx$$
$$\geq (f, w - u) \qquad \forall w \in K. \tag{5}$$

The existence of a solution u of (5) is well known if $z \in L^{q'}(\Omega)$ (e.g., see Lions [5, p.247]). Therefore let $(z_n)_{n=1}^\infty$ be a sequence of functions in $L^{q'}(\Omega)$ with $z_n \to z$ in $L^1(\Omega)$, and let $u_n \in K$ denote the corresponding solutions

$$\tilde{a}(v, u_n, w - u_n) + \int_\Omega \gamma(u_n)(w - u_n)\, dx + \int_\Omega z_n(w - u_n)\, dx$$
$$\geq (f, w - u_n) \qquad \forall w \in K. \tag{6}$$

Setting $w = 0$ in (6) we get

$$\tilde{a}(v, u_n, u_n) \leq d_1 + d_2 \|u_n\|_{1,q} \qquad \forall n$$

(d_k being constants). Thus by (A3), $\|u_n\|_{1,q} \leq \text{const}\ \forall n$, and we may pass to a subsequence such that

$$u_n \to u \qquad \text{weakly in } V, \quad \text{weak* in } L^\infty(\Omega).$$

Obviously $u \in K$.

Now setting $w = u$ in (6), we obtain

$$\limsup_{n\to\infty} \tilde{a}(v, u_n, u_n - u) \leq 0.$$

By a well-known result on mappings of type $(S)^+$ (e.g., see Browder [2, Appendix to Section 1]) it follows that

$$u_n \to u \qquad \text{strongly in} \quad V.$$

Passing to the limit in (6) we obtain (5).

The uniqueness of the solution of (5) is an immediate consequence of hypothesis (A2).

(ix) Next we define a mapping $S\colon K \to K$ as follows: to each element $v \in K$ we associate the unique solution $u = Sv$ of the variational inequality

$$\tilde{a}(v, Sv, w - Sv) + \int_\Omega \gamma(Sv)(w - Sv)\,dx + \int_\Omega (P \circ Tv)(w - Sv)\,dx$$
$$\geq (f, w - Sv) \qquad \forall w \in K. \tag{7}$$

Setting $K_\rho = K \cap B_\rho$, where $B_\rho = \{w \in V : \|w\|_{1,q} \leq \rho\}$, we note that there exists $\rho > 0$ such that $S\colon K_\rho \to K_\rho$. This follows readily from the inequality

$$\begin{aligned}\|Sv\|_{1,q}^q &\leq d_3\|Sv\|_{1,q} + d_4\|P \circ Tv\|_{L^1(\Omega)}\\ &\leq d_3\|Sv\|_{1,q} + d_5\|v\|_{1,q}^{q-\varepsilon} + d_6,\end{aligned}$$

which we obtain from (7) by setting $w = 0$.

(x) We now show that S is a compact mapping of K_ρ into itself with respect to the norm-topology in V (i.e., $S(K_\rho)$ is relatively compact in V and S is continuous).

In order to prove the relative compactness of $S(K_\rho)$ in V, let $(v_n)_{n=1}^\infty$ be a sequence in K_ρ. We may assume that for some $v \in K_\rho$

$$v_n \to v \begin{cases} \text{weakly in } V, \\ \text{weak* in } L^\infty(\Omega), \\ \text{strongly in } L^r(\Omega) \qquad \forall 1 \leq r < \infty. \end{cases}$$

Since $Sv_n \in K_\rho$ $\forall n$, there exists further $\chi \in K$ such that (for a subsequence)

$$Sv_n \to \chi \begin{cases} \text{weakly in } V, \\ \text{weak* in } L^\infty(\Omega), \\ \text{strongly in } L^r(\Omega) \qquad \forall 1 \leq r < \infty. \end{cases}$$

We have

$$\tilde{a}(v_n, Sv_n, Sv_n - \chi) + \int_\Omega \gamma(Sv_n)(Sv_n - \chi)\,dx + \int_\Omega (P \circ Tv_n)(Sv_n - \chi)\,dx$$
$$\leq (f, Sv_n - \chi) \qquad \forall n. \tag{8}$$

By hypothesis (1),

$$\left|\int_\Omega (P \circ Tv_n)(Sv_n - \chi)\,dx\right| \le \int_\Omega |k_1|\,|Sv_n - \chi|\,dx$$

$$+ c_1 \int_\Omega |\nabla(Tv_n)|^{q-\varepsilon}|Sv_n - \chi|\,dx. \qquad (9)$$

Since $|Sv_n - \chi| \to 0$ weak* in $L^\infty(\Omega)$, the first term on the right of (9) tends to 0 as $n \to \infty$. Moreover, $|\nabla(Tv_n)|^{q-\varepsilon}$ remains bounded in $L^{1+\eta}(\Omega)$, $\eta > 0$ suitable, while $Sv_n - \chi \to 0$ strongly in $(L^{1+\eta}(\Omega))^*$. Thus the second term on the right of (9) tends to 0 too, and we conclude from (8) that

$$\limsup_{n\to\infty} \tilde{a}(v_n, Sv_n, Sv_n - \chi) \le 0.$$

Hence $Sv_n \to \chi$ strongly in V (by the previously mentioned result on mappings of type $(S)^+$), which shows that $S(K_\rho)$ is relatively compact in V.

For proving the continuity of S, let $(v_n)_{n=1}^\infty$ be a sequence in K_ρ such that

$$v_n \to v \qquad \text{strongly in } V.$$

We note that then

$$P \circ Tv_n \to P \circ Tv \qquad \text{strongly in } L^1(\Omega).$$

(This follows from Lebesque's theorem on dominated convergence and the fact that there exist $L^q(\Omega)$-functions $\eta_0, \eta_1, \ldots, \eta_N$ such that, for suitable subsequences,

$$|Tv_n(x)| \le \eta_0(x),$$

$$\left|\frac{\partial(Tv_n)}{\partial x_i}(x)\right| \le \eta_i(x), \qquad i = 1, \ldots, N,$$

for a.a. $x \in \Omega$, $\forall n$.) Further we already know that

$$Sv_n \to \chi \qquad \text{strongly in } V, \quad \text{with } \chi \in K_\rho.$$

Passing to the limit in the variational inequalities that determine Sv_n, we obtain

$$\tilde{a}(v, \chi, w - \chi) + \int_\Omega \gamma(\chi)(w - \chi)\,dx + \int_\Omega (P \circ Tv)(w - \chi)\,dx$$

$$\ge (f, w - \chi) \qquad \forall w \in K.$$

Hence by uniqueness $\chi = Sv$, i.e.,

$$Sv_n \to Sv \qquad \text{strongly in } V.$$

This proves the continuity of S.

(xi) The Schauder fixed point theorem now guarantees the existence of a fixed point of S: there exists $u \in K_\rho$ such that $u = Su$. For this u we have (2).

Remark If we make the stronger assumption that $k_1 \in L^{q'}(\Omega)$ and $\varepsilon = 1$, the assertion of the theorem also holds in *unbounded* domains (Hess [4]).

References

1. H. Amann, On the existence of positive solutions of nonlinear elliptic boundary value problems. *Indiana Univ. Math. J.* **21** (1971), 125–146.
2. F. E. Browder, Existence theorems for nonlinear partial differential equations. *Proc. Symp. Pure Math.* **16**, 1–60. Amer. Math. Soc., Providence, Rhode Island, 1970.
3. J. Deuel and P. Hess, A criterion for the existence of solutions of nonlinear elliptic boundary value problems. *Proc. Roy. Soc. Edinburgh* **74A** (1974–1975), 49–54.
4. P. Hess, Nonlinear elliptic problems in unbounded domains. *Proc. Conf. Nonlinear Operators, Berlin, 1975* (to appear in *Abh. Akad. Wiss. DDR*).
5. J.-L. Lions, "Quelques Méthodes de Résolution des Problèmes aux Limites Non Linéaires." Dunod, Paris, 1969.
6. J. P. Puel, Existence, comportement à l'infini et stabilité dans certains problèmes quasilinéaires elliptiques et paraboliques d'ordre 2. *Ann. Scuola Norm. Sup. Pisa Sci. Fis. Mat.* **3** (1976), 89–119.

AMS (MOS) 1970 Subject Classification: 35B45, 35J60, 47H15.

Tikhonov Regularization and Nonlinear Problems at Resonance—Deterministic and Random

*R. Kannan**

Department of Mathematical Sciences
University of Missouri—St. Louis
St. Louis, Missouri

1. Introduction

This paper is motivated by the recent paper of Bakhushinskii and Apartsin [2] and Bakhushinskii [1], where the questions of stochastic approximation and regularization algorithms are considered for random linear operator equations. Our interest is in extending their studies to investigate the statistical properties of random solutions of nonlinear differential equations of the type

$$Lu + N(\omega, u) = f(\omega), \tag{1}$$

*Present address: Mathematics Department, The University of Texas, Arlington, Texas.

where

(i) $\omega \in \Omega$, $(\Omega, \mathscr{B}, \mu)$ being a probability space,

(ii) L is a linear self-adjoint differential operator over a real Hilbert space S such that L has a nontrivial null space,

(iii) $N: \Omega \times S \to S$ is a random nonlinear operator, and

(iv) $f(\omega)$ is a given random variable with values in S.

In earlier papers we considered the question of existence of random solutions and a naturally important role was played by the existence analysis of the deterministic analog of problem (1). However in order to study the statistical properties of random solutions of (1) we need further information on the methods of solving (1) and construction of approximate solutions. As an example of problems of type (1) we can consider the case of random nonlinear perturbations of harmonic oscillators at resonance:

$$x'' + m^2x + g(\omega, t, x) = p(\omega, t),$$
$$x(0) = x(2\pi), \qquad x'(0) = x'(2\pi).$$

In [1], Bakushinskii studies the linear analog of (1) (i.e., $N \equiv 0$) and applies the method of Tikhonov regularization to obtain an approximation scheme, and in [2], methods of stochastic approximation type are derived for solving improperly posed problems. In this paper we point out similar techniques for solving the deterministic analog of the nonlinear problem (1), thereby raising the question of obtaining results similar to those of Bakhushinskii for the random case.

In Section 2 we study the deterministic linear analog and point out the behavior of the solutions of the problems of the type $Lu - \lambda u = f$ as $\lambda \to 0$. We then state some results due to Krjanev [20] on methods of successive approximation for such problems.

An outline of the method used to establish the existence of solutions of the deterministic nonlinear analog of (1) is presented in Section 3. This method, referred to as the method of "alternative problems" in recent literature [6], has been studied extensively in recent years and applied to a wide variety of situations. For a detailed survey see Cesari [6].

After this outline we present in Section 4 two recent results on the question of existence of solutions of deterministic nonlinear problems of type (1) due to Hess [13], De Figureido [8], and Brézis [4]. These authors employ a "perturbation" technique similar to the linear case as in Section 2, except that in one case the perturbation is linear while in the other it is nonlinear.

In Section 5 we study the deterministic analog of (1), where N is maximal monotone and L is positive for methods of successive approximation. We discuss the proximal point algorithm as introduced by Moreau [24] (also see Rockafellar [29]) and state some results on the behavior of approximate solutions.

Section 6 is devoted to a brief outline of the question of the existence of random solutions to problem (1).

We finally consider the random linear problem of type (1). After presenting the Tikhonov regularization method that is closely related to Sections 4 and 5, we present the work of Khalfin and Sudakov [19] on a probabilistic approach to improperly posed problems (cf. Lavrent'ev [22]) and then we discuss the work of Bakhushinskii [1]. The similarity of these techniques for the linear case to those of the deterministic nonlinear analog of (1) will hopefully lead to further study of the statistical properties of random solutions of (1).

2. Linear Case

In this section we discuss the linear boundary value problem at resonance and state some results whose nonlinear analogs will be discussed in the later sections. Beforc we consider the linear boundary value problem we recall the notion of "condition number" for finite dimensional problems. Thus, let $Ax = b$ be a linear equation in R^n with nonsingular $A \in L(R^n)$, and $Bx = c$ be a perturbation of the linear equation where $B \in L(R^n)$ is close to A in the sense that $\|A^{-1}\| \, \|B - A\| < 1$. Then B is also nonsingular, and for $b \neq 0$ the solutions $x^* = A^{-1}b$ and $y^* = B^{-1}c$ satisfy the estimate

$$\frac{\|x^* - y^*\|}{\|x^*\|} \leq \frac{\alpha(A)}{1 - \alpha(A)\|B - A\|/\|A\|} \left[\frac{\|B - A\|}{\|A\|} + \frac{\|b - c\|}{\|b\|} \right],$$

where $\alpha(A) = \|A\| \, \|A^{-1}\|$. $\alpha(A)$ is called the condition number of A under the particular norm. It is an indicator of the sensitivity of the solution of $Ax = b$ under small changes of the matrix on the right-hand side. Rheinboldt presents a concept of condition number for nonlinear equations in [27].

Now let S be the real Hilbert space $L^2[a, b]$ with inner product $(\cdot)$ and norm $\|\cdot\|$ and let $H^n[a, b]$ be the subspace consisting of all functions $u(t)$ in $C^{n-1}[a, b]$ with $u^{(n-1)}$ absolutely continuous on $[a, b]$ and $u^{(n)}$ in S. Then $H^n[a, b]$ is a Banach space under the norm

$$|u|_n = \sum_{i=0}^{n-1} \max_{a \leq t \leq b} |u^{(i)}(t)| + \|u^{(n)}\|, \qquad u \in H^n[a, b].$$

The topology induced by this norm will be referred to as the H^n topology. Let $\tau = \sum_{i=0}^{n} a_i(t)(d/dt)^i$ be an nth order formal differential operator with coefficients $a_i(t)$ in $C^\infty[a, b]$ and $a_n(t) \neq 0$ on $[a, b]$, and let

$$B_i(u) = \sum_{j=0}^{n-1} \alpha_{ij} u^{(j)}(a) + \sum_{j=0}^{n-1} \beta_{ij} u^{(j)}(b), \qquad i = 1, \ldots, n,$$

be a set of linearly independent boundary values and let L be the differential operator in S defined by

$$\mathscr{D}(L) = \{u \in H^n[a, b] \mid B_i(u) = 0,\ i = 1, \ldots, n\}, \qquad Lu = \tau u.$$

We assume that L is self-adjoint and let ϕ_i, $i = 1, 2, \ldots$, be an orthonormal basis for S made up of eigenfunctions of L with corresponding eigenvalues λ_i, $i = 1, 2, \ldots$, so that $L\phi_i = \lambda_i \phi_i$ and $|\lambda_i| \to \infty$. For a fixed function $f \in S$ and a real number λ we consider the linear boundary value problem (BVP)

$$Lu - \lambda u = f.$$

If $L_\lambda = L - \lambda I$ and P_λ is the orthogonal projection of S onto the null space $\mathscr{N}(L_\lambda)$, then $(I - P_\lambda)$ is the orthogonal projection of S onto the range $\mathscr{N}(L_\lambda)^\perp$ and we know the following:

THEOREM 2.1 The BVP $Lu - \lambda u = f$ is solvable if and only if $P_\lambda f = 0$. Thus, if $\lambda \neq \lambda_i$, $i = 1, 2, \ldots$, then $Lu - \lambda u = f$ is uniquely solvable, while if $\lambda = \lambda_i$ for some $i \geq 1$, then $Lu - \lambda u = f$ has either no solutions or infinitely many solutions.

In [16] we obtained the following results:

THEOREM 2.2 The BVP $Lu - \lambda_0 u = f$ is solvable if and only if $\|u_\lambda\|$ remains bounded (where u_λ is the solution of $Lu - \lambda u = f$) as $\lambda \to \lambda_0$, in which case $P_{\lambda_0} u_\lambda = 0$ and further,

$$|u_\lambda - u_{\lambda_0}|_n \leq (2\gamma/\delta)\|f\| \, |\lambda - \lambda_0|,$$

where $\delta = \inf_{\lambda_i \neq \lambda_0} |\lambda_i - \lambda_0|$ and γ is a constant arising from the generalized inverse of $L - \lambda_0 I$.

THEOREM 2.3 If the BVP $Lu - \lambda u = f$ is not solvable for $\lambda = \lambda_0$, then $\|P_{\lambda_0} u_\lambda\| = |\lambda - \lambda_0|^{-1} \|P_{\lambda_0} f\| \to \infty$ as $\lambda \to \lambda_0$ and $(I - P_{\lambda_0}) u_\lambda = w_\lambda \xrightarrow{H^n} u_{\lambda_0}$ (the least squares solution of $Lu - \lambda_0 u = f$) with the error estimate

$$|w_\lambda - u_{\lambda_0}|_n \leq (2\gamma/\delta)\|f\| \, |\lambda - \lambda_0|$$

for all $|\lambda - \lambda_0| < \delta/2$.

Krjanev [20] considers the question of methods of successive approximation for incorrectly posed problems as in Theorem 2.3 and obtains the following results. Let us consider the equation $Ax = f$ where $A: H \to H$ is a self-adjoint nonnegative linear operator, H is a Hilbert space, and $f \in H$. Then we have

THEOREM 2.4 Let B: $H \to H$ be a self-adjoint positive definite linear operator and let $\rho(C)$ be the spectral radius of the operator $C = (A + B)^{-1}B$. Then $\rho(C) = 1$.

We now consider the iterative process

$$Ax_n + \varepsilon x_n = \varepsilon x_{n-1} + f,$$

where $x_0 \in H$ and $\varepsilon > 0$. This problem is equivalent to

$$x_n = (A + \varepsilon I)^{-1}\varepsilon x_{n-1} + (A + \varepsilon I)^{-1}f$$

(because $(A + \varepsilon I)^{-1}$ exists and is bounded). Clearly for the operator $C = (A + \varepsilon I)^{-1}\varepsilon I$, by Theorem 2.4, we have $\rho(C) = 1$. Then we obtain

THEOREM 2.5 For each $x_0 \in H$ and each $\varepsilon > 0$, the sequence $\{x_n\}$ defined above converges to a solution of $Ax = f$.

More generally one could consider the problem

$$Ax_n + Bx_n = Bx_{n-1} + f, \tag{1'}$$

where $x_0 \in H$ and B satisfies the hypotheses of Theorem 2.4. One can obtain the same conclusion as in Theorem 2.5. Clearly by virtue of the results mentioned earlier we obtain that in order that $\{x_n\}$ obtained from (1′) be convergent, it is necessary and sufficient that (1′) is solvable. There exists, of course, a unique solution x^* of $Ax = f$ with $Px^* = 0$ where P is the projection onto the null space of A, and this is the solution of minimum norm if $Ax = f$ is solvable. If $Ax = f$ has more than one solution, we denote by x_1 the element for which $\|x_1 - x_g\| = \min\|x - x_g\|$, where x_g is a given element and the minimum is taken over all solutions of $Ax = f$. We can then see that in order for x_n obtained from (1′) to converge to x_1 is is necessary and sufficient that $Px_0 = Px_g$.

In practice, A and f may not be known exactly. Krjanev [20] then obtains the following: Consider the sequence of equations given by

$$(A + \Delta A)x_n' + Bx_n' = Bx_{n-1}' + f + \Delta f, \tag{2}$$

where ΔA and Δf are perturbations of A and f, respectively. Let $\|\Delta f\| \leq \delta$, $\|\Delta A\| \leq M\delta$, $\|f\| \leq N$ and $\|B^{-1}\,\Delta A\| < 1$. Then there exists a constant N_0 such that $\|x_n\| \leq N_0$ where x_n is obtained from (1′). Let $q = \|(A + B)^{-1}B\|$.

THEOREM 2.6 Let x^* be a solution of (1′) for which $x_n \to x^*$. Then

$$\|x^* - x_n'\|$$
$$\leq M_0(n) + \delta(q^n - 1)/(q - 1)\|B^{-1}\|[MN_0 q + MN\|B^{-1}\| + 1 + 0(\delta)],$$

where $0 \leq M_0(n) \to 0$ as $n \to \infty$.

3. Alternative Method

Let L be a linear self-adjoint differential operator with homogeneous boundary conditions such that L has a countable system of eigenvalues λ_i, $|\lambda_i| \to \infty$ and the corresponding eigenfunctions form a complete orthonormal basis for the real Hilbert space $\mathcal{S}$. Further, let $0 < \dim \mathcal{N}(L) < \infty$. Let $N: \mathcal{D}(N) \subset S \to S$ be a nonlinear operator and let $f \in S$. We consider the nonlinear problem

$$Lu + Nu = f \tag{3}$$

for the existence of solutions. We denote by $P: S \to \mathcal{N}(L)$ the linear projection operator of $\mathcal{S}$ onto $\mathcal{N}(L)$, so that $(I - P): S \to \mathcal{R}(L)$ and $S = \mathcal{N}(L) \oplus \mathcal{R}(L)$. Finally let H be the generalized inverse of L, i.e., $H = [L | \mathcal{D}(L) \cap \mathcal{N}(L)^{\perp}]^{-1}$.

If $u \in S$ is a solution of (3), then we can write $u = v + w$, $v \in \mathcal{R}(L)$ and $w \in \mathcal{N}(L)$. Then (3) reduces to

$$Lv + N(v + w) = f \qquad \text{or} \qquad v + H(I - P)N(v + w) = H(I - P)f. \tag{4}$$

Conversely, if there exists $w \in \mathcal{N}(L)$ and $v \in \mathcal{N}(L)^{L}$ which satisfy (4), then by applying L to (4)

$$Lv + (I - P)N(v + w) = (I - P)f$$

or

$$L(v + w) + N(v + w) - f = PN(v + w) - Pf.$$

Thus a solution of (4) is a solution of (3) if and only if

$$PN(v + w) - Pf = 0. \tag{5}$$

Hence, solving problem (3) is equivalent to finding $v \in \mathcal{R}(L)$ and $w \in \mathcal{N}(L)$ such that

$$v - H(I - P)[N(v + w) - f] = 0$$
$$PN(v + w) - Pf = 0.$$

This decomposition of (3) into an equivalent system of a pair of equations dates back to Lyapunov and Schmidt and was put into a functional-analytic setting by Cesari [5]. Since then there has been a large literature on the various extensions and applications of this scheme to nonlinear differential equations—ordinary and partial. See Cesari [6] for an extensive survey of the literature.

We restrict ourselves to outlining briefly two of the methods that have been used to solve the system of Eqs. (4) and (5).

The first method consists of solving (4) for each w in the ball $|w| \le R$ in the space $\mathcal{N}(L)$ and obtaining a function $v(w)$ which is continuously dependent on w by restricting ourselves to a suitable closed convex set in $\mathcal{R}(L)$ and applying the Banach fixed point theorem. Equation (5) then reduces to a finite dimensional equation of the type $Tw = 0$ where $T\colon |w| \leq R \subset \mathcal{N}(L) \to \mathcal{N}(L)$ defined by $Tw = PN[v(w) + w] - Pf$. Any one of the several tools available for finite dimensional problems, e.g., the implicit function theorem, the Brouwer fixed point theorem, or any of its several modifications may be utilized to solve $Tw = 0$. An important point to be noted here is that since $|\lambda_i| \to \infty$ we can always select P to be the projection operator from S onto the subspace generated by the first m eigenfunctions where $m > \dim \mathcal{N}(L)$ is so chosen that $\|H(I - P)\|$ can be made very small and one is in a setting to apply the Banach fixed point theorem for (4) (by assuming N to satisfy a local Lipschitz condition).

The second method consists of utilizing the fact that $H(I - P)\colon S \to \mathcal{R}(L)$ is bounded and compact, and $P\colon S \to \mathcal{N}(L)$, being a finite dimensional projection, is also compact so that if $N\colon \mathcal{D}(N) \subset S \to S$ is continuous and maps bounded sets into bounded sets, we can reduce the system of Eqs. (4) and (5) to a single operator equation of the type $(I + T_1)\, u = 0$, where $u = (u, w) \in \mathcal{R}(L) \oplus \mathcal{N}(L)$ and $T_1\colon \mathcal{R}(L) \oplus \mathcal{N}(L) \to \mathcal{R}(L) \oplus \mathcal{N}(L)$ is defined by

$$T_1\colon \begin{pmatrix} v \\ w \end{pmatrix} \to \begin{pmatrix} -H(I - P)[N(v + w) - f] \\ PN(v + w) - Pf - w \end{pmatrix}.$$

Since T_1 is compact, we are now in a position to apply either the Leray–Schauder degree theory or Schauder's fixed point theorem.

We conclude this section with the remark that the first method has also been applied to handle nonlinear hyperbolic problems for which P is no longer a finite dimensional projection (for details see Cesari [6], Hale [10], and Hall [11]).

4. Perturbation Method

We now consider two nonlinear boundary value problems in which problems of the type considered in Section 3 have been dealt with by the use of perturbation methods. The first result we discuss is due to De Figureido:

Let S be a real Hilbert space, $A\colon S \to S$ be a linear, bounded, self-adjoint operator with closed range $R(A)$ and a finite dimensional null space $N(A)$. Let $N\colon S \to S$ be a compact nonlinear mapping (i.e., N is continuous and takes bounded sets into relatively compact sets) such that

$$\|Nu\| \leq c\|u\|^{\alpha} + d$$

for all $u \in S$, where $c > 0$ and $0 \leq \alpha < 1$ are fixed constants.

THEOREM 4.1 Let $\{\lambda_n\}$ be a sequence of positive real numbers such that $\lambda_n \to 0$. Also, let the following condition hold:

Given $y \in N(A)$, $\|y\| = 1$, and sequences $t_n \to \infty$, $y_n \to y$, $y_n \in N(A)$, $z_n \in R(A)$, $\|z_n\| \leq K_1$, where K_1 is a constant, we have $(h, y) > \lim_{n\to\infty} \inf\langle N(t_n y_n + t_n{}^\alpha z_n), y\rangle$ where h is a given element of S. Then the nonlinear problem

$$Au + Nu = h$$

has a solution $u \in S$.

Proof We consider the sequence of perturbed problems

$$Au_n - \lambda_n u_n + Nu_n = h.$$

This equation is equivalent to

$$u_n = (A - \lambda_n)^{-1}(h - Nu_n).$$

The operator $T\colon S \to S$ defined by $Tu = (A - \lambda_n)^{-1}(h - Nu_n)$ is compact, and the growth hypothesis on N enables us to apply Schauder's fixed point theorem. The fact that $\alpha < 1$ implies that T maps a ball of sufficiently large radius into itself, and thus T has a fixed point which is a solution of the perturbed problems.

We first show that if the sequence $\{u_n\}$ is bounded, then Theorem 4.1 is proved. Each u_n can be written as $v_n + w_n$, where $v_n \in N(A)$ and $w_n \in R(A)$, by virtue of the hypotheses on A, since S can be decomposed as $N(A) \oplus R(A)$. By hypothesis N maps bounded sets into relatively compact sets, and thus there exists $g \in S$ and a subsequence of $\{u_n\}$, which will be referred to as $\{u_n\}$ henceforth, such that $Nu_n \to g$. But the boundedness of $\{u_n\}$ implies that $\{v_n\}$ and $\{w_n\}$ are also bounded. Hence we can extract subsequences of $\{v_n\}$ and $\{w_n\}$ such that $v_n \to v$ and $w_n \to w$ where, $v, w \in S$. Passing to the limit in the perturbed problem, we obtain $Aw_n \to h - g$. Since A restricted to $R(A)$ is a linear homeomorphism, we obtain that $w_n \to w$. Let $u = v + w$. It can then be seen that

$$Au + Nu = h,$$

or u is a solution of the given nonlinear problem.

This theorem, due to De Figureido, is a generalization of the proof due to Hess [13] of the following result of Landesman and Lazer [21]: Let Ω be a bounded domain in R^N and let L be a uniformly elliptic, formally self-adjoint differential operator defined on Ω such that $\mathcal{N}(L) \neq \Phi$. Also let $g\colon R \to R$ be a bounded continuous function such that the limits $\lim_{s\to\infty} g(s) = g(\infty)$ and $\lim_{s\to\infty} g(s) = g(-\infty)$ exist and suppose that $g(-\infty) < g(s) < g(\infty)$. Let $\Omega_+ = \{x \in \Omega\colon w(x) > 0\}$ and $\Omega_- = \{x \in \Omega\colon w(x) < 0\}$, where w spans $\mathcal{N}(L)$.

Then the inequalities

$$g(-\infty)\int_{\Omega_+}|w|\,dx - g(\infty)\int_{\Omega_-}|w|\,dx$$

$$< \int_{\Omega} fw\,dx < g(\infty)\int_{\Omega_+}|w|\,dx - g(-\infty)\int_{\Omega_-}|w|\,dx$$

are necessary and sufficient for the existence of a weak solution of the nonlinear problem

$$(Lu)(x) = g[u(x)] - f(x), \qquad x \in \Omega$$

$$u(x) = 0, \qquad x \in \partial\Omega.$$

It is interesting to note the connection between the proof of the above theorem and the results for the linear case discussed in Section 2. Another situation in which the perturbation method has been utilized is the following result due to Brézis [4].

THEOREM 4.2 Let A, B be two maximal monotone operators over a Hilbert space S and let $\langle Au, B_\lambda u\rangle \geq 0$ for every $u \in \mathscr{D}(A)$, where $B_\lambda = (I - (I + \lambda B)^{-1})/\lambda$. Then $\mathscr{R}(A + B) = \mathscr{R}(A) + \mathscr{R}(B)$ and

$$\text{Int } \mathscr{R}(A + B) = \text{Int } \mathscr{R}(A) + \text{Int } \mathscr{R}(B).$$

Proof The proof is carried out in three stages. We refer to the paper of Brézis for details. The first step is to show that $A + B$ is maximal monotone. This implies that the nonlinear operator equation

$$\varepsilon u + Au + Bu \ni f, \qquad f \in S,$$

has a solution $u_\varepsilon \in S$ for every $\varepsilon > 0$. We then show that $\|Au_\varepsilon\| \leq C(f)$, $\|Bu_\varepsilon\| \leq C(f)$. In order to prove this we consider the sequence $u_{\varepsilon\lambda}$, where $u_{\varepsilon\lambda}$ is the solution of

$$\varepsilon u_{\varepsilon\lambda} + Au_{\varepsilon\lambda} + B_\lambda u_{\varepsilon\lambda} \ni f.$$

For each ε, $u_{\varepsilon\lambda} \to u_\varepsilon$ as $\lambda \to 0$, where u_ε is the solution of $\varepsilon u + Au + Bu \ni f$. Finally we use the bounds on Au_ε and Bu_ε to prove the theorem.

As an application of the theorem we have: Let Ω be a bounded smooth domain in R^N and let $\beta\colon R \to \mathscr{R}$ be a continuous monotone increasing function. Then the equation

$$-\Delta u + \beta(u) = f(x) \qquad \text{in} \quad \Omega,$$

$$\frac{\partial u}{\partial n} = 0 \qquad \text{on} \quad \partial\Omega,$$

has a solution if $(1/|\Omega|)\int_\Omega f\,dx \in \text{Int } \mathscr{R}(\beta)$.

5. The Proximal Point Algorithm

In this section we study nonlinear boundary value problems of the type considered in Section 3 for methods of successive approximation. In particular the results of Krjanev for the linear case as discussed in Section 2 can be extended to the nonlinear case when N and L are both monotone. We survey some of the results in this direction and the details of the unpublished results may be seen in forthcoming papers.

Thus we consider the nonlinear boundary value problem (3), where L is a positive self-adjoint operator with nontrivial finite dimensional kernel and $N: \mathscr{D}(N) \subset S \to S$ is a maximal monotone operator. Then operator $H(I - P): S \to \mathscr{R}(L)$ can be written as J^*J [15] and thus if we require the operator N to be defined over $\mathscr{R}(J^*)$, then Eqs. (4) and (5) can be written as

$$v_1 + JN(J^*v_1 + w) = 0$$

$$PN(J^*v_1 + w) = 0,$$

where $u = J^*v_1 + w$, $v_1 \in \mathscr{R}(L)$. For details see [15]. We now assume for the sake of simplicity that the operator N is continuous and maps bounded sets into bounded sets under appropriate topologies. If we define the operator $T: \mathscr{R}(L) \oplus \mathscr{N}(L) \to \mathscr{R}(L) \oplus \mathscr{N}(L)$ by

$$T: \begin{pmatrix} v_1 \\ w \end{pmatrix} \to \begin{pmatrix} v_1 + JN(J^*v_1 + w) \\ PN(J^*v_1 + w) \end{pmatrix},$$

it can be seen that problem (3) can be written in the form $Tz = 0$, where T is a maximal monotone operator over $\mathscr{R}(L) \oplus \mathscr{N}(L)$.

Many problems in the theory of variational inequalities, minimax problems, and minimizations of lower semicontinuous proper convex functions can be reduced to operator equations of this type. Thus for example, let $\mathscr{S}$ be the product of two Hilbert spaces S_1 and S_2 and let $\phi: S \to [-\infty, \infty]$ be such that $\phi(x, y)$ is convex in $x \in S_1$ and concave in $y \in S_2$. For each $z = (x, y)$ let Tz be the set of all $w = (u, v)$ such that

$$\begin{aligned} \phi(x_1, y) - \langle x_1, u\rangle + \langle y_1, u\rangle &\geq \phi(x, y) - \langle x, u\rangle + \langle y, u\rangle \\ &\geq \phi(x, y_1) - \langle x, u\rangle + \langle y_1, u\rangle \\ &\qquad \text{for all} \quad x_1 \in S_1,\ y_1 \in S_2. \end{aligned}$$

Under suitable hypotheses on ϕ, Rockafellar [29] proved that T is maximal monotone. And the global saddle points of L are those $z = (x, y)$ such that $0 \in Tz$.

A fundamental algorithm for solving $0 \in Tz$, where T is maximal monotone, was introduced by Moreau [24] in the context of minimization of lower semicontinuous proper convex functions. This algorithm which is referred to as the proximal point algorithm may be described as follows: for a lower semicontinuous proper convex function $f(z)$ we define the sequence $\{z_k\}$ by selecting $\{z_k\}$ to be the minimizer of $f(z) + (1/2c_k)\| z - z_k \|^2$, where $c_k > 0$. In the setup of operator equations of the type $0 \in Tz$ the sequence $\{z_k\}$ is generated as follows: $z_{k+1} = (I + c_k T)^{-1} z_k$. It must be noted here that by Minty's theorem $(I + c_k T)^{-1}$ is everywhere defined. Martinet [23] proved that in the case $c_k \equiv c > 0$ the sequence $\{z_k\}$ converges weakly to a solution of $0 \in Tz$.

The equation defining z_k may be rewritten as follows: z_{k+1} is the solution of

$$Tz_{k+1} + (z_{k+1}/c_k) = z_k/c_k.$$

The connection with the work of Krjanev [20] for the linear case as was discussed in Section 2 is now obvious.

In this section we now state some properties of $\{z_k\}$. The details may be seen elsewhere [14].

PROPOSITION 5.1 $\{\|z^n - z^{n+1}\|/c_n\}$ is a monotone decreasing sequence.

PROPOSITION 5.2 If $T^{-1}(0) \neq \Phi$, then the sequence $\{Mz_n\}$ where M is the projection of S on $T^{-1}(0)$ (which is a nonempty closed convex set) is a Cauchy sequence.

PROPOSITION 5.3 If $T^{-1}(0) \neq \Phi$, then $(z^n - z^{n+1})/c_n \to 0$.

PROPOSITION 5.4 The solution $\{x_n\}$ of the equation $Tx + \alpha_n x = 0$ converges as $\alpha_n \to 0$ to an element of $T^{-1}(0)$ if and only if $Tx = 0$ has a solution. (This result may be seen partially in the work of Dolph and Minty [9].)

PROPOSITION 5.5 The solution $\{x_n\}$ of $Tx + (1/n)x = p$, $p \in S$ being such that $\|p\| \leq \delta$, converges to a solution of $0 \in Tx$ if $n\delta \to 0$.

Remarks In Propositions 5.4 and 5.5 one can show that the limit is precisely the solution of minimum norm. The comparison with the linear case and the work of Krjanev is now obvious. These results are of important significance in attempts to define the concept of "condition number" for nonlinear problems. In our paper [14] we obtain a result on the strong convergence of the ergodic sequence generated by $\{z^n\}$.

6. Existence of Random Solutions

In this section we discuss the general method of proving the existence of random solutions of random analogs of the nonlinear problems considered in Section 3. We start with a few preliminary definitions from the theory of random operator equations [3]. Let $(\Omega, \mathscr{B}, \mu)$ be a probability space and X and Y be two metric spaces. A mapping T from $\Omega \times X$ into Y is called a random operator if for each fixed $x \in X$ the function $T(\cdot, x)$ is a Y valued random variable. A random operator T is said to be continuous if for each $\omega \in \Omega$, $T(\omega, \cdot)$ is continuous. Any X-valued random correspondence $x(\omega)$ which satisfies $T(\omega, x(\omega)) = y(\omega)$, when $y(\omega) \in Y$, is said to be a random solution of the equation $T(\omega, x) = y(\omega)$. An important result on the randomness of the inverse of a random operator T is the following:

PROPOSITION 6.1 ([12,25]) Let $(\Omega, \mathscr{B}, \mu)$ be a complete probability space, X be a separable metric space and Y a metric space. Let T be a separable random operator from $\Omega \times X$ into Y such that $T(\omega, \cdot)$ is continuous. Then T^{-1} is also a random operator.

Debreu [7] proved the following important relation between the graph of an operator and its randomness.

PROPOSITION 6.2 Let g be a correspondence of $(\Omega, \mathscr{B}, \mu)$ into a complete separable metric space X. If the graph of g belongs to the σ-algebra generated by $\mathscr{B} \times \mathscr{B}_x$, then g is measurable.

We now consider the random analog of (3). Let S and L be the same as in Section 3 and let N be a random nonlinear operator. We consider the question of the existence of random solutions of

$$Lu + N(\omega, u) = 0.$$

As in Section 3, we consider the equivalent system of Eqs. (4) and (5). If we consider the case in which L and N are monotone, then one can proceed to prove the randomness in two ways. We outline them both because of their applicability to a more general class of nonlinear boundary value problems.

The Iterative Approach

By the results indicated in Sections 4 and 5, it is sufficient to rewrite the random problem in the form $T(\omega, u) = 0$, where T is the maximal monotone operator and prove the randomness of $\{z_k\}$ where z_0 is arbitrary and random. The convergence of z_k would then establish the randomness of the solution of the random boundary value problem. Each z_{k+1} can be written in the form $(I + c_k T)^{-1} z_k$. But since T is maximal monotone and $c_k > 0$, it follows that $T + (1/c_k)I$ is strictly monotone. Hence $(T + (1/c_k)I)^{-1}$ is well-defined for each ω and is also continuous. Hence, if $T + (1/c_k)I$ is random, by Proposition 6.1 it follows that $(T + (1/c_k)I)^{-1}$ is random and the randomness of z_k implies the randomness of z_{k+1}. The randomness of T follows immediately from the randomness of N. In fact it is easy to see that we can also let L be random, which is natural in many applications.

The "Measurability of Graph" Method

We indicate this method also because when iterative techniques are not available, as in the situations when we apply the Schauder fixed-point theorem or the Leray–Schauder principle, this technique is most suitable. By Section 3, the random problem can be rewritten in the form $(I + T)u = 0$, where $u = (v, w) \in \mathscr{R}(L) \oplus \mathscr{N}(L)$ and

$$T\colon \begin{pmatrix} v \\ w \end{pmatrix} \to \begin{pmatrix} H(I - P)N(v + w) \\ PN(v + w) - w \end{pmatrix}.$$

Clearly $T\colon (\Omega, \mathscr{R}(L) \oplus \mathscr{N}(L)) \to (\mathscr{R}(L) \oplus \mathscr{N}(L))$ is random if L and N are so. Let us assume that the hypotheses on N are such that, for each $\omega \in \Omega$, the deterministic problem (3) is solvable. Thus we generate a correspondence $g\colon \omega \to \{\tilde{u}(\omega)\}$, where for each ω, $u(\omega)$ is the set of solutions of the deterministic problem (3). If we now show that the graph of g belongs to the appropriate σ-algebra as in Proposition 6.2, then g is a random correspondence. The details of these steps may be seen in [17], [18]. Finally, if g is closely valued for each ω, then by the Kuratowski–Ryll–Nardzewski selection theorem there exists a single-valued random solution to the random analog of problem (3).

7. Tikhonov Regularization and Random Problems

We now survey some of the results obtained for approximate solutions of random linear problems by using the method of regularization due to Tikhonov. Let A be a closed linear operator from a Hilbert space S to another Hilbert space S_1 such that the domain of A is dense in S and A does not have a bounded inverse. Let f be an element in the range of A. By a regularization algorithm we mean a family of bounded linear operators $\{R\}$ from S_1 to S such that for every f in the range of A there is a $x \in A^{-1}f$ such that for every $\varepsilon > 0$ there exists $\delta = \delta(f, \varepsilon, A)$ and an operator R from the family such that $\| R\bar{f} - x \| < \varepsilon$ if $\|\bar{f} - f\| < \delta$. This concept of the regularization algorithm stems from the concept of "properly posed" problems as defined by Tikhonov (cf. [31]). According to Tikhonov the linear operator equation $Ax = f$ is properly posed if there exists a closed set $M \subset S$ such that

(1) it is known a priori that a solution x exists for some class of data and $x \in M$,

(2) the solution is unique in a class of functions belonging to M,

(3) the solution depends continuously on f.

Khalfin and Sudakov [19] consider the linear operator equation

$$Ax = f$$

in which the right-hand side f is not known accurately but a certain statistical distribution of errors in the definition of f is known. Thus let us assume that S and S_1 are measurable spaces and the operator A is random. Let $(\Omega, \mathscr{B}, \mu)$ be a probability space and we assume that x, f are random variables with values in S and S_1. Let f be a random variable with values in S_1 and let B be an operator with domain contained in S_1 and range in S such that $\bar{x} = B\bar{f}$, where $\bar{x}$ is a random variable with values in S. The problem of constructing an approximate solution to $Ax = f$ consists in finding an operator B such that x and $\bar{x}$ are close to each other—both in the sense of S and Ω. Khalfin and Sudakov obtain estimates on $\mu\{\|B_\lambda \bar{f} - x\| > \varepsilon\}$, where B_λ is a regularizer of the equation $Ax = f$ and then these estimates are applied to a particular regularizer for the Cauchy problem for the Laplace equation.

Bakushinskii [1] continues the above study to the problem of construction of regularizing algorithms. Thus we are considering the random linear operator equation

$$Ax = f.$$

Let $B_\alpha = \psi(A^*A, \alpha)A^*$, $\alpha > 0$ be a family of operators where $\psi(A^*A, \alpha)$ is generated by functions $\psi(\lambda, \alpha)(\lambda \in S(A^*A))$ with the following properties:

$$\lim_{\alpha \to 0} \lambda\psi(\lambda, \alpha) = 1, \qquad \lambda \neq 0,$$

$$\sup|\psi(\lambda, \alpha)|\sqrt{\lambda} = K_\alpha < \infty,$$

$$\lambda \in S(A^*A).$$

Further let $\psi(\lambda, \alpha)$ be continuous for $\lambda \geq 0$, $\alpha > 0$ and

$$\sup|\psi(\lambda, \alpha)|\sqrt{\lambda} = K_\alpha, \qquad 0 \leq \lambda < \infty,$$

$$\lambda\psi(\lambda, \alpha) \geq \lambda\psi(\lambda, \alpha'), \qquad \alpha' > \alpha.$$

Then we may form the expression

$$B_\alpha(\omega) = \psi(A^*(\omega)A(\omega), \alpha)A^*(\omega).$$

For each $w \in \Omega$, the above is Bakushinskii's [1] construction of a regularization algorithm. For the above randomized version, Bakushinskii proves that if $\bar{A}$ is close to A, then $\bar{B}_\alpha$ is close to B for fixed α. By $\bar{A}$ being close to A we mean $[\int_\Omega |\bar{A}(\omega) - A(\omega)|^p \, d\mu]^{1/p} \to 0$ and a suitable meaning is assigned to $\bar{B}_\alpha$ being close to B. It must be noted that just as we proceeded in the previous section it must be shown that the naturally generated mapping from Ω into the space $B_\alpha(\omega)$ of linear operators from S_1 into S should be proved to be measurable.

Bakhushinskii and Apartsin [2] continue these investigations to obtain methods of stochastic approximation type for solving linear incorrectly posed problems. Thus let us consider the linear problem

$$Ax = f,$$

where A is a positive noninvertible operator defined in a dense domain of a Hilbert space. Let us assume that f is such that the above equation is solvable, even though the solution may not be unique. Let $f_1, f_2, \ldots, f_n$ be independent realizations of the random variable $\bar{f}$ (whose values are in $\mathcal{S}$), where $E(\bar{f}) = f$. Let $\|\bar{f} - f\|^2_{L^2(S)} = \alpha^2 < \infty$, where the norm is in the usual sense of $\mathcal{S}$-valued random variables, as defined for $L^p(S)$ above. Then under suitable hypotheses on the function $\psi(A, n)$ Bakhushinskii and Apartsin [2] prove that $\{x_n\}$, generated by

$$x_{n+1} - x_n = -\psi(A, n)x_n + \xi_n,$$

where $\xi_n = (1/n)\hat{f}_n + ((n-1)/n)\xi_{n-1}$, $\xi_0 = x_0 = 0$ converges in $L_2(S)$ to the normal solution of $Ax = f$. The above scheme of approximation is patterned after the Robbins–Monro [28] method.

We conclude with the remark that the results for the deterministic nonlinear case as obtained in Sections 4 and 5 provide the natural setting for generalizations of Tikhonov regularization to random nonlinear problems and apply the methods of stochastic approximation to such situations. It must be mentioned here that the Robbins–Monro method has been applied by Nevel'son and Has'minskii [26] in the study of stochastic differential equations.

References

1. A. B. Bakushinskii, On construction of regularizing algorithms in the presence of random noise. *Dokl. Akad. Nauk SSSR* **189** (1969), 231–233.
2. A. B. Bakushinskii and A. S. Apartsin, Methods of the stochastic approximation type for solving linear incorrectly posed problems. *Siberian Math. J.* **16** (1975), 9–14.
3. A. T. Bharucha-Reid, "Random Integral Equations." Academic Press, New York, 1972.
4. H. Brézis, Quelques propriétés des operateurs monotones et des semigroupes nonlinéaires. Lecture Notes in Mathematics No. 543. Springer-Verlag, Berlin and New York, 1976.
5. L. Cesari, Functional analysis and Galerkin's method. *Michigan Math. J.* **11** (1964), 385–414.
6. L. Cesari, Functional analysis, nonlinear differential equations and the alternative method, *in* "Nonlinear functional analysis and differential equations." Lect. Notes Pure Appl. Math. Marcel Dekker, New York, 1976.
7. G. Debreu, Integration of correspondences. *Berkeley Symp. Stat. Prob. 5th* **2** 351–372.
8. D. G. De Figureido, On the range of nonlinear operators with linear asymptotes which are not invertible. *Comment. Math. Univ. Carolinae* **15** (1974), 415–428.

9. C. L. Dolph and G. J. Minty, On nonlinear integral equations of the Hammerstein type, *in* "Integral Equations," pp. 99–154. Univ. of Wisconsin Press, Madison, Wisconsin, 1964.
10. J. K. Hale, Applications of alternative problems. Brown Univ. Lect. Notes. Brown Univ. Press, Providence, Rhode Island, 1971.
11. W. S. Hall, Periodic solutions of a class of weakly nonlinear evolution equations. *Arch. Rat. Mech. Anal.* **39** (1970), 294–322.
12. O. Hans, Random operator equations, *Fourth Berkeley Symp. Stat. Prob.* **2** (1961), 185–202.
13. P. Hess, On a theorem by Landesman and Lazar. *Indiana Univ. Math. J.* **23** (1974), 827–829.
14. R. Kannan, On the proximal point algorithm (to appear).
15. R. Kannan and J. Locker, Nonlinear boundary value problems and operators TT^*. *J. Differential Equations* (to appear).
16. R. Kannan and J. Locker, Continuous dependence of least squares solutions of linear boundary value problems. *Proc. Amer. Math. Soc.* **59** (1976), 107–110.
17. R. Kannan and H. Salehi, Measurability of solutions of nonlinear equations, *in* "Nonlinear functional analysis and differential equations." Lect. Notes Pure Appl. Math. Marcel Dekker, New York, 1976.
18. R. Kannan and H. Salehi, Random nonlinear equations and monotonic nonlinearities, *J. Math. Anal. Appl.* **58** (1977),
19. L. A. Khalfin and V. N. Sudakov, A statistical approach to improperly posed problems of mathematical physics. *Dokl. Akad. Nauk SSSR* **5** (1957),
20. A. V. Krjanev, The solution of incorrectly posed problems by methods of successive approximations. *Dokl. Akad. Nauk. SSSR* **210** (1973), 20–22.
21. E. M. Landesman and A. C. Lazer, Nonlinear perturbation of linear elliptic boundary value problems at resonance. *Indiana Univ. Math. J.* **19** (1970), 609–623.
22. M. M. Lavrent'ev, "Some improperly posed problems of mathematical physics." Springer-Verlag, Berlin and New York, 1967.
23. B. Martinet, Regularisation d'inéquations variationelles par approximations successives. *Rev. Française Automat. Informat. Recherche Opérationelle* (1970), 154–159.
24. J. J. Moreau, Proximité et dualité dans un espace Hilbertien. *Bull. Soc. Math. France* **93** (1965), 273–299.
25. M. Z. Nashed and H. Salehi, Measurability of generalized inverses of randon linear operators. *SIAM J. Appl. Math.* **25** (1973), 681–692.
26. M. B. Nevel'son and R. Z. Has'minskii, Stochastic approximation and recursive estimation. *Translations Amer. Math. Soc.* **47** (1977).
27. W. C. Rheinboldt, On measures of ill-conditioning for nonlinear equations. *Math. Comput.* **30** (1976), 104–111.
28. H. Robbins and S. Monro, A stochastic approximation method. *Ann. Math. Statist.* **22** (1951), 400–407.
29. R. T. Rockafellar, Monotone operators associated with saddle points and minimax problems, Nonlinear Functional Analysis Part I. "Symposium On Pure Mathematics," Vol. 18, pp. 397–407. American Math. Soc., Providence, Rhode Island, 1970.
30. R. T. Rockafellar, Monotone operators and the proximal point algorithm. *SIAM J. Control Optimization* **14** (1976), 877–898.
31. A. N. Tikhonov, Solution of incorrectly formulated problems and the regularization method. *Dokl. Akad. Nauk. SSSR* **151** (1963), 501–504.

The Eigenvalue Problem for Variational Inequalities and a New Version of the Ljusternik–Schnirelmann Theory

M. Kučera, J. Nečas, and J. Souček

Czechoslovak Academy of Sciences

Introduction: Motivation

Although the theory of variational inequalities has been widely developed in the last years (see, e.g., Lions [1]), little attention has been devoted to spectral questions of variational inequalities. Nevertheless, it seems reasonable to formulate for these inequalities problems analogous to those studied in the usual (linear or nonlinear) theory of operators. This chapter deals with one of the main problems of this type—the existence of eigenvalues.

First, let us explain a physical example that led us to the notion of an eigenvalue of a variational inequality. Consider a beam simply supported at its ends and compressed by a force P (see Fig. 1). Denote by H the space of

Figure 1

all functions $u \in W_2^{\,2}(\langle 0, 1\rangle)$ (the well-known Sobolev space) satisfying the conditions $u(0) = u(1) = 0$. It is a Hilbert space with inner product

$$\langle u, v\rangle = \int_0^1 u''(x)v''(x)\,dx.$$

Denote by A an operator from H into H such that

$$\langle Au, v\rangle = \int_0^1 u'(x)v'(x)\,dx \qquad \text{for each} \quad u, v \in H.$$

The case of the beam mentioned above can be described by the equation

$$(IE/P)u - Au = 0, \tag{0.1}$$

where E is Young's modulus of elasticity, I is the moment of inertia, and P is the force compressing the beam. More precisely, the situation is described by the boundary value problem

$$(IE/P)u^{(4)} + u'' = 0 \qquad \text{on} \quad (0, 1), \tag{0.2}$$

$$u(0) = u(1) = u''(0) = u''(1) = 0; \tag{0.3}$$

but it is usual to introduce a weak solution of this problem as a function $u \in H$ satisfying (0.1). The beam can bend if IE/P is an eigenvalue of the operator A (i.e., of problem (0.2), (0.3)) and this bending is given by the corresponding eigenfunction. Now, let us consider that the beam is supported by fixed obstacles from one side at some points $x_1, x_2, \ldots, x_n \in (0, 1)$. If IE/P is an eigenvalue of the operator A, then the beam can realize only an eigenfunction satifying the conditions $u(x_i) \geq 0$, $i = 1, \ldots, n$ (Fig. 2).

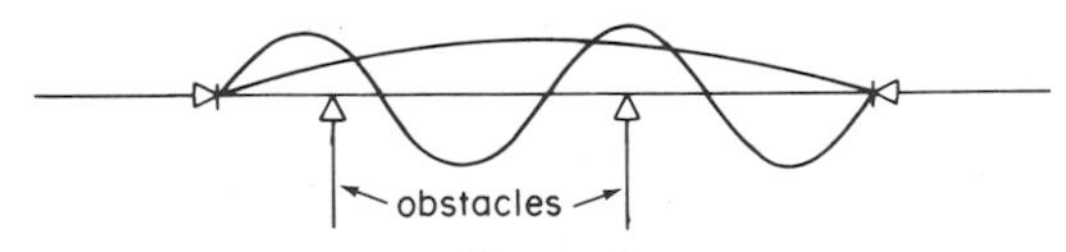

Figure 2

Moreover, it can be proved that the beam can bend for some "new eigenvalues" and some "new eigenfunctions" can be realized. These are not eigenvalues and eigenfunctions of the operator A but of the corresponding variational inequality given by the convex closed cone

$$K = \{u \in H;\ u(x_i) \geq 0,\ i = 1, \ldots, n\}.$$

Precisely, if λ is a real number such that there exists a nontrivial $u \in H$, satisfying the conditions

$$u \in K, \tag{0.4}$$

$$\langle \lambda u - Au, v - u \rangle \geq 0 \qquad \text{for each} \quad v \in K, \tag{0.5}$$

then we say that λ is an eigenvalue and u is an eigenfunction of the variational inequality (0.5). It can be shown that the supported beam can bend if IE/P is an eigenvalue of the variational inequality (0.5) and the bending is given by the corresponding eigenfunction of inequality (0.5).

Hence, it seems to be interesting to examine the structure of the set of all eigenvalues of a variational inequality. For illustration, we shall give a full description of this set for an inequality of the type just explained; but for the sake of simplicity, we shall consider an inequality of the second order instead of the fourth order (see Section 1). Further, in analogy with the spectral theory of nonlinear operators (see, for example, Fučík *et al.* [2]), we shall study a more general problem (Section 2). We shall consider a general Hilbert space H with the inner product $\langle \cdot, \cdot \rangle$, two even nonnegative functionals f, g on H, and a convex closed cone K in H. For a given positive r we shall investigate the problem of finding $u \in H$ and λ real such that

$$u \in M_r(f) = \{u \in H; f(u) = r\}, \tag{0.6}$$

$$u \in K, \tag{0.7}$$

$$\lambda \langle f'(u), v - u \rangle - \langle g'(u), v - u \rangle \geq 0 \qquad \text{for each} \quad v \in K, \tag{0.8}$$

where f', g' denote the Fréchet derivatives of the functionals f, g. If $u \in H$ satisfies conditions (0.6)–(0.8) with some real λ, then we say that u is a critical point of problem (0.6)–(0.8), the number λ is the corresponding eigenvalue, and the number $\gamma = g(u)$ is the corresponding critical level of problem (0.6)–(0.8). For the aforementioned case of the beam

$$f(u) = \tfrac{1}{2} \int_0^1 (u''(x))^2 \, dx, \qquad g(u) = \tfrac{1}{2} \int_0^1 (u'(x))^2 \, dx.$$

Let us remark that in special cases (with f, g positive homogeneous) the set of all critical levels equals the set of all eigenvalues except for a multiplicative constant (see Section 2). In the case $K = H$ (i.e., in the case of an equation

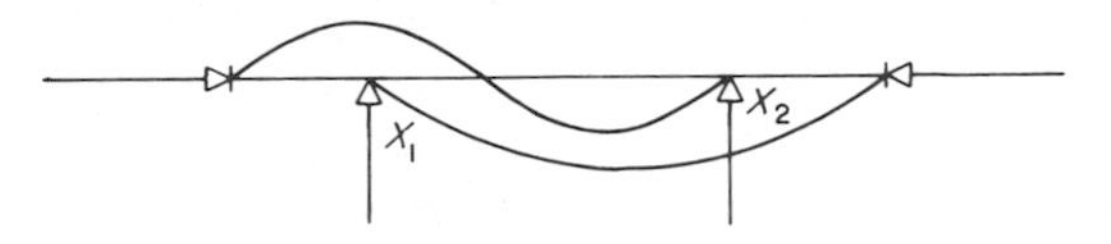

Figure 3

instead of an inequality) the set of critical levels has been investigated by many workers. The well-known Ljusternik–Schnirelmann theory gives (under some assumptions) the existence of a sequence of critical levels converging to zero (see references [2–5], special versions of the Morse–Sard theorem give upper bounds for the "size" of the set of critical levels (see, for example, references [2,6]). A complete survey of these methods is contained in Lecture Notes by Fučík *et al.* [2].

The classical Ljusternik–Schnirelmann theory is based on the notion of the category of a set defined for compact subsets of the symmetric manifold $M_r(f)$. In the case of variational inequalities, it would be necessary to investigate only subsets of a nonsymmetric set $M_r(f) \cap K$, which represents a serious difficulty. Therefore we do not use the basic ideas of Ljusternik–Schnirelmann theory directly but together with a penalty method. We take a penalty functional Φ (for the properties of Φ see Section 2) and consider the problem

$$u \in M_r(f + \varepsilon\Phi), \tag{0.9}$$

$$\lambda(f'(u) + \varepsilon\Phi'(u)) - g'(u) = 0. \tag{0.10}$$

We want to prove the existence of critical levels of problem (0.9)–(0.10) for each nonnegative ε and, letting $\varepsilon \to +\infty$, to obtain the existence of critical levels of problem (0.6)–(0.8). The classical ideas of Ljusternik–Schnirelmann cannot be used in the case of problem (0.9)–(0.10) either because the manifold $M_r(f + \varepsilon\Phi)$ is not symmetric. But it proves to be possible to follow the basic areas of Ljusternik–Schnirelmann by using special classes of subsets of $M_r(f + \varepsilon\Phi)$ instead of classical categories. These classes are given by certain basic sets, and their deformations along suitable trajectories are considered, for example, in the paper by Nečas [5]. The previously mentioned trajectories define, in a certain sense, the smallest class of homotopies that are necessary for using the considerations of the Ljusternik–Schnirelman type.

Although we shall have infinitely many classes and a critical level will correspond to any of them, it will not be proved whether these critical levels are different. A better situation occurs in the special case of a half-space, where the set of critical levels is in fact a sequence converging to zero (see Section 3).

1. A Simple Example

Let $x_1, x_2, \ldots, x_n$ be real numbers, $0 < x_1 < x_2 < \cdots < x_n < 1$. We shall solve the eigenvalue problem (0.6)–(0.8) for

$$H = \mathring{W}_2^1(\langle 0, 1\rangle), \qquad K = \{u \in \mathring{W}_2^1(\langle 0, 1\rangle);\ u(x_i) \geq 0, i = 1, \ldots, n\}$$

and for the functionals f, g defined as

$$f(u) = \tfrac{1}{2} \int_0^1 (u')^2\, dx, \qquad g(u) = \tfrac{1}{2} \int_0^1 u^2\, dx \qquad \text{for each} \quad u \in H.$$

In this special case, conditions (0.7), (0.8) can be written as

$$u \in K, \tag{1.1}$$

$$\lambda \int_0^1 u'(v' - u')\, dx - \int_0^1 u(v - u)\, dx \geq 0 \qquad \text{for each} \quad v \in K. \tag{1.2}$$

By the usual considerations concerning weak solutions of differential equations or inequalities (an integration by parts and comparisons by test functions satisfying the conditions $v(x_i) = 0$ and then $v(x_i) \geq 0$), it is easy to show that our problem is equivalent to the following one:

$$\lambda u'' + u = 0 \qquad \text{on} \quad (x_i, x_{i+1}), \quad i = 0, \ldots, n, \tag{1.3}$$

where we set $x_0 = 0$, $x_{n+1} - 1$,

$$u(0) = u(1) = 0, \tag{1.4}$$

$$u(x_i) \geq 0, \qquad i = 1, \ldots, n, \tag{1.5a}$$

$$u'(x_i - 0) - u'(x_i + 0) \geq 0, \qquad i = 1, \ldots, n, \tag{1.5b}$$

$$u(x_i)[u'(x_i - 0) - u'(x_i + 0)] = 0, \qquad i = 1, \ldots, n. \tag{1.5c}$$

Here condition (1.3) is meant in the classical sense; i.e., the function u is supposed to have its second derivative continuous on (x_i, x_{i+1}).

We shall say that a function u is a fundamental solution of the problem (1.1)–(1.2), (i.e., (1.3)–(1.5)) if u satisfies (1.1), (1.2) (i.e., (1.3)–(1.5)) with some real λ and $u(x_i) = 0$ for each $i = 1, 2, \ldots, n$. First, we shall seek fundamental solutions only. Let u be such a solution. Then we obtain by calculation that for each interval (x_i, x_{i+1}) $(i = 0, 1, \ldots, n)$ on which u does not vanish identically, we have

$$u(x) = \sin K_i \pi \frac{x - x_i}{x_{i+1} - x_i} \qquad \text{for all} \quad x \in (x_i, x_{i+1}), \tag{1.6}$$

$$\lambda = \frac{(x_{i+1} - x_i)^2}{K_i^2 \pi^2}, \tag{1.7}$$

where K_i is a suitable integer (different from zero). The number λ must be the same for all the intervals. Hence if there are two intervals (x_i, x_{i+1}), (x_j, x_{j+1}) on which u is not identically zero, then we have

$$\frac{K_i}{K_j} = \frac{x_{j+1} - x_j}{x_{i+1} - x_1}. \tag{1.8}$$

First, let us consider the case in which

(*) $(x_{j+1} - x_j)/(x_{i+1} - x_i)$ is irrational for each $i \neq j$ $(i, j = 0, 1, \ldots, n)$.

In this case condition (1.8) cannot be fulfilled for $i \neq j$ and therefore an arbitrary fundamental solution is different from zero on one interval (x_i, x_{i+1}) only. Using (1.5b) we obtain for this interval that

$$\text{if} \quad x_i \neq 0, \quad \text{then} \quad u'(x_i + 0) \leq 0, \tag{1.9a}$$

$$\text{if} \quad x_{i+1} \neq 1, \quad \text{then} \quad u'(x_{i+1} - 0) \geq 0. \tag{1.9b}$$

On the whole we see that in the case (*) a function $u \in \mathring{W}_2^1(\langle 0, 1\rangle)$ is the fundamental solution of problem (1.1)–(1.2) if and only if u satisfies (1.6) on a certain interval (x_i, x_{i+1}) with some integer $K_i \neq 0$, u equals zero on the other intervals (x_j, x_{j+1}), $j \neq i$, $j = 0, 1, \ldots, n$, and (1.9a) and (1.9b) are valid (see Fig. 4). The corresponding eigenvalues are given by formula (1.7). The multiplication of a fundamental solution by a positive constant gives a fundamental solution, too. On the other hand, two solutions corresponding to the same eigenvalue are equal to each other except for a positive multiplicative constant. In other words, we can say that these eigenvalues are simple.

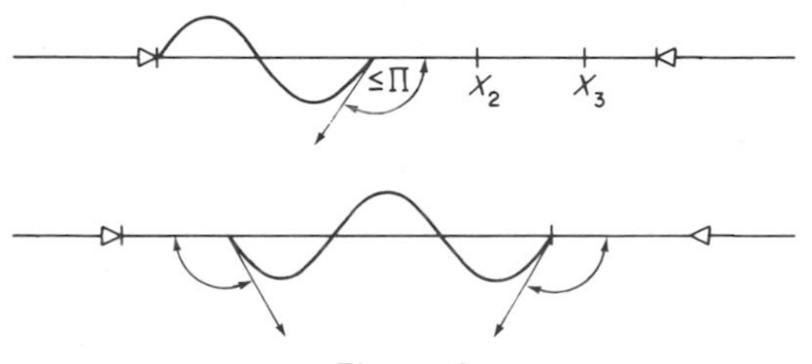

Figure 4

Now let us suppose that

(**) $(x_{j+1} - x_j)/(x_{i+1} - x_i) = p/q$ with positive integers p, q for some $i \neq j$.

In addition to solutions of the type just described, we can obtain some of their combinations that are fundamental solutions, too. Precisely, consider positive integers K_i, K_j satisfying (1.8). The functions $u_i(x)$ (or $u_j(x)$) satisfying (1.6) (or (1.6) with j instead of i) and defined as zero on the remainder of the interval $\langle 0, 1\rangle$ are two fundamental solutions and formula (1.7) gives the same λ for i and j instead of i. It is clear that an arbitrary linear combination $au_i + bu_j$, $a \geq 0$, $b \geq 0$ is a fundamental solution. We can say that the multiplicity of the eigenvalue λ is equal to two. Moreover, if $x_i = x_{j+1}$, then we have only the condition (1.5b) instead of (1.9a), (1.9b) at the point x_i. Therefore, we can obtain some further new eigenvalues and fundamental solutions which are not given by the procedure just described (see Fig. 5). Obviously, these eigenvalues are also given by formula (1.7) for some i and some integer K_i, and these fundamental solutions satisfy (1.6) on an arbitrary interval (x_i, x_{i+1}) with some integer K_i.

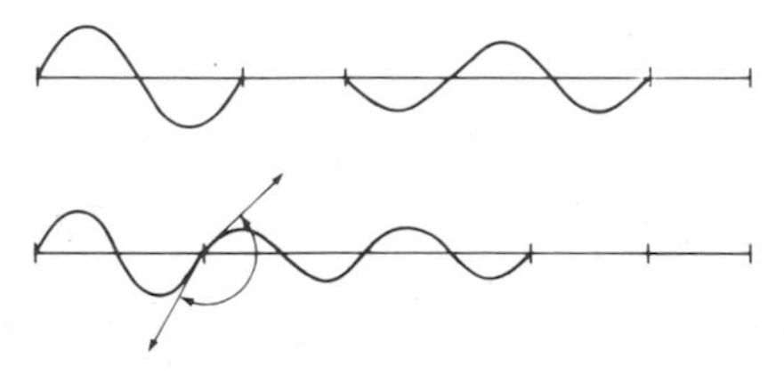

Figure 5

A similar situation occurs in the case in which there exist several intervals such that their lengths are in a rational relation to each other. In the case of r such intervals we obtain eigenvalues of multiplicity r (in a sense analogous to that which was explained above).

We have now described all the fundamental solutions corresponding to the obstacle $B = \{x_1, \ldots, x_n\}$. If u is an arbitrary solution, then we can take a subset B_0 of all the points from B at which u equals zero. (It is possible that $B_0 = \varnothing$.) Hence, u is a fundamental solution with respect to this new obstacle B_0 (i.e., in the case $B_0 = \varnothing$, u is a classical solution of the problem (1.3), (1.4), where we write (0, 1) instead of (x_i, x_{i+1})). But such solutions can be described analogously as above. Hence, it is sufficient to consider all subsets $B_0 \subset B$ (including the empty set), take all fundamental solutions with respect to these sets, and select those of them that are nonnegative at all the remaining points from B. In this way, we obtain all the solutions of the problem (1.1)–(1.2).

Remark 1.1 It may happen that the eigenvalues corresponding to the fundamental solutions of our problem (i.e., fundamental with respect to B) are simple and that some eigenvalues corresponding to the solutions fundamental with respect to some subset B_0 of B are multiple.

2. General Theory. The Penalty Method and a New Version of the Ljusternik–Schnirelmann Theory

Let H be a Hilbert space with the inner product $\langle \cdot, \cdot \rangle$ and with the corresponding norm $\|\cdot\|$. We shall consider two even nonnegative functionals f, g on H. We shall suppose that f, g have the Fréchet derivatives

f', g' on H and satisfy the following assumptions:

(2a) $\lim_{\|u\| \to +\infty} f(u) = +\infty$;
(2b) $f(0) = g(0) = 0$;
(2c) $f', g'\colon H \to H$ are continuous and bounded mappings;
(2d) $\langle f'(u) - f'(v), u - v\rangle \geq C\|u - v\|^2$ for all $u, v \in H$;
(2e) if $u_n \rightharpoonup u$ (weak convergence in H), then $g'(u_n) \to g'(u)$ (strong convergence in H);
(2f) if $\langle g'(u), u\rangle = 0$, then $u = 0$.

Further, let K be a closed convex cone with the vertex at the origin of the space H, i.e., a closed convex set satisfying the conditions

(*) if $u \in K$, then $tu \in K$ for each $t \geq 0$;
(**) if $u, v \in K$, then $u + v \in K$.

Consider a Fréchet differentiable functional Φ (a penalty functional) on H such that

(2g) $\Phi(u) = 0$ for each $u \in K$, $\Phi(u) > 0$ for each $u \in H \backslash K$;
(2h) $\Phi'\colon H \to H$ is a continuous and bounded mapping, $\Phi'(u) = 0$ if and only if $u \in K$;
(2i) $\langle \Phi'(u) - \Phi'(v), u - v\rangle \geq 0$ for all $u, v \in H$.

For arbitrary positive ε, r we set

$$\varphi_\varepsilon(u) = f(u) + \varepsilon\Phi(u) \quad \text{for each} \quad u \in H,$$

$$M_r(\varphi_\varepsilon) = \{u \in H;\ \varphi_\varepsilon(u) = r\}, \qquad M_r(f) = M_r(\varphi_0).$$

For given positive ε and r, we shall consider the eigenvalue problem

$$u \in M_r(f), \tag{2.1}$$

$$\lambda\varphi_\varepsilon'(u) - g'(u) = 0. \tag{2.2}$$

In accord with the program announded in the Introduction, we shall define the classes V_k^ε $(k = 1, 2, \ldots)$ replacing the Ljusternik–Schnirelmann categories in our next considerations. Consider the abstract ordinary differential equation

$$\dot{u}(t) = g'(u(t)) - \varphi_\varepsilon'(u(t)) \frac{\langle \varphi_\varepsilon'(u(t)), g'(u(t))\rangle}{\langle \varphi_\varepsilon'(u(t)), \varphi_\varepsilon'(u(t))\rangle} \quad \text{for} \quad t \geq 0, \tag{2.3}$$

$$u(0) = u_0, \tag{2.4}$$

where $u_0 \in H$ and $\dot{u}$ denotes the derivative of an abstract function u with respect to the variable t. For each $u_0 \in M_r(\varphi_\varepsilon)$, there exists a solution $u(t)$ of problem (2.3)–(2.4) and $u(t)$ lies on $M_r(\varphi_\varepsilon)$ for all $t \geq 0$ (because $\dot{u}(t)$ is a tangent vector to $M_r(\varphi_\varepsilon)$). Let us denote this solution by $u(t, u_0)$. If C is a subset of H and $\delta > 0$, then we set $u(\delta, C) = \{u \in H;\ u = u(\delta, u_0),\ u_0 \in C\}$. Now, let us define V_k^ε as the class of all the sets $u(\delta, C)$, where δ is an arbitrary nonnegative number and C is an arbitrary set of the type $C = H_k \cap M_r(\varphi_\varepsilon)$, H_k is a finite-dimensional linear subspace of H such that $\dim H_k \geq k$. In other words, the class V_k^ε contains all the sets obtained as an intersection of $M_r(\varphi_\varepsilon)$ with an arbitrary finite-dimensional subspace H_k of H with $\dim H_k \geq k$, and at the same time it contains arbitrary deformations of these sets along trajectories given by Eq. (2.3).

Analogously to the usual Ljusternik–Schnirelmann theory, we set

$$\gamma_k^\varepsilon = \sup_{C \in V_k^\varepsilon} \min_{u \in C} g(u).$$

Remark 2.1 It follows from assumption (2a) that the sets $M_r(\varphi_\varepsilon)$ are uniformly bounded with respect to $\varepsilon \geq 0$ (for a given $r > 0$).

Remark 2.2 The functional f is nonnegative and $f(0) = 0$; therefore $f'(0) = 0$. In particular, setting $v = 0$ in (2d), we obtain

$$\langle f'(u), u\rangle \geq C\|u\|^2 \qquad \text{for each} \quad u \in H.$$

By a similar consideration for the functional Φ (using (2i)) we obtain $\langle \Phi'(u), u\rangle \geq 0$. Hence, we have

$$\langle \varphi_\varepsilon'(u), u\rangle \geq C\|u\|^2 \qquad \text{for each} \quad u \in H, \quad \varepsilon \geq 0.$$

Remark 2.3 For an arbitrary $\varepsilon \geq 0$, the functional φ_ε satisfies the following condition:

(S) if $u_n \rightharpoonup u$, $\varphi_\varepsilon'(u_n) \to v$ for some $v \in H$, then $u_n \to u$, $\varphi_\varepsilon'(u) = v$.

Indeed, with respect to (2d), (2i) we have

$$C\|u_n - u\|^2 \leq \langle \varphi_\varepsilon'(u_n) - \varphi_\varepsilon'(u), u_n - u\rangle,$$

where the right-hand side converges to zero under the assumption $u_n \rightharpoonup u$, $\varphi_\varepsilon'(u_n) \to v$.

THEOREM 2.1 Suppose that assumptions (2a)–(2i) are satisfied. Then for each positive integer k there exists $u_k^\varepsilon \in H$ satisfying (2.1)–(2.2) with some real λ_k^ε and such that

$$g(u_k^\varepsilon) = \gamma_k^\varepsilon.$$

Proof Let k be a given positive integer. There exist sets $K_n \in V_k^\varepsilon$ $(n = 1, 2, \ldots)$ such that

$$\mu_n = \min_{u \in K_n} g(u) \xrightarrow[n \to \infty]{} \gamma_k^\varepsilon.$$

For each positive δ we write

$$Q_n(\delta) = \{u \in K_n;\ \mu_n \le g(u) \le \mu_n + \delta\}.$$

We shall prove that for each $\delta > 0$ we have

$$\lim_{n\to\infty} \min_{u \in Q_n(\delta)} \|T(u)\| = 0, \tag{2.5}$$

where we write

$$T(u) = g'(u) - \varphi_\varepsilon'(u)\frac{\langle \varphi_\varepsilon'(u), g'(u)\rangle}{\langle \varphi_\varepsilon'(u), \varphi_\varepsilon'(u)\rangle}.$$

Suppose the contrary. Then we can suppose that

$$\min_{u \in Q_n(\delta)} \|T(u)\|^2 \ge v \tag{2.6}$$

with some $v > 0$. Using Eq. (2.3) we obtain by calculation that

$$\begin{aligned} \frac{d}{dt}(g(u(t, u_0))) &= \langle g'(u(t, u_0)), \dot{u}(t)\rangle \\ &= \langle g'(u(t, u_0)), g'(u(t, u_0)) \\ &\quad - \frac{\langle g'(u(t, u_0)), \varphi_\varepsilon'(u(t, u_0))\rangle^2}{\langle \varphi_\varepsilon'(u(t, u_0)), \varphi_\varepsilon'(u(t, u_0))\rangle} \\ &= \|T(u(t, u_0))\|^2. \end{aligned} \tag{2.7}$$

The operator T is continuous on $M_r(\varphi_\varepsilon)$ ($u(t, u_0)$ is continuous as the mapping of the variables t, u_0), and the set $Q_n(\delta)$ is compact. It follows from this and (2.6) that there exists $\eta > 0$ such that

$$\|T(u(t, u_0))\|^2 \ge v/2 \qquad \text{for each} \quad u_0 \in Q_n(\delta), \quad t \in \langle 0, \eta\rangle.$$

Hence, using (2.7) we obtain

$$g(u(\eta, u_0)) \ge \mu_n + (v/2)\eta \qquad \text{for each} \quad u_0 \in Q_n(\delta).$$

Simultaneously, we have

$$g(u(\eta, u_0)) \ge \mu_n + \delta \qquad \text{for each} \quad u_0 \in K_n \backslash Q_n(\delta).$$

This means that

$$\begin{aligned} &\min_{u \in u(\eta, K_n)} g(u) \\ &\ge \min(\mu_n + (v/2)\eta, \mu_n + \delta) \xrightarrow[n\to\infty]{} \min(\gamma_k^\varepsilon + (v/2)\eta, \gamma_k^\varepsilon + \delta) > \gamma_k^\varepsilon, \end{aligned}$$

which is impossible with respect to the definition of the number γ_k^ε. Hence (2.5) is proved.

Now let us consider a sequence δ_n ($n = 1, 2, \ldots$) of positive numbers $\delta_n \to 0$. It follows from (2.5) that there exists an increasing sequence of indices m_n ($n = 1, 2, \ldots$) such that

$$\lim_{n \to +\infty} \left(\min_{u \in Q_{m_n}(\delta_n)} \|T(u)\| \right) = 0. \tag{2.8}$$

Denote by u_n a point from Q_{m_n} such that

$$\|T(u_n)\| = \min_{u \in Q_{m_n}(\delta_n)} \|T(u)\|. \tag{2.9}$$

The set $M_r(\varphi_\varepsilon)$ is bounded (see Remark 2.1) and therefore we can suppose $u_n \rightharpoonup u$. Further, by the assumptions (2c) and (2h) and Remark 2.2 we can suppose that

$$\frac{\langle \varphi_\varepsilon'(u_n), g'(u_n) \rangle}{\langle \varphi_\varepsilon'(u_n), \varphi_\varepsilon'(u_n) \rangle} \to C_1. \tag{2.10}$$

(Otherwise we could consider a suitable subsequence.) First suppose $C_1 = 0$. Conditions (2e) and (2.8), together with the definition of $T(u)$, imply $g'(u) = 0$. Assumptions (2f) and (2b) give $u = 0$ and $g(u) = 0$. This is impossible because g is continuous in the weak topology (this is a consequence of (2e)) and hence

$$g(u) = \lim_{n \to \infty} g(u_{m_n}) \geq \lim_{n \to \infty} \mu_{m_n} = \gamma_k^{\,\varepsilon} > 0.$$

This means that $C_1 \neq 0$. It follows from (2.10), (2.9), (2.8), (2e), and from the definition of $T(u)$ that $\varphi_\varepsilon'(u_n) \to y$ for some $y \in H$. By Remark 2.3 we have $u_n \to u$. Then $u \in M_r(\varphi_\varepsilon)$ and $g'(u) - C_1 \varphi_\varepsilon'(u) = 0$ by (2.8) and (2.9). Setting $\lambda_k^{\,\varepsilon} = 1/C_1$, $u_k^{\,\varepsilon} = u$, we have (2.2), and the proof is completed.

Up to now, an arbitrary but fixed nonnegative ε was considered. Letting $\varepsilon \to +\infty$ in the sequel, we shall pass to the eigenvalue problem for the variational inequality:

$$u \in M_r(f) \cap K, \tag{2.11}$$

$$\lambda \langle f'(u), v - u \rangle - \langle g'(u), v - u \rangle \geq 0 \qquad \text{for each} \quad v \in K. \tag{2.12}$$

For such considerations the following additional assumptions on the cone K and the penalty functional Φ will be useful:

(2j) $C_1 \langle \Phi'(u), u \rangle \leq \Phi(u) \leq C_2 \|\Phi'(u)\|^{1+\alpha}$ for each $u \in H$, where C_1, C_2, and α are positive constants;

(2k) the set K contains an infite-dimensional subspace of the space H.

Remark 2.4 If condition (2k) is valid, then for each positive integer k there exists $v > 0$ such that $\gamma_k^{\,\varepsilon} \geq v$ for all $\varepsilon \geq 0$. Indeed, if H_k is a k-

dimensional subspace of H, $H_k \subset K$, $C = H_k \cap M_r(f)$, then $C \in V_k^\varepsilon$ for each $\varepsilon \geq 0$ and therefore $\gamma_k^\varepsilon \geq \min_{u \in C} g(u)$. The last number is positive because the functional g is positive on the compact set C by (2b) and (2f).

THEOREM 2.2 Let the assumptions (2a)–(2k) be fulfilled. Let us consider a fixed positive integer k and the corresponding u_k^ε, λ_k^ε ($\varepsilon \geq 0$) from Theorem 2.1. Then there exists a sequence ε_n such that $\varepsilon_n > 0$ ($n = 1, 2, \ldots$), $\varepsilon_n \to +\infty$, $\lambda_k^{\varepsilon_n} \to \lambda_k$, $u_k^{\varepsilon_n} \to u_k$ (if $n \to \infty$), where (2.11), (2.12) hold (with $u = u_k$, $\lambda = \lambda_k$).

Proof Consider a sequence ε_n such that $\varepsilon_n > 0$ ($n = 1, 2, \ldots$), $\varepsilon_n \to +\infty$ and write $\lambda_k{}^n = \lambda_k^{\varepsilon_n}$, $u_k{}^n = u_k^{\varepsilon_n}$, $\gamma_k{}^n = \gamma_k^{\varepsilon_n}$. By the uniform boundedness of M_r (φ_ε) (see Remark 2.1), we can suppose that $u_k{}^n \rightharpoonup u_k$. Furthermore, we have

$$\lambda_k{}^n \langle f'(u_k{}^n) + \varepsilon_n \Phi'(u_k{}^n), u_k{}^n \rangle = \langle g'(u_k{}^n), u_k{}^n \rangle. \tag{2.13}$$

It follows from this by Remark 2.2 and the uniform boundedness of $M_r(\varphi_\varepsilon)$ that the sequence $\lambda_k{}^n$ ($n = 1, 2, \ldots$) is bounded. Hence we can suppose $\lambda_k{}^n \to \lambda_k$ for some λ_k. Let us show that $\lambda_k \neq 0$. Assumption (2j) implies

$$C_1 \langle \varepsilon_n \Phi'(u_k{}^n), u_k{}^n \rangle \leq \varepsilon_n \Phi(u_k{}^n) \leq f(u_k{}^n) + \varepsilon_n \Phi(u_k{}^n) = r.$$

Since the sequence $\langle f'(u_k{}^n), u_k{}^n \rangle$ is bounded, in the case $\lambda_k = 0$ we would obtain $\langle g'(u_k), u_k \rangle = \lim \langle g'(u_k{}^n), u_k{}^n \rangle = 0$ and $u_k = 0$ by virtue of (2.13), (2e), and (2f). Hence, $g(u_k{}^n) \to g(u_k) = 0$. Simultaneously, we have $g(u_k{}^n) \geq \nu > 0$ (see Remark 2.4), and this not possible. This means that $\lambda_k \neq 0$.

Now we shall show that $u_k \in K$. We have

$$\langle \Phi'(v) - \Phi'(u_k{}^n), v - u_k{}^n \rangle \geq 0 \qquad \text{for each} \quad v \in H \tag{2.14}$$

because of (2i). It follows from Eq. (2.2) that the sequence $\varepsilon_n \Phi'(u_k{}^n)$ is bounded and therefore $\Phi'(u_k{}^n) \to 0$. Hence, letting $n \to \infty$ and setting $v = u_k + tw$, $t > 0$, $w \in H$, we obtain from (2.14)

$$\langle \Phi'(u_k + tw), w \rangle \geq 0 \qquad \text{for each} \quad t > 0, \quad w \in H.$$

Letting $t \to 0+$, we obtain

$$\langle \Phi'(u_k), w \rangle \geq 0 \qquad \text{for each} \quad w \in H,$$

which is equivalent to the equation $\Phi'(u_k) = 0$, i.e., $u_k \in K$ (see (2h)).

Now we have

$$\lambda_k{}^n \langle f'(u_k{}^n) + \varepsilon_n \Phi'(u_k{}^n), v - u_k{}^n \rangle = \langle g'(u_k{}^n), v - u_k{}^n \rangle \tag{2.15}$$

for each $v \in H$. Moreover, assumptions (2h), (2i) imply that for $v \in K$

$$\langle \Phi'(u_k{}^n), v - u_k{}^n \rangle = \langle \Phi'(u_k{}^n) - \Phi'(v), v - u_k{}^n \rangle \leq 0, \tag{2.16}$$

and this together with (2.15) gives

$$\lambda_k{}^n\langle f'(u_k{}^n), v - u_k{}^n\rangle \geq \langle g'(u_k{}^n), v - u_k{}^n\rangle \qquad \text{for each} \quad v \in K. \tag{2.17}$$

Let us show that $u_k{}^n \to u_k$. It follows from (2.17) (for $v = u_k$) that

$$\lambda_k{}^n\langle f'(u_k{}^n) - f'(u_k), u_k{}^n - u_k\rangle \leq \langle g'(u_k{}^n), u_k{}^n - u_k\rangle$$
$$- \lambda_k{}^n\langle f'(u_k), u_k{}^n - u_k\rangle.$$

The right-hand side converges to zero (as $n \to \infty$) and the left-hand side is not less than $C_1 \|u_k{}^n - u_k\|^2$ with a certain positive constant C_1 because of $\lambda_k{}^n \to \lambda_k > 0$ and (2d). Hence $u_k{}^n \to u_k$. Now, letting $n \to \infty$ in (2.17), we obtain (2.12).

It remains to show that $u_k \in M_r(f)$. We have

$$f(u_k{}^n) + \varepsilon_n \Phi(u_k{}^n) = r. \tag{2.18}$$

By virtue of (2j)

$$\varepsilon_n \Phi(u_k{}^n) \leq C_2 \|\varepsilon_n \Phi'(u_k{}^n)\| \cdot \|\Phi'(u_k{}^n)\|^\alpha,$$

where the right-hand side converges to zero because $\varepsilon_n \Phi'(u_k{}^n)$ is bounded and $\Phi'(u_k{}^n) \to 0$, as we proved above. Letting $n \to \infty$ in (2.18), we obtain $f(u_k) = r$. The proof is complete.

Remark 2.5 Let us consider that the functionals f and Φ are positive two-homogeneous and the functional g is positive $(b + 1)$-homogeneous with $b > 0$ (i.e., $f(tu) = t^2 f(u)$, $\Phi(tu) = t^2\Phi(u)$, $g(tu) = t^{b+1}g(u)$ for each $u \in H$, $t > 0$). If u is a critical point of problem (2.11)–(2.12) and λ is the corresponding eigenvalue, then we have

$$\lambda = \frac{\langle g'(u), u\rangle}{\langle f'(u), u\rangle} = \frac{(b + 1)g(u)}{2f(u)} = \frac{b + 1}{2r}\, g(u).$$

In other words, the set of all critical levels equals the set of all eigenvalues except for a multiplicative constant (cf. Fučík *et al.* [2]).

Remark 2.6 The critical level $g(u_k)$ from Theorem 2.2 can be denoted by γ_k and called the Ljusternik–Schnirelmann critical level (of order k) of our problem. Unfortunately, we do not know whether the sequence γ_k $(k = 1, 2, \ldots)$ really contains infinitely many distinct numbers (cf. Section 3).

Remark 2.7 There is a problem of the existence of the penalty functional satisfying our conditions. For each $u \in H$ there exists a unique point $P_K u \in K$ such that

$$\|u - P_K u\| = \inf_{v \in K} \|u - v\|.$$

Further, this point $P_K u$ is the unique point satisfying the condition

$$\langle u - P_K u, v - P_K u\rangle \le 0 \qquad \text{for each} \quad v \in K. \tag{2.19}$$

The mapping P_K is called the projection on the cone K. It is known that P_K is a monotone Lipschitz mapping (cf. Zarantonello [7]) and that the mapping $I - P_K$ is also a projection on a certain convex cone with vertex at the origin (see Zarantonello [7]).

LEMMA 2.1 (cf. Zarantonello [7]) The functional Φ, defined by the formula

$$\Phi(v) = \tfrac{1}{2}\|(I - P_K)v\|^2, \qquad v \in H,$$

is differentiable in the sense of Fréchet and

$$\Phi'(u) = (I - P_K)(u).$$

(This lemma is contained in Zarantonello [7], but for the sake of completeness we give a proof.)

Proof It is well known that the continuous Gateaux derivative of some functional is also the derivative in the sense of Fréchet. Thus it is sufficient to prove the existence of the Gateaux derivative because P_K is clearly continuous. For $u, h \in H$ and t real we have

$$\begin{aligned}(1/t)[\tfrac{1}{2}\|(I - P_K)(u + th)\|^2 - \tfrac{1}{2}\|(I - P_K)(u)\|^2]\\ = (1/t)[\tfrac{1}{2}\|(I - P_K)(u + th) - (I - P_K)(u)\|^2\\ - \|(I - P_K)(u)\|^2 + \langle (I - P_K)(u + th), (I - P_K)(u)\rangle].\end{aligned}$$

The first term on the right-hand side converges to zero for $t \to 0$ because $(I - P_K)$ is a Lipschitz mapping. The remainder of the right-hand side can be written as

$$\begin{aligned}&- (1/\mathrm{t})\|(I - P_K)(u)\|^2 + \langle h, (I - P_K)u\rangle\\ &\quad + (1/t)\langle u - P_K(u + th), (I - P_K)(u)\rangle\\ &= \langle h, (I - P_k)(u)\rangle + (1/t)\langle P_K u - P_K(u + th), (I - P_K)(u)\rangle.\end{aligned}$$

It is sufficient to show that the last expressions converge to zero as $t \to 0$. They can be written as

$$\begin{aligned}&(1/t)\langle P_K u - P_K(u + th), (I - P_K)(u + th)\rangle\\ &\quad + (1/t)\langle P_K u - P_K(u + th), (I - P_K)(u) - (I - P_K)(u = th)\rangle.\end{aligned} \tag{2.20}$$

The second term converges to zero because the mappings P_K, $I - P_K$ are Lipschitz. We shall show that

$$\lim_{t\to 0} \frac{\langle P_K u - P_K(u + th), (I - P_K)(u + th)\rangle}{\|P_K u - P_K(u + th)\|} = 0. \tag{2.21}$$

It will follow that the first expression in (2.20) also converges to zero. Since $P_K u \in K$,

$$\langle P_K u - P_K(u + th), (I - P_K)(u + th)\rangle \leq 0.$$

(see (2.19)). If (2.21) is not true, then there exists a sequence t_n, $t_n \to 0$, and a constant $C < 0$ such that

$$\frac{\langle P_K(u) - P_K(u + t_n h), (I - P_K)(u + t_n h)\rangle}{\|P_K u - P_K(u + t_n h)\|} \leq C.$$

We can suppose that

$$\frac{P_K u - P_K(u + t_n h)}{\|P_K u - P_K(u + t_n h)\|} \rightharpoonup y$$

for some $y \in H$. We have

$$\langle y, (I - P_K)(u)\rangle \leq C,$$

but simultaneously

$$\langle P_K u - P_K(u + t_n h), (I - P_K)u\rangle \geq 0$$

(by (2.19)) and therefore

$$\langle y, (I - P_K)(u)\rangle \geq 0.$$

This is a contradiction and (2.21) is proved.

Remark 2.8 Because of Lemma 2.1 and Remark 2.7, it is easy to see that the functional Φ fulfills assumptions (2g)–(2i).

Moreover, it is proved by Zarantonello [7] that

$$\langle u, P_K u\rangle = \|P_K u\|^2. \tag{2.22}$$

(This follows from the fact that an arbitrary point $u \in H$ can be written as $u = P_K u + P_{\tilde{K}} u$, where $\tilde{K}$ is some dual cone and $\langle P_{\tilde{K}} u, P_K u\rangle = 0$.) But the mapping $I - P_K$ is a projection too (see Remark 2.7). Hence, we can write $I - P_K$ instead of P_K in (2.22). The left-hand part of inequality (2j) follows from this and the right-hand part of (2j) is clear too. Thus, the functional Φ defined in Lemma 2.1 satisfies all our assumptions.

3. Special Case of a Halfspace

In this section, we shall suppose that K is a halfspace. Precisely, we shall suppose that K satisfies the assumptions introduced in Section 2 and

$$u \in K^0 \qquad \text{if and only if} \qquad -u \notin K, \tag{3.1}$$

where K^0 denotes the interior of the set K. Let us consider nonnegative even functionals f, g satisfying the usual assumptions of the Ljusternik–Schnirelmann theory. Precisely, in addition to conditions (2a)–(2f), we shall suppose that for each $\eta > 0$ there exists $\delta > 0$ such that the inequalities

$$\|f'(x+h) - f'(x)\| \leq \eta, \qquad \|g'(x+h) - g'(x)\| \leq \eta \tag{3.2}$$

hold for each $x \in M_r(f)$ and all $h \in X$ with $\|h\| \leq \delta$ (cf. Fučík *et al.* [2]). The classical Ljusternik–Schnirelmann theory gives the existence of points $u_k \in H$ ($k = 1, 2, \ldots$) such that

$$g(u_k) = \sup_{D \in W_k} \min_{u \in D} g(u). \tag{3.3}$$

Here W_k is the class of all the compact sets $D \subset M_r(f)$ satisfying the condition cat $D \geq k$, where cat D is the usual Ljusternik–Schnirelmann category (or order, see Fučík *et al.* [2]) of the set D in $M_r(f)$. These points u_k ($k = 1, 2, \ldots$) satisfy the conditions

$$u_k \in M_r(f), \tag{3.4}$$

$$\lambda_k f'(u_k) - g'(u_k) = 0 \tag{3.5}$$

with some real λ_k. Moreover,

$$\lim_{k \to \infty} \gamma_k{}^0 = 0, \tag{3.6}$$

where $\gamma_k{}^0 = g(u_k)$. The elements $-u_k$ also satisfy conditions (3.4), (3.5). Moreover, because of assumption (3.1), at least one of the points u_k, $-u_k$ lies in K.

Hence, this point is a critical point of the problem

$$u \in M_r(f) \cap K, \tag{3.7}$$

$$\lambda\langle f'(u), v - u\rangle - \langle g'(u), v - u\rangle \geq 0 \qquad \text{for each} \quad v \in K, \tag{3.8}$$

and λ_k is the corresponding eigenvalue of this problem. In particular, problem (3.7)–(3.8) has all the eigenvalues and critical levels of problem (3.4)–(3.5) and these critical levels converge to zero. The aforementioned critical points of problem (3.7)–(3.8) can lie in K^0 or in ∂K. Moreover, we shall prove the existence of an infinite sequence of critical points $\tilde{u}_k \in \partial K$ with the corresponding critical levels converging to zero.

For the case in which problem (3.4)–(3.5) does not have infinitely many critical points on ∂K, this gives the existence of the sequence of infinitely many new critical points of the variational inequality (3.7), (3.8), which are not critical points of the equation (3.4), (3.5). The proof of the last assertion is based on the penalty method as in Section 2, but instead of the functional Φ we use an even functional Ψ defined by

$$\Psi(u) = \Phi(u) + \Phi(-u) \qquad \text{for each} \quad u \in H.$$

The functional Φ is supposed to possess properties (2g)–(2i). Analogously to Section 2, we set

$$\Psi_\varepsilon(u) = f(u) + \varepsilon\Psi(u), \quad u \in H; \qquad M_r(\Psi_\varepsilon) = \{u \in H;\ \Psi_\varepsilon(u) = r\}.$$

For given positive ε and r, let us consider the problem

$$u \in M_r(\Psi_\varepsilon), \tag{3.9}$$

$$\lambda\Psi_\varepsilon'(u) - g'(u) = 0. \tag{3.10}$$

Under our assumptions the usual Ljusternik–Schnirelmann theory is applicable to the functionals Ψ_ε, g as well as to f, g. Thus, for each positive integer k there exists $u_k{}^\varepsilon$ such that

$$g(u_k{}^\varepsilon) = \sup_{D \in W_k{}^\varepsilon} \min_{u \in D} g(u), \tag{3.11}$$

where $W_k{}^\varepsilon$ is the class of all the compact sets $D \subset M_r(\Psi_\varepsilon)$ with cat $D \geq k$; cat D denotes the usual Ljusternik–Schnirelmann category of the set D in $M_r(\Psi_\varepsilon)$. These points $u_k{}^\varepsilon$ are the solutions of problem (3.9)–(3.10) with some $\lambda_k{}^\varepsilon$ and

$$\lim_{k\to\infty} \gamma_k{}^\varepsilon = 0 \tag{3.12}$$

(for ε fixed), where $\gamma_k{}^\varepsilon = g(u_k{}^\varepsilon)$. Because of the symmetry and assumption (3.1), we can choose $u_k{}^\varepsilon \notin K^0$. Now we can use the method of the proof of Theorem 2.2. This gives the existence of a sequence ε_n $(n = 1, 2, \ldots)$ such that $\varepsilon_n > 0$, $\varepsilon_n \to \infty$, $\lambda_k^{\varepsilon_n} \to \tilde{\lambda}_k$, $u_k^{\varepsilon_n} \to \tilde{u}_k$, where $\tilde{\lambda}_k$, $\tilde{u}_k$ satisfy (3.7), (3.8) (with $u = \tilde{u}_k$, $\lambda = \tilde{\lambda}_k$), Moreover, we obtain $\tilde{u}_k \in \partial K$ because $u_k^{\varepsilon_n} \notin K^0$, $\tilde{u}_k \in K$. Let us suppose that the functional g satisfies the additional assumption

$$g(\mathscr{H}u) \leq g(u) \qquad \text{for each} \quad u \in H, \quad 0 < \mathscr{H} \leq 1. \tag{3.13}$$

For a set $D_\varepsilon \subset M_r(\Psi_\varepsilon)$ we have cat $D_\varepsilon = m$ if and only if cat $D = m$, where $D \subset M_r(f)$, is a projection of D_ε into $M_r(f)$ constructed with the help of the halflines originating at the origin. This follows from the usual definition of the Ljusternik–Schnirelmann categories.

It follows from this and from (3.3), (3.11), (3.13) that $\gamma_k^{\varepsilon} \leq \gamma_k^{0}$. Thus, we have $\tilde{\gamma}_k \leq \gamma_k^{0}$, where $\tilde{\gamma}_k = g(\tilde{u}_k)$. In particular $\lim_{k\to\infty} \tilde{\gamma}_k = 0$ because of (3.6).

Let us remark that the situation described in this section corresponds to the one-point obstacle in the case of the supported beam in the Introduction or in the example from Section 1. If this one-point obstacle is placed at an irrational number, then the equation has no critical points on ∂K and therefore we obtain infinitely many new critical points of problem (3.7)–(3.8) with the corresponding critical levels converging to zero. These critical points lie in ∂K (i.e., the functions $\tilde{u}_k$ touch the obstacle) and hence they are not the critical points of the corresponding equation.

If the functionals f and Φ are two-homogeneous and the functional g is $(b + 1)$-homogeneous with $b > 0$ (i.e., $f(tu) = t^2 f(u)$, $g(tu) = t^{b+1} g(u)$ for each $u \in H$, $t > 0$), then we have

$$\lambda_k^{\varepsilon} = \frac{b+1}{2r} \gamma_k^{\varepsilon}$$

(see Fučik *et al.* [2]). It follows from this that

$$\tilde{\lambda}_k = \frac{b+1}{2r} \tilde{\gamma}_k$$

and therefore $\lim_{k\to\infty} \tilde{\lambda}_k = 0$.

4. An Open Problem

Let us consider the situation from Section 2 again. We have proved that for each $\varepsilon \geq 0$ there exists u_k^{ε} $(k = 1, 2, \ldots)$ satisfying (2.1), (2.2) with some λ_k^{ε}. Let us consider the following additional assumption:

the critical point u_k^{ε} is unique for each λ_k^{ε} and u_k^{ε} is differentiable as an abstract function of the variable ε. (4.1)

Then the function $\gamma_k^{\varepsilon} = g(u_k^{\varepsilon})$ is nonincreasing on $< 0, +\infty)$ (for k fixed). Under the assumptions of the usual Ljusternik–Schnirelmann theory we have $\gamma_k^{0} \to 0$ (cf. Section 3) and therefore we obtain $\gamma_k \to 0$ too. In particular, we obtain infinitely many critical levels and critical points of problem (2.11)–(2.12).

Proof Denote by $\dot{u}_k^{\varepsilon}$ the derivative of u_k^{ε} with respect to ε. Then

$$\frac{d}{d\varepsilon}(\gamma_k^{\varepsilon}) = \langle g'(u_k^{\varepsilon}), \dot{u}_k^{\varepsilon}\rangle = \lambda_k^{\varepsilon}\langle f'(u_k^{\varepsilon}) + \varepsilon\Phi'(u_k^{\varepsilon}), \dot{u}_k^{\varepsilon}\rangle.$$

We have

$$f(u_k^\varepsilon) + \varepsilon\Phi(u_k^\varepsilon) = r,$$

and therefore

$$\langle f'(u_k^\varepsilon) + \varepsilon\Phi'(u_k^\varepsilon), \dot{u}_k^\varepsilon\rangle + \Phi(u_k^\varepsilon) = 0.$$

Substituting into (4.1), we obtain

$$\frac{d}{d\varepsilon}(\gamma_k^\varepsilon) = -\lambda_k^\varepsilon\Phi(u_k^\varepsilon).$$

From assumptions (2d), (2f) it follows that $\lambda_k^\varepsilon > 0$ and therefore the last expression is not positive. Hence the function γ_k^ε is nonincreasing.

Open Problems

(1) Whether assumption (4.1) is fulfilled in concrete examples.
(2) Whether assumption (4.1) can be weakened or removed.

References

1. J. L. Lions, "Quelques Methodes de Résolution des Problèmes aux Limites non Linéaires." Dunod, Paris, 1969.
2. S. Fučík, J. Nečas, J. Souček, and V. Souček, "Spectral Analysis of Nonlinear Operators," Lecture Notes in Mathematics, Vol. 346. Springer-Verlag, Berlin and New York, 1973.
3. L. A. Ljusternik and L. G. Schnirelmann, Application of topology to variational problems (Russian). *Trudy Vsesojuz. Mat. Sjezda* **1** (1935), 224–237.
4. S. Fučík and J. Nečas, Ljusternik–Schnirelmann theorem and nonlinear eigenvalue problems. *Math. Nachr.* **53** (1972), 277–289.
5. J. Nečas, The numerical method of finding critical points of even functionals (Russian). *Trudy Mat. Inst. Stoklov.* **134** (1975), 235–239.
6. S. Fučík, J. Nečas, J. Souček, and V. Souček, Upper bound for the number of eigenvalues for nonlinear operators. *Ann. Scuola Norm. Sup. Pisa Sci. Fis. Mat.* **27** (1973), 53–71.
7. E. H. Zarantonello, Projections on convex sets in hilbert space and spectral theory. *In* "Contributions to Nonlinear Functional Analysis" (E. H. Zarantonello, ed.). Academic Press, New York, 1971.

AMS (MOS) 1970 Subject Classification: 58E05, 34B25.

Nonlinear Boundary Value Problems for Ordinary Differential Equations: From Schauder Theorem to Stable Homotopy

Jean Mawhin

Université de Louvain, Belgium

1. Introduction

Since 1920 topological methods have played a fundamental role in proving existence theorems for nonlinear differential and integral equations. The pioneering paper of Birkhoff and Kellogg [2] extending the Brouwer fixed point theorem to some function spaces was undoubtedly motivated by existence theorems in analysis and contains applications to nonlinear boundary value problems for ordinary differential equations. Also the importance of the theory of nonlinear elliptic equations in the genesis of the work of Schauder [23] and Leray and Schauder [8] is well known.

The fundamental contributions of Professor Erich Rothe to degree theory in infinite dimensional spaces [17–21], and in particular his fixed point theorem and his generalization of the Hopf theorem, are still influencing deeply this domain of nonlinear functional analysis. It seemed therefore appropriate, in a paper dedicated to him, to show how much activity today still involves the interaction between topological methods and differential equations in the specific domain of nonlinear boundary value problems for ordinary differential equations.

All the results are stated and proved for second-order differential equations with bounded right-hand sides, but similar results can obviously be obtained by the same approach for higher-order differential equations with quasibounded right-hand sides (Mawhin [14]) of sufficiently small quasinorm.

2. Formulation of the Problem

We shall consider the following class of nonlinear boundary value problems for ordinary differential equations. Let $I = [0, 1]$,

$$f\colon I \times R^n \times R^n \to R^n, \qquad (t, x, y) \mapsto f(t, x, y)$$

be a continuous mapping such that for some $M \geq 0$ and all $(t, x, y) \in I \times R^n \times R^n$, one has

$$|f(t, x, y)| \leq M, \tag{2.1}$$

where $|\cdot|$ denotes a norm in R^n. Denote by $C^1(I)$ the Banach space of continuously differentiable mappings $x\colon I \to R^n$ with the usual norm

$$\|x\| = \max\left(\max_{t \in I} |x(t)|, \qquad \max_{t \in I} |x'(t)|\right)$$

$(x'(t) = dx/dt)$, and let

$$g\colon C^1(I) \to R^m, \qquad x \mapsto g(x)$$

be a continuous mapping with $0 \leq m \leq 2n$ an integer. We shall be interested in determining the solutions x of the vector ordinary differential equation

$$x''(t) = f(t, x(t), x'(t)) \tag{2.2}$$

which are defined over I and satisfy the boundary conditions

$$g(x) = 0. \tag{2.3}$$

Without loss of generality we can assume that there exists an integer $0 \leq p \leq m$ such that $g = (\tilde{l}, k)$ where, if $q = m - p$,

(a) $\tilde{l}\colon C^1(I) \to R^p$, $x \mapsto \tilde{l}(x) = (\tilde{l}_1(x), \ldots, \tilde{l}_p(x))$ with the $\tilde{l}_i$ linear continuous functionals over $C^1(I)$;

(b) $k: C^1(I) \to R^q$ with $k = l - r$, $l = (l_1, \ldots, l_q)$ with the l_i linear continuous functionals over $C^1(I)$ and $r: C^1(I) \to R^q$ continuous not necessarily linear and taking bounded sets into bounded sets.

Let now $X \subset C^1(I)$ be the closed vector subspace of $x \in C^1(I)$ such that $\tilde{l}(x) = 0$ with the induced norm. If $C(I)$ denotes the Banach space of continuous mappings $x: I \to R^n$ with the usual uniform norm

$$|x|_0 = \max_{t \in I} |x(t)|,$$

let Z denote the Banach space

$$Z = C(I) \times R^q$$

with the norm

$$\|z\| = \|(x, a)\| = \max(|x|_0, |a|).$$

If we define

$$\begin{aligned} &\text{dom } L = \{x \in X: x \text{ is of class } C^2 \text{ on } I\}, \\ &L: \text{dom } L \subset X \to Z, \qquad x \mapsto (x'', l(x)), \\ &N: X \to Z, \qquad x \mapsto (f(\cdot, x(\cdot), x'(\cdot)), r(x)), \end{aligned} \tag{2.4}$$

it is clear that the boundary value problem (2.2)–(2.3) is equivalent to the operator equation

$$Lx = Nx \tag{2.5}$$

because the linear part of the boundary conditions (2.3) is included in the definition of X. We shall see in the following sections how topological methods in function spaces can be used to prove the existence of a solution for (2.5).

3. The Case Where L Has an Inverse

We shall assume in this section that $\tilde{l}$ and l are such that $L^{-1}: Z \to X$ exists and is a compact linear mapping. In this case Eq. (2.5) is equivalent to

$$x = L^{-1}Nx = Tx \tag{3.1}$$

with $T: X \to X$ compact, i.e., continuous and taking bounded sets into compact sets. A powerful tool for proving the existence of one solution to (2.6), i.e., for showing that T has a fixed point, is the Leray–Schauder continuation theorem [8] which implies for (2.5) the following result.

PROPOSITION 3.1 Let X, Z be normed spaces, L: dom $L \subset X \to Z$ linear and such that $L^{-1}: Z \to X$ exists and is compact, and $N: X \to X$ continuous and taking bounded sets into bounded sets. If there exists an open ball $B(0, R)$ such that for each $\lambda \in]0, 1[$ and each possible solution of

$$Lx = \lambda Nx \tag{3.2}$$

one has

$$x \notin \partial B(0, R),$$

then equation (2.5) has at least one solution in $\bar{B}(0, R)$.

This result contains as a special case the Schauder fixed point theorem.

In the case of the boundary value problem (2.2)–(2.3), the conditions of Proposition 3.1 will be fulfilled, for example, if we assume moreover that there exists $M' \geq 0$ such that for all $x \in X$,

$$|r(x)| \leq M'. \tag{3.3}$$

In fact, one has then for all $x \in X$ and $\lambda \in]0, 1[$,

$$\|L^{-1}Nx\| \leq \|L^{-1}\| \max(M, M')$$

and the conditions of Proposition 3.1 are then satisfied if $R > \|L^{-1}\| \max(M, M')$. One has thus proved the following:

THEOREM 3.1 If the mapping L defined in (2.4) has a compact inverse L^{-1} and if conditions (2.1) and (2.8) are satisfied, then the boundary value problem (2.2)–(2.3) has at least one solution.

In the case where $n = 1$ (scalar equation), $m = q = 2$ and

$$\begin{aligned} g_i(x) \equiv l_i(x) - c_i \equiv \int_0^1 (p_{i0}(t)x(t) + p_{i1}(t)x'(t))\, dt \\ + q_{i0}\, x(0) + q_{i1}\, x(1) + q_{i2}\, x'(0) \\ + q_{i3}\, x'(1) - c_i \qquad (i = 1, 2), \end{aligned} \tag{3.4}$$

where the c_i are constants and the continuous functions p_{ij} and the constants q_{ij} $(i, j = 1, 2; j = 0, 1)$ are such that L^{-1} exists, Theorem 3.1 (and even its extension to differential equations of any order) was proved in 1922 by Birkhoff and Kellogg [2]. It was an application of their fixed-point theorem, a prototype of the Schauder fixed-point theorem in some particular function spaces.

On the other hand, the existence and compactness of L^{-1}, as well as condition (2.8), are always insured in the case of the *first boundary value problem* or *Picard problem* for (2.2), i.e., when $q = m = 2n$ and

$$\begin{aligned} g_i(x) &\equiv x_i(0) - b_i \qquad (i = 1, \ldots, n), \\ g_{n+i}(x) &\equiv x_i(1) - c_i \qquad (i = 1, \ldots, n), \end{aligned} \tag{3.5}$$

which is obviously a special case of (3.4) when $n = 1$. It is easy to verify that if $z = (y, a) \in Z$, the equation

$$Lx = z$$

has the unique solution (with $b = (b_1, \ldots, b_n)$, $c = (c_1, \ldots, c_n)$)

$$x(t) = b + t(c - b) + \int_0^t \int_0^u y(s)\, ds\, du - t \int_0^1 \int_0^u y(s)\, ds\, du,$$

which can still be written, integrating by parts,

$$x(t) = b + t(c - b) + \int_0^1 G(t, s) y(s)\, ds \tag{3.6}$$

introducing the *Green function*

$$G(t, s) = \begin{cases} (t - 1)s & \text{for} \quad 0 \le s \le t \le 1, \\ (s - 1)t & \text{for} \quad 0 \le t \le s \le 1. \end{cases}$$

It follows easily from (3.6) and Arzela–Ascoli theorem that L^{-1} is compact so that Theorem 3.1 implies the following

THEOREM 3.2 If (2.1) is satisfied, then for each $(b, c) \in R^{2n}$, the Picard problem (2.2)–(2.3) with g defined in (3.5) has at least one solution.

As we have noticed in the preceding, Theorem 3.2 is in the scalar case contained in Birkhoff and Kellog's result and an "elementary" proof (i.e., avoiding the use of a fixed point theorem in a function space) was given by Scorza-Dragoni [24,25] when $n = 1$ and $n = 2$. It was Bass [1] who remarked that for any n the result was a direct consequence of the Schauder fixed-point theorem, a fact used later by many authors. Theorem 3.2 is usually called the "Scorza-Dragoni lemma" (see, e.g., Hartman [5]). It can be generalized at once to the case where the constants b_i and c_i are replaced by continuous functionals r_i $(i = 1, \ldots, 2n)$ such that $r = (r_1, \ldots, r_{2n})$ satisfies (3.3) and to various other situations. Also condition (2.1) can be weakened in many different ways (see, e.g., Mawhin [12] and Rouche and Mawhin [22].

4. The Case of Linear Boundary Conditions Such That L Has No Inverse

The situation described in Section 3 no longer holds when we consider the *second boundary value problem* or the *Neumann problem* for (2.2), i.e., when $q = m = 2n$ and

$$\begin{aligned} g_i(x) &\equiv x_i'(0) - b_i \qquad (i = 1, \ldots, n), \\ g_{n+i}(x) &\equiv x_i'(1) - c_i \qquad (i = 1, \ldots, n). \end{aligned} \tag{4.1}$$

In this case, L^{-1} does not exist because

$$\ker L = \{x \in \operatorname{dom} L : x(t) = x(0),\ t \in I\}$$

is isomorphic to R^n, and the equation

$$Lx = z$$

($z = (y, a) \in Z$, $a = (b, c) \in R^{2n}$) is solvable if and only if

$$c - b = \int_0^1 y(t)\, dt.$$

Therefore

$$\dim \ker L = \operatorname{codim} \operatorname{Im} L = n$$

and L is a Fredholm mapping of index zero [9]. Condition (2.1) is no more sufficient to ensure the existence of a solution as shown by the example

$$x''(t) = 1, \qquad x'(0) = x'(1) = 0.$$

A way to obtain sufficient conditions of existence for this problem of Neumann is the use of the following generalized continuation theorem proved by Mawhin [9] in a more general setting.

Proposition 4.1 Let X, Z be normed spaces, L: $\operatorname{dom} L \subset X \to Z$ linear, Fredholm of index zero and having compact right inverses, $N: X \to X$ continuous and taking bounded sets into bounded sets. Assume that there exists an open ball $B(0, R)$ such that the following conditions hold:

(a) for each $\lambda \in]0, 1[$ and each possible solution of

$$Lx = \lambda N x, \tag{4.2}$$

one has $x \notin \partial B(0, R)$;

(b) for each $x \in \ker L \cap \partial B(0, R)$ one has $QNx \neq 0$, where $Q: Z \to Z$ is a continuous projector such that $\operatorname{Im} L = \ker Q$;

(c) $d_B[JQN \mid \ker L, B(0, R) \cap \ker L, 0] \neq 0$, where d_B denotes the Brouwer degree and J: $\operatorname{Im} Q \to \ker L$ is any isomorphism.

Then the equation $Lx = Nx$ has at least one solution in $\bar{B}(0, R)$.

In the case of the Neumann problem (2.2)–(2.3) with a g given by (4.1), one may verify easily that one can take for Q the mapping defined by

$$Qz = Q(y, (b, c)) = \left(\int_0^1 y(t)\, dt + b - c, (0, 0)\right),$$

where $y \in C(I)$, $b \in R^n$, $c \in R^n$. We shall prove the following:

THEOREM 4.1 Assume that (2.1) is satisfied and that there exists $S > 0$ such that

$$\int_0^1 f(t, u(t), u'(t))\, dt + b - c \neq 0$$

for all $u \in C^2(I)$ such that $\inf_{t \in I} |u(t)| \geq S$ and $|u''|_0 \leq M$. Then problem (2.2)–(2.3) with g defined in (4.1) has at least one solution if (c) holds.

Proof Let $\lambda \in]0, 1[$ and x be any possible solution to (4.2). Then,

$$x''(t) = \lambda f(t, x(t), x'(t)), \qquad t \in I, \tag{4.3}$$

$$x'(0) = \lambda b, \tag{4.4}$$

$$x'(1) = \lambda c, \tag{4.5}$$

which also implies that

$$c - b = \int_0^1 f(t, x(t), x'(t))\, dt. \tag{4.6}$$

Therefore, using (4.3), (4.4), and (2.1), we have for all $t \in I$,

$$x'(t) = x'(0) + \int_0^t x''(s)\, ds \leq |b| + M. \tag{4.7}$$

On the other side it follows from (4.6) and the assumptions of the theorem that there must exist $\tau \in I$ such that

$$|x(\tau)| < S.$$

Therefore, we have for all $t \in I$,

$$\begin{aligned} |x(t)| &= \left| x(\tau) + \int_\tau^t x'(s)\, ds \right| \\ &= \left| x(\tau) + \int_\tau^t \left[b + \int_0^s x''(u)\, du \right] ds \right| \\ &< S + |b| + M/2. \end{aligned} \tag{4.8}$$

(4.7) and (4.8) imply that condition (a) of Proposition 4.1 is satisfied if we take

$$R > \max(S + |b| + M/2,\ |b| + M).$$

Now $QN|\ker L$, with $\ker L$ naturally identified with R^n, is explicitly given by

$$a \mapsto \int_0^1 f(t, a, 0)\, dt + b - c,$$

and by assumption this mapping has no zero for $|a| \geq S$ and hence for $a \geq R$. Since condition (c) of Proposition 4.1 is assumed to hold, the result follows. For a more general result, see Mawhin [10].

As a simple example, suppose that $n = 1$, $b = c = 0$, f does not depend explicitly upon x' and

$$\lim_{x \to \pm\infty} f(t, x) = f_{\pm}(t)$$

uniformly in $t \in I$. If

$$\left(\int_0^1 f_+(t)\, dt\right)\left(\int_0^1 f_-(t)\, dt\right) < 0, \tag{4.9}$$

it is not difficult to show that all conditions of Theorem 4.1 are satisfied for all sufficiently great $R > 0$ and the existence of at least one solution follows. On the other hand if we assume moreover that for all $t \in I$ and $x \in R$ either

$$f_-(t) < f(t, x) < f_+(t) \tag{4.10}$$

or

$$f_+(t) < f(t, x) < f_-(t), \tag{4.11}$$

then (4.9) is also a necessary condition for the solvability of the Neumann problem

$$x''(t) = f(t, x(t)), \qquad x'(0) = x'(1) = 0. \tag{4.12}$$

In fact, if x is a solution to (4.12), then necessarily,

$$\int_0^1 f(t, x(t))\, dt = 0$$

and hence, using (4.10) or (4.11) one has either

$$\int_0^1 f_+(t)\, dt < 0 < \int_0^1 f_-(t)\, dt \qquad \text{or} \qquad \int_0^1 f_-(t)\, dt > 0 > \int_0^1 f_+(t)\, dt,$$

which is equivalent to (4.9). Necessary and sufficient conditions of this type were first introduced for the Dirichlet problem in elliptic equations by Landesman and Lazer [7]. For more results in this line, see references [3,10,13,14,15,16].

The situation we have just described is quite analogous in the case of *periodic boundary conditions* (the *Poincaré problem*), i.e., when $p = m = 2n$ and

$$\begin{aligned} g_i(x) &\equiv x_i(0) - x_i(1) \qquad (i = 1, \ldots, n), \\ g_{n+i}(x) &\equiv x_i'(0) - x_i^1(1) \qquad (i = 1, \ldots, n). \end{aligned} \tag{4.13}$$

In this case in fact all the boundary conditions are included in the definition of X and in this space it is immediate that the linear mapping L defined by $Lx = x''$ is Fredholm of index zero with a kernel isomorphic to R^n.

5. The Case of Nonlinear Boundary Conditions Such That Ind $L = 0$

In this section and in the following we shall consider problems in which the nonlinear part of the boundary conditions will play a crucial role. We shall assume, with the notations of Section 2, that $l = 0$, which does not strictly mean that the part

$$k(x) = 0$$

of the boundary conditions (2.3) does not contain a linear part but that we do not want this linear part, if any, playing a special role in formulating the problem as an abstract equation, and so we include it in $r(x)$. We therefore write the considered boundary value problem in the form

$$x''(t) = f(t, x(t), x'(t)), \tag{5.1}$$

$$\tilde{l}(x) = 0, \tag{5.2}$$

$$0 = r(x), \tag{5.3}$$

where l: $C^1(I) \to R^p$ is linear and continuous, r: $C^1(I) \to R^q$ is not necessarily linear, continuous and takes bounded sets into bounded sets. We also assume that

$$p + q = m = 2n. \tag{5.4}$$

We shall denote by $\tilde{x}''$ the linear mapping

$$\tilde{x}''\colon C^2(I) \cap X \subset X \to C(I), \qquad x \mapsto x'',$$

where X is defined as in Section 2, and we shall assume that the following condition holds:

(A) dim ker $\tilde{x}'' = 2n - p$ and codim Im $\tilde{x}'' = 0$ (i.e., $\tilde{x}''$ is onto).

Therefore, with the notations of Section 2,

$$Lx = (\tilde{x}'', 0),$$

which implies that

$$\dim \ker L = \dim \ker \tilde{x}'' = q, \qquad \operatorname{codim} \operatorname{Im} L = q,$$

and hence L is Fredholm of index zero, and N is given by (2.4). Also L has compact right inverses by the Arzela–Ascoli theorem. Using Proposition 4.1 we shall prove the following:

THEOREM 5.1 Let us assume that (5.4) and condition (A) hold, as well as (2.1). If the following conditions hold:

(B) There exists $R > 0$ such that each possible solution x of class C^2 of

$$r(x) = 0$$

such that $|x''|_0 \leq M$ verifies the relation $\|x\| \neq R$.

(C) $d_B[r\,|\ker L, B(0, R) \cap \ker L, 0] \neq 0$, then problem (5.1)–(5.3) has at least one solution.

Proof Let $\lambda \in]0, 1[$ and x be a possible solution of

$$Lx = \lambda Nx.$$

Then, $x \in X \cap C^2(I)$,

$$x''(t) = \lambda f(t, x(t), x'(t)), \qquad t \in I, \tag{5.5}$$

$$0 = r(x). \tag{5.6}$$

Thus, using (5.5) and (2.1), $|x''|_0 \leq M$, and condition (B) implies that hypothesis (a) of Proposition 4.1 is verified. It is now clear that $Q: Z \to Z$, defined by

$$Qz = Q(y, a) = (0, a),$$

is a continuous projector such that $\ker Q = \operatorname{Im} L$, and hence

$$QN(x) = r(x) \neq 0$$

for $x \in \ker L$ such that $\|x\| = R$, because $x'' = 0$ if $x \in \ker L$. Condition (C) just corresponds to assumption (c) of Proposition 4.1 and the proof is complete.

As an example, let us consider the following scalar case ($n = 1$) with $p = 0$, $q = 2$ and

$$\begin{aligned} r_1(x) &= x^2(0) - [x'(0)]^2 - b, \\ r_2(x) &= 2x'(1)[x(1) - x'(1)] - c. \end{aligned} \tag{5.7}$$

This example was considered by Gaines and Mawhin [4] where other results in the spirit of this section can be found. Let us first consider condition (B) of Theorem 5.1, and let x be a solution of

$$r(x) = 0 \tag{5.8}$$

such that $|x''|_0 \leq M$. Therefore, by the Taylor formula,

$$x'(1) = x'(0) + a, \quad \text{with} \quad |a| \leq M,$$

$$x(1) = x(0) + x'(0) + d, \quad \text{with} \quad |d| \leq M/2.$$

Hence (5.8) can be written

$$x^2(0) - [x'(0)]^2 - b$$
$$= 0, \; 2x(0)x'(0) + 2ax(0) + 2(d-a)x'(0) + 2a(d-a) - c = 0. \tag{5.9}$$

Hence, writing $z = x(0) + ix'(0)$, we obtain from (5.9)

$$|z|^2 \leq K_1 |z| + K_2,$$

where K_i only depends upon M, b, and c. Therefore,

$$|(x(0), x'(0))| = |z| \leq K_3$$

with the same dependence for K_3. This together with $|x''|_0 \leq M$ implies the existence of some $R > 0$ such that $\|x\| < R$. Now $\ker L = \{x: t \mapsto u_1 + tu_2, (u_1, u_2) \in R^2\}$ is isomorphic to R^2 and up to this isomorphism

$$r(u_1, u_2) = (u_1{}^2 - u_2{}^2 - b, 2u_1u_2 - c),$$

which implies that

$$d_B[r \,|\ker L, B(0, R) \cap \ker L, 0] = 2$$

and condition (B) of Theorem 5.1 is verified.

It follows therefore that if condition (2.1) holds for f, the boundary value problem (5.1) (with $n = 1$) through (5.8) (with r given by (5.7)) has at least one solution.

Let us now come back to the periodic boundary value problem (2.2)–(2.3) with g given by (4.13). As was noticed by Zezza (personal communication) this problem is equivalent to the following one:

$$x''(t) = f(t, x(t), x'(t)), \tag{5.10}$$

$$x(0) - x(1) = 0, \tag{5.11}$$

$$\int_0^1 f(t, x(t), x'(t))\, dt = 0, \tag{5.12}$$

which is a special case of (5.1)–(5.3). Moreover, if

$$X = \{x \in C^1(I) : x(0) - x(1) = 0\},$$

one verifies easily that

$$\dim \ker \tilde{x}'' = n, \qquad \operatorname{codim} \operatorname{Im} \tilde{x}'' = 0,$$

so that condition (A) holds. The reader can then easily deduce from Theorem 5.1 the following:

COROLLARY 5.1 Assume that (2.1) is satisfied and that there exists $S > 0$ such that

$$\int_0^1 f(t, u(t), u'(t))\, dt \neq 0$$

for all $u \in C^2(I)$ such that $\inf_{t \in I} |u(t)| \geq S$ and $d_B[F, B(0, R), 0] \neq 0$, where

$$F: R^n \to R^n, \quad a \mapsto \int_0^1 f(t, a, 0)\, dt.$$

Then the periodic problem for (5.10) has at least one solution.

For a derivation of this result in the line of Section 4, see, for example, Mawhin [11].

Similar considerations can be applied to the Neumann problem.

6. The Case of Nonlinear Boundary Conditions Such That Ind $L > 0$

Let us come back to the boundary value problem (5.1)–(5.3) with the conditions of Section 5 but assume now that

$$p + q = m < 2n. \tag{6.1}$$

Because $\dim \ker x'' = 2n$ in $C^2(I)$ and because of the definition of X, one necessarily has

$$\dim \ker \tilde{x}'' \geq 2n - p$$

and hence, using (6.1)

$$\dim \ker L = \dim \ker \tilde{x}'' \geq 2n - p > q = \operatorname{codim} \operatorname{Im} L,$$

i.e., Ind $L > 0$. In this case degree theory, which is the basic tool used in proving Propositions 3.1 and 4.1, cannot be used because the corresponding degrees are necessarily zero (see, for example, references [9,13,15]). One is therefore led to look for other topological invariants and the following theorem, given by Mawhin [14], is based upon an important topological result due to Nirenberg [15,16].

PROPOSITION 6.1 Let X, Z be normed spaces, L: dom $L \subset X \to Z$ linear, Fredholm such that dim ker $L = d > d^* =$ codim Im L and having compact right inverses, $N: X \to X$ continuous and taking bounded sets into bounded sets. Assume that there exists $R > 0$ such that the following conditions hold, with S^n the unit sphere in R^{n+1}:

(a) for each $\lambda \in]0, 1[$ and each possible solution x of $Lx = \lambda Nx$, one has $x \notin \partial B(0, R)$.

(b) for each $x \in \ker L \cap B(0, R)$ one has $QNx \neq 0$, where $Q: Z \to Z$ is a continuous projector such that Im $L = \ker Q$.

(c) $\Lambda: R^d \to \ker L$ and Λ': Im $Q \to R^{d^*}$ being isomorphisms, the mapping

$$\psi: S^{d-1} \to S^{d^*-1}, \qquad u \mapsto \frac{\Lambda' QN\Lambda(Ru)}{|\Lambda' QN\Lambda(Ru)|}$$

has nontrivial stable homotopy.

Then the equation $Lx = Nx$ has at least one solution in $\bar{B}(0, R)$.

Let us recall (see, for example, Hilton [6] and Nirenberg [16]) that if ψ: $S^{d-1} \to S^{d^*-1}$ is such that $\psi = F/|F|$ with F: $\bar{B}(0, 1) \subset R^d \to R^{d^*}$ and

$$F(S^{d-1}) \subset R^{d^*}\backslash\{0\},$$

one can define analytically the suspension $F_1 = \sum F$ of F by

$$F_1: B(0, 1) \subset R^{d+1} \to R^{d^*+1}, \qquad (x, t) \mapsto (F(x), t)$$

with $x \in R^d$ and $t \in R$. Then the suspension $\psi_1 = \sum \psi$ of ψ can be defined by the restriction to S^d of $F_1/|F_1|$. Similarly the jth suspension $F_j = \sum^j F$ of F is defined by

$$F_j: B(0, 1) \subset R^{d+j} \to R^{d^*+j}, \qquad (x, t) \mapsto (F(x), t)$$

with $x \in R^d$ and $t \in R^j$. The jth suspension $\psi_j = \sum^j \psi$ of ψ is then $F_j/|F_j|$ restricted to S^{d+j-1}. One then says that ψ has nontrivial stable homotopy if $\sum^j \psi$ for j sufficiently great is not homotopic to a constant mapping from S^{d+j-1} into S^{d^*+j-1}.

Let us now come back to the boundary value problem (5.1)–(5.3) when (6.1) holds. By a proof quite similar to that of Theorem 5.1, but based upon Proposition 6.1 instead of Proposition 4.1, one obtains the following:

THEOREM 6.1 Let us assume that (2.1), (6.1), and condition (A) of Section 5 hold, as well as condition (B) of Theorem 5.1. Let Λ: $R^q \to \ker L$ be any isomorphism and assume that the mapping

$$\psi: S^{2n-p-1} \to S^{q-1}, \qquad u \mapsto \frac{r(R\Lambda u)}{|r(R\Lambda u)|},$$

has nontrivial stable homotopy. Then problem (5.1)–(5.3) has at least one solution.

As an example let us consider the following system of equations ($n = 2$):

$$\begin{aligned} x_1''(t) &= f_1(t, x_1(t), x_2(t), x_1'(t), x_2'(t)), \\ x_2''(t) &= f_2(t, x_1(t), x_2(t), x_1'(t), x_2'(t)), \end{aligned} \tag{6.2}$$

$$\begin{aligned} x_1^2(0) + x_1'(1)^2 - x_2^2(0) - x_2'(1)^2 - c_1 &= 0, \\ 2(x_1(0)x_2(0) + x_1'(1)x_2'(1)) - c_2 &= 0, \\ 2(x_1'(1)x_2(0) - x_1(0)x_2'(1)) - c_3 &= 0, \end{aligned} \tag{6.3}$$

where $f = (f_1, f_2)$ satisfies (2.1). Here $p = 0$, $n = 2$, $q = 3$ so that Ind $L = 1$. We proceed as in the example of Section 5 and get, if $x = (x_1, x_2) \in C^2(I)$ satisfies

$$r(x) = 0 \tag{6.4}$$

and $|x''|_0 \leq M$, where r is defined by the left-hand side of (6.3),

$$\left.\begin{aligned} x_i'(1) &= x_i'(0) + a_i, && \text{with } |a_i| \leq M \\ x_i(1) &= x_i(0) + x_i'(0) + b_i, && \text{with } |b_i| \leq M/2 \end{aligned}\right\} \quad (i = 1, 2). \tag{6.5}$$

Therefore, letting

$$z^j = x_j(0) + ix_j'(0) \qquad (j = 1, 2)$$

and using (6.4) and (6.5), one gets after elementary computations

$$\begin{aligned} \big||z^1|^2 - |z^2|^2\big| &\leq 2|a_1||z^1| + 2|a_2||z^2| + k_1, \\ 2|z^1||z^2| &\leq 2|a_1||z^2| + 2|a_2||z^1| + k_2, \end{aligned}$$

where $k_1 \geq 0$ and $k_2 \geq 0$ only depend upon a_i ($i = 1, 2$) and c_i ($i = 1, 2, 3$). Hence letting $u = |z^1| + i|z^2|$, we obtain

$$|u|^4 \leq k_3 |u|^2 + k_4 |u| + k_5,$$

where the k_i ($i = 3, 4, 5$) only depend upon the a_i and the c_i. Consequently, there exists a constant C depending only upon the a_i and c_i such that

$$|u| = (x_1^2(0) + x_1'(0)^2 + x_2^2(0) + x_2'(0)^2)^{1/2} \leq C,$$

and the existence of one $R > 0$ verifying condition (B) of Theorem 5.1 easily follows, using the Taylor formula. Now it is easy to show that the map, with $\Lambda(u + tv) = (u_1, v_1, u_2, v_2)$,

$$\psi: S^3 \to S^2, \; u \mapsto \frac{r(R\Lambda u)}{r(R\Lambda u)}$$

is homotopic to the Hopf map

$$\phi: S^3 \to S^2, \; (u_1, v_1, u_2, v_2) \mapsto$$

$$(u_1{}^2 + v_1{}^2 - u_2{}^2 - v_2{}^2, 2(u_1 u_2 + v_1 v_2), 2(u_2 v_1 - u_1 v_2)),$$

where S^3 is considered as the unit sphere in R^4, and it is known [6,16] that for any integer $j \geq 1$, the jth suspension $\sum^j \phi$ of the Hopf map is the generator of the homotopy group $\pi_{3+j}(S^{2+j})$ and hence is not trivial because for $n \geq 3$, $\pi_{n+1}(S^n)$ is cyclic of order 2. Thus ψ has nontrivial stable homotopy, and by Theorem 6.1 the boundary value problem (6.2)–(6.3) has at least one solution.

References

1. R. W. Bass, On non-linear repulsive forces. *In* "Contributions to Nonlinear Oscillations", Annals of Mathematical Studies, Vol. 4, pp. 201–211. Princeton Univ. Press, Princeton, New Jersey, 1958.
2. G. D. Birkhoff and O. D. Kellogg, Invariant points in function spaces. *Trans. Amer. Math. Soc.* **23** (1922), 96–115.
3. S. Fučík and J. Mawhin, Periodic solutions of some nonlinear differential equations of higher order. *Časopis Pěst. Mat.* **100** (1975), 276–283.
4. R. E. Gaines and J. Mawhin, Ordinary differential equations with nonlinear boundary conditions. *J. Differential Equations* **26** (1977), 200–222.
5. P. Hartman, "Ordinary Differential Equations." Wiley, New York, 1964.
6. P. J. Hilton, "An Introduction to Homotopy Theory." Cambridge Univ. Press, London and New York, 1953.
7. E. M. Landesman and A. C. Lazer, Nonlinear perturbations of linear elliptic boundary value problems at resonance. *J. Math. Mech.* **19** (1970), 609–623.
8. J. Leray and J. Schauder, Topologie et équations fonctionnelles. *Ann. Sci. École Norm. Sup.* **51** (3) (1934), 45–78.
9. J. Mawhin, Equivalence theorems for nonlinear operator equations and coincidence degree theory for some mappings in locally convex topological vector spaces. *J. Differential Equations* **12** (1972), 610–636.
10. J. Mawhin, Problèmes aux limites du type de Neumann pour certaines équations différentielles ou aux dérivées partielles non linéaires. *In* "Equations Différentielles et Fonctionnelles Non Linéaires," pp. 124–134. Hermann, Paris, 1973.
11. J. Mawhin, Periodic solutions of some vector retarded functional equations. *J. Math. Anal. Appl.* **45** (1974), 588–603.
12. J. Mawhin, Boundary value problems for nonlinear second-order vector differential equations. *J. Differential Equations* **16** (1974), 257–269.
13. J. Mawhin, Nonlinear perturbations of Fredholm mappings in normed spaces and applications to differential equations. *Univ. Brasilia Trab. Mat.* **61** (May 1974).

14. J. Mawhin, Topology and nonlinear boundary value problems. *In* "Dynamical Systems: An International Symposium" (L. Cesari, J. Hale, and J. La Salle, eds.), vol. 1, pp. 51–82, Academic Press, New York, 1976.
15. L. Nirenberg, An application of generalized degree to a class of nonlinear problems, *In Troisième Coll. C.B.R.M. d'Analyse Fonctionnelle*, pp. 57–74. Vander, Louvain, 1971.
16. L. Nirenberg, "Topics in Nonlinear Functional Analysis," Courant Institute Lecture Notes. New York University, New York, 1973–74.
17. E. Rothe, Uber Abbildungsklassen von Kugeln des Hilbertschen Raumes. *Compositio Math.* **4** (1937), 294–307.
18. E. Rothe, Uber den Abbildungsgrad bei Abbildungen von Kugeln des Hilbertschen Raumes. *Compositio Math.* **5** (1937), 166–176.
19. E. Rothe, Zur Theorie des topologischen Ordnung und der Vektorfelder in Banach Raumen. *Compositio Math.* **5** (1937), 177–197.
20. E. Rothe, Topological proofs of uniqueness theorems in the theory of differential and integral equations. *Bull. Amer. Math. Soc.* **45** (1939), 606–613.
21. E. Rothe, The theory of topological order in some linear topological spaces. *Iowa State Coll. J. Sci.* **13** (1939), 373–390.
22. N. Rouche and J. Mawhin, "Equations Différentielles Ordinaires," Vol. II. Masson, Paris, 1973.
23. J. Schauder, Der Fixpunktsatz in Funktionalräumen. *Studia Math.* **2** (1930), 171–180.
24. G. Scorza-Dragoni, Il problema dei valori ai limiti studiato in grande per le equazioni differenziali del secondo ordine. *Math. Ann.* **105** (1931), 133–143.
25. G. Scorza-Dragoni, Sul problema dei valori ai limiti per i sistemi di equazioni differenziali del secondo ordine. *Boll. Un. Mat. Ital.* **14** (1935), 225–230.

AMS (MOS) 1970 Subject Classification: 34B15, 47H15.

Some Minimax Theorems and Applications to Nonlinear Partial Differential Equations

Paul H. Rabinowitz

University of Wisconsin

Introduction

The purpose of this paper is to prove some existence theorems for critical points of a real-valued function on a real Banach space and to apply these results to elliptic and hyperbolic partial differential equations. The abstract results on critical points will be presented in Section 1. They are obtained using minimax arguments. Applications to elliptic equations are given in Section 2. In particular, we give a new proof of a recent result of Ahmad *et al.* [1] as well as some variants of their result. Lastly, in Section 3, the work of Section 1 is applied to hyperbolic problems.

Our research was largely motivated by trying to understand the relationship of the work of Ahmed, Lazer, and Paul [1] to the earlier work of

Ambrosetti and Rabinowitz [2]. We thank Louis Nirenberg for calling the work of Ahmad *et al.* [1] to our attention.

1. The Abstract Theorems

Let E be a real Banach space and $I \in C^1(E, \mathbb{R})$. The Fréchet derivative of I at u will be denoted by $I'(u)$. In this section we shall give sufficient conditions for I to have a critical point in E. For the case in which I is even so that 0 is a critical point, criteria will be given for I to have multiple critical points.

A standard compactness condition will be required of I. We say I satisfies the Palais–Smale (PS) condition if every sequence (u_m) on which I is bounded and $I' \to 0$, possesses a convergent subsequence. The following standard "deformation theorem" will be used repeatedly. For $c \in \mathbb{R}$, $A_c = \{x \in \mathrm{E} \mid I(x) \le c\}$ and $K_c = \{x \in E \mid I(x) = c \text{ and } I'(x) = 0\}$.

LEMMA 1.1 Let $I \in C^1(E, \mathbb{R})$ and satisfy (PS). If $c \in \mathbb{R}$, $\bar{\varepsilon} > 0$, and $\mathscr{U}$ is any neighborhood of K_c, there exists an $\varepsilon \in (0, \bar{\varepsilon})$ and $\eta \in C([0, 1] \times E, E)$ such that

(1) $\eta(0, x) = x$ for all $x \in E$,
(2) $\eta(t, x) = x$ if $x \notin I^{-1}[c - \bar{\varepsilon}, c + \bar{\varepsilon}]$, $t \in [0, 1]$,
(3) $\eta(t, \cdot)$ is a homeomorphism of E onto E for all $t \in [0, 1]$,
(4) $I(\eta(t, x)) \le I(x)$ for all $t \in [0, 1]$ and $x \in E$,
(5) $\eta(1, A_{c+\varepsilon} \backslash \mathscr{U}) \subset A_{c-\varepsilon}$,
(6) if $K_c = \varnothing$, $\eta(1, A_{c+\varepsilon}) \subset A_{c-\varepsilon}$,
(7) if I is even, $\eta(t, \cdot)$ is odd for all $t \in [0, 1]$.

See, e.g., Rabinowitz [3] for a proof of Lemma 1.1. Our main abstract result is the following:

THEOREM 1.2 Let $I \in C^1(E, \mathbb{R})$ and satisfy (PS). Suppose there is a finite dimensional subspace $E_1 \subset E$ with a (topologically) complementary subspace E_2 such that

(1) $I|_{E_2}$ is bounded from below by $b_2 > -\infty$;
(2) there is a bounded neighborhood Ω of 0 in E_1 and a constant $b_1 < b_2$ so that $I(u) \le b_1$ for $u \in \partial\Omega$.

Then I possesses a critical point $u \in E$ with $I(u) = b \ge b_2$.

Proof We will give a minimax characterization of b. Suppose first that $E_1 \ne \{0\}$. Noting that $E = E_1 \oplus E_2$, we define

$$\mathscr{S} = \{\chi(z) = (\varphi(z), \psi(z)) \in C(\bar{\Omega}, E_1 \oplus E_2) \mid \varphi(z) = z \text{ and } \psi(z) = 0 \text{ for } z \in \partial\Omega\}.$$

Next define

$$b = \inf_{\chi \in \mathscr{S}} \max_{z \in \bar{\Omega}} I(\chi(z)). \tag{1.3}$$

We will show that b is a critical value of I. First observe that if $\chi \in \mathscr{S}$, since $\varphi(z) = z$ for $z \in \partial\Omega$ and $\varphi \in C(\bar{\Omega}, E_1)$, by the homotopy invariance of Brouwer degree there is a $\tilde{z} \in \Omega$ such that $\varphi(\tilde{z}) = 0$ [4]. Therefore

$$\max_{z \in \bar{\Omega}} I(\chi(z)) \geq I(\chi(\tilde{z})) = I(\psi(\tilde{z})) \geq \inf_{u \in E_2} I(u) \geq b_2$$

so $b \geq b_2 > -\infty$.

If b is not a critical value of I, we invoke Lemma 1.1 with $b = c$ and $\bar{\varepsilon} = \frac{1}{2}(b_2 - b_1)$. With ε and η as given by the Lemma, choose $\chi \in \mathscr{S}$ so that

$$\max_{z \in \Omega} I(\chi(z)) \leq b + \varepsilon. \tag{1.4}$$

Consider $\eta(1, \chi(z))$. For $z \in \partial\Omega$, $\chi(z) = z$ and $I(z) \leq b_1 < b_2 - \bar{\varepsilon}$. Therefore $\eta(1, \chi(z)) \in \mathscr{S}$. But by (1.4) and (6) of Lemma 1.1,

$$\max_{z \in \bar{\Omega}} I(\eta(1, \chi(z))) \leq b - \varepsilon$$

contrary to the definition of b. Thus b is a critical value of I and Theorem 1.2 is proved for this case.

Lastly, if $E_1 = \{0\}$, $E_2 = E$, a simple application of Lemma 1.1 shows $\inf_{u \in E} I(u)$ is a critical value of I.

Remark 1.5 If there is a neighborhood $\mathcal{O}$ of Ω in E so that $\eta(1, \cdot): \mathcal{O} \to \mathcal{O}$, then we can replace $\mathscr{S}$ by $\mathscr{S}_{\mathcal{O}} = \{\chi \in \mathscr{S} \mid \chi(\bar{\Omega}) \subset \mathcal{O}\}$ and the proof of Theorem 1.2 applies equally well to give a critical point of I in $\mathcal{O}$. Indeed, for this procedure to work it suffices to have $I \in C^1(\mathcal{O}, \mathbb{R})$ satisfy the (PS) condition in $\mathcal{O}$, and have an analog of Lemma 1.1 relative to $\mathcal{O}$.

A "dual" maximin characterization of b can be given in the following sense: Suppose $E_1 \neq \{0\}$ and $\mathscr{S}^* = \{A \subset E \mid A \text{ is closed and } A \cap \chi(\bar{\Omega}) \neq \varnothing$ for all $\chi \in \mathscr{S}\}$. Then $\mathscr{S}^* \neq \varnothing$ since $E_2 \in \mathscr{S}^*$. Define

$$b^* = \sup_{A \in \mathscr{S}^*} \inf_{u \in A} I(u).$$

COROLLARY 1.6 Under the hypotheses of Theorem 1.2, $b^* = b$.

Proof Let $A \in \mathscr{S}^*$ and $\chi \in \mathscr{S}$. Then there exists $y \in A \cap \chi(\bar{\Omega})$. Consequently,

$$\max_{z \in \bar{\Omega}} I(\chi(z)) \geq I(y) \geq \inf_{u \in A} I(u)$$

from which we conclude that $b \geq b^*$. Moreover, since we can take $A = E_2$, $b^* > -\infty$. To show that $b^* = b$, let $\chi \in \mathscr{S}$. Since $\bar{\Omega}$ is compact, there exists

$\tilde{z} = \tilde{z}(\chi) \in \Omega$ such that

$$I(\chi(\tilde{z})) = \max_{z \in \bar{\Omega}} I(\chi(z)).$$

Of course $\tilde{z}$ need not be unique, but for each $\chi \in \mathscr{S}$, choose any such $\tilde{z}$. Let $S = \{\chi(\tilde{z}) \mid \chi \in \mathscr{S}\}$ and $A = \bar{S}$. Then A is closed and $A \cap \chi(\bar{\Omega}) \neq \varnothing$ for all $\chi \in \mathscr{S}$. Hence $A \in \mathscr{S}^*$ and

$$\inf_{u \in A} I(u) \leq b^*,$$

Clearly we can find $(u_m) \subset S$ so that $I(u_m) \to \inf_{u \in A} I(u)$. But $I(u_m) = I(\chi(\tilde{z}_m)) \geq b$ so $b^* \geq \inf_{u \in A} I(u) \geq b$. Thus we must have equality.

Remark 1.7 A similar argument settles some open questions in references [2] and [3], see, e.g., Remark 3.40 of Rabinowitz [3].

For our later purposes we give one more characterization of a critical value of I.

COROLLARY 1.8 Under the hypotheses of Theorem 1.2, let $\mathscr{T} = \{h \in C(E, E) \mid h$ is a homeomorphism of E onto E and $h(u) = u$ for $u \in \partial\Omega\}$. If

$$c = \sup_{h \in \mathscr{T}} \inf_{u \in E_2} I(h(u)),$$

then $c \leq b$ and is a critical value of I.

Proof Choosing $h(u) = u$, we see from (1) of Theorem 1.2 that $c > -\infty$. Since $h(E_2)$ is closed for each $h \in \mathscr{T}$, $h(E_2) \in \mathscr{S}^*$ provided that $h(E_2) \cap \chi(\bar{\Omega}) \neq \varnothing$ for all $\chi \in \mathscr{S}$. This last statement follows since it is equivalent to $E_2 \cap h^{-1} \circ \chi(\bar{\Omega}) \neq \varnothing$, which is immediate from the proof of Theorem 1.2 on observing that $h^{-1} \circ \chi \in \mathscr{S}$. Therefore $c \leq b < \infty$. To show that c is a critical value of I, on replacing I by $-I$ and arguing as in the proof of Theorem 1.2, we need only verify that $\eta(1, \cdot) \circ h \in \mathscr{T}$ whenever $h \in \mathscr{T}$. But by (3) and (2) of Lemma 1.1, $\eta(1, \cdot) \circ h$ is a homeomorphism of E onto E and $\eta(1, h(u)) = u$ for $u \in \partial\Omega$ so the proof is complete.

Remark We do not know if $c = b$ in general.

If I has a known critical value, (1.3) may still provide a new one. This is the case, for example, if 0 is a critical point of I, $I(0) = 0$, and $b \neq 0$. If I is even, 0 is a critical point of I and critical points occur in antipodal pairs. Under further conditions on I which we will study next, I possesses additional critical points.

Let $A \subset E \backslash \{0\}$ be closed and symmetric (with respect to the origin). The genus of A, $\gamma(A)$, is defined to be the least integer k such that there is an odd

$f \in C(A, \mathbb{R}^k \backslash \{0\})$. For the properties of genus see, e.g., Rabinowitz [3, Section 1]. Let $\Sigma = \{A \subset E\backslash\{0\} \mid A \text{ is closed and symmetric}\}$.

THEOREM 1.9 Let $I \in C^1(E, \mathbb{R})$ be even with $I(0) = 0$ and satisfy the (PS) condition. Suppose further that

(1) there is a closed subspace $\tilde{E} \subset E$ of codimension j and a constant $\tilde{b}$ such that $I|_{\tilde{E}} \geq \tilde{b}$, and
(2) there is an $A \in \Sigma$ having $\gamma(A) = m > j$ and $\sup_A I < 0$.

Then I possesses at least $m - j$ distinct pairs of nonzero critical points.

Proof Let $\Sigma_k = \{K \in \Sigma \mid \gamma(K) \geq k\}$. Define

$$c_k = \inf_{K \in \Sigma_k} \sup_{u \in K} I(u).$$

Clearly $c_k \leq c_{k+1}$ and for $1 \leq k \leq m$, $c_k < 0$ via (2). By a theorem of Clark ([5] or Rabinowitz [3, Theorem 3.1]), if c_k also satisfies $c_k > -\infty$, then c_k is a critical value of I and if $c_k = \cdots = c_{k+p} \equiv c$, then $\gamma(K_c) \geq p + 1$. The definition of genus implies that K contains infinitely many distinct points if $\gamma(K) > 1$. Thus Theorem 1.9 will be proved once we show that $c_{j+1} > -\infty$. It is easy to see that if $\gamma(K) > j$ and $\tilde{E}$ is a closed subspace of E of codimension j, then $K \cap \tilde{E} \neq \varnothing$ (by (7) of Lemma 1.1, and Rabinowitz [3]). Therefore by hypothesis (1),

$$\sup_{u \in K} I(u) \geq \sup_{u \in K \cap \tilde{E} \neq \varnothing} I(u) \geq \inf_{u \in \tilde{E}} I(u) \geq \tilde{b} > -\infty.$$

Thus $c_{j+1} \geq \tilde{b}$ and the theorem follows.

Our final result in this section gives another criterion for the existence of multiple critical points in the symmetric case and is based on work by Ambrosetti and Rabinowitz [2]. Let $B_\rho = \{x \in E \mid \|x\| < \rho\}$ and let $\tilde{A}_c = \{x \in E \mid I(x) \geq c\}$.

THEOREM 1.10 Let $I \in C^1(E, \mathbb{R})$ be even with $I(0) = 0$ and satisfy the (PS) condition. Suppose that

(1) there are constants $\rho, \alpha > 0$ and a j dimensional subspace E_1 of E with (topologically) complementary subspace E_2 such that $I > 0$ on $(B_\rho \cap E_2)\backslash\{0\}$ and $I \geq \alpha$ on $\partial B_\rho \cap E_2$;
(2) there is an m ($> j$) dimensional subspace $\tilde{E}$ of E and a constant $R > 0$ such that $I(u) < 0$ if $u \in \tilde{E}$ and $\|u\| > R$.

Then I possesses at least $m - j$ distinct pairs of nonzero critical points.

Proof As in Ambrosetti and Rabinowitz [2], define

$$\Lambda^* = \{h \in C(E, E) \mid h \text{ is an odd homeomorphism of } E \text{ onto } E \text{ with } h(B_1) \subset \tilde{A}_0 \cup B_\rho\}.$$

Then $h(u) = \rho u \in \Lambda^* \neq \varnothing$. Next define

$$\Lambda_k = \{K \subset E \mid K \text{ is compact, symmetric, and } \gamma(K \cap h(\partial B_1)) \geq k \text{ for all } h \in \Lambda^*\}.$$

The following lemma based on Lemma 2.18 of Ambrosetti and Rabinowitz [2] lists the properties of these sets we require:

LEMMA 1.11 If I satisfies the hypotheses of Theorem 1.10, then for $1 \leq k \leq m$

(1) $\Lambda_{k+1} \subset \Lambda_k$,
(2) $\Lambda_k \neq \varnothing$,
(3) if $K \in \Lambda_k$ and $Y \in \Sigma_r$ for $r < k$, then $\overline{K \backslash Y} \in \Lambda_{k-r}$,
(4) if f is an odd homeomorphism of E onto E satisfying $f(u) = u$ where $I(u) < 0$ and $f^{-1}(\tilde{A}_0) \subset \tilde{A}_0$, then $f(K) \in \Lambda_k$ whenever $K \in \Lambda_k$.

Proof The definition of Λ_k implies (1). To prove (2), let $K_R = \overline{B_R} \cap \tilde{E}$. Then $K_R \supset (\tilde{A}_0 \cup B_\rho) \cap \tilde{E} \supset h(B_1) \cap \tilde{E}$ for all $h \in \Lambda^*$ via hypothesis (2) of Theorem 1.10. Hence $K_R \cap h(\partial B_1) = \tilde{E} \cap h(\partial B_1)$. Since h is a homeomorphism of E onto E and $h(0) = 0$, $h(B_1)$ is a neighborhood of 0 in E. Therefore, $\tilde{E} \cap h(B_1)$ is a symmetric bounded open neighborhood of 0 in $\tilde{E}$ whose boundary lies in $\tilde{E} \cap h(\partial B_1)$. By the monotonicity property of genus ((2) of Lemma 1.1 [3]),

$$\gamma(K_R \cap h(\partial B_1)) = \gamma(\tilde{E} \cap h(B_1)) \geq \gamma(\partial(\tilde{E} \cap h(B_1))). \tag{1.12}$$

Since $\tilde{E}$ is m dimensional, the right-hand side of (1.12) equals m (Theorem 1.2 of [3]). Thus (2) is established for $k = m$ and by (1) for $k < m$.

To get (3), observe that $\overline{K \backslash Y}$ is compact and symmetric and for $K \in \Lambda^*$,

$$\overline{K \backslash Y} \cap h(\partial B_1) = \overline{(K \cap h(\partial B_1)) \backslash Y}.$$

Therefore

$$\gamma(\overline{K \backslash Y} \cap h(B_1)) = \gamma(\overline{(K \cap h(\partial B_1)) \backslash Y}) \geq \gamma(K \cap h(\partial B_1)) - \gamma(Y) \geq m - r$$

(using (5) of Lemma 1.1 [3]). Lastly to prove (4), observe that $f(K)$ is compact and symmetric. If $h \in \Lambda^*$,

$$\gamma(f(K) \cap h(\partial B_1)) = \gamma(K \cap f^{-1} \circ h(\partial B_1))$$

(via (3) of Lemma 1.1 [3]). If $f^{-1} \circ h \in \Lambda^*$, the result is immediate. Clearly $f^{-1} \circ h$ is an odd homeomorphism of E onto E. Since $h \in \Lambda^*$, $h(B_1) \subset \tilde{A}_0 \cup \bar{B}_\rho$ and therefore

$$\begin{aligned} f^{-1}(h(B_1)) &\subset f^{-1}(\tilde{A}_0) \cup f^{-1}(\{x \in \bar{B}_\rho \mid I(x) < 0\}) \\ &\subset \tilde{A}_0 \cup \{x \in \bar{B}_\rho \mid I(x) < 0\} \subset \tilde{A}_0 \cup \bar{B}_\rho. \end{aligned}$$

Now we can complete the proof of Theorem 1.10. Define

$$c_k = \inf_{A \in \Lambda_k} \max_{u \in A} I(u), \qquad 1 \le k \le m.$$

Clearly $c_k \le c_{k+1}$ via (1) of Lemma 1.11. We will show c_k is a critical value of I with $c_k \ge \alpha > 0$ for $j < k \le m$ and if $c_{k+1} = \cdots = c_{k+p} \equiv c$, $\gamma(K_c) \ge p$. It suffices to prove the latter statement. First observe that for $A \in \Lambda_k$ and $h(u) = \rho u \;\varepsilon\; \Lambda^*$, $\gamma(A \cap h(\partial B_1)) = \gamma(A \cap \partial B_\rho) \ge k$. Hence by hypothesis (1), for $k > j$, $A \cap \partial B_\rho \cap E_2 \ne \varnothing$ as in Theorem 1.9. Therefore, $\max_{u \in A} I(u) \ge \alpha$ and $c_k \ge \alpha$ for $k > j$. Next suppose $\gamma(K_c) < p$. Then we can find a neighborhood $\mathscr{U}$ of K_c such that $\gamma(\mathscr{U}) = \gamma(K_c) < p$ ((6) of Lemma 1.1 [3]). By Lemma 1.1 with $\bar{\varepsilon} = \alpha$, there is an $\varepsilon > 0$ and $\eta \in C([0, 1] \times E, E)$ with $\eta(t, x)$ odd in x and

$$\eta(1, A_{c+\varepsilon} \backslash \mathscr{U}) \subset A_{c-\varepsilon}. \tag{1.13}$$

Choose $K \in \Lambda_{k+p}$ so that

$$\max_{u \in K} I(u) \le c + \varepsilon. \tag{1.14}$$

Then by (3) of Lemma 1.11, $\overline{K \backslash \mathscr{U}} \in \Lambda_{k+1}$. Since $\eta(1, \cdot)^{-1} \tilde{A}_0 \subset \tilde{A}_0$ and $\eta(1, u) = u$ for $x \in A_0$ by (4) and (2) of Lemma 1.1, it follows that $\eta(1, K \backslash \mathscr{U}) \in \Lambda_{k+1}$. But this is contrary to (1.13), (1.14), and the definition of c.

2. Applications to Elliptic Partial Differential Equations

In this section several applications of the results of Section 1 will be given to elliptic partial differential equations. In particular a simple proof of the result of Ahmad *et al.* [1] which motivated our work will be presented. A perturbed version of their theorem will also be proved and the symmetric case will be studied briefly.

Let $\mathscr{D} \subset \mathbb{R}^n$ be a bounded domain with a smooth boundary and let

$$Lu \equiv - \sum_{i, j=1}^{n} (a_{ij}(x) u_{x_j})_{x_i} + c(x) u,$$

where $a_{ij}(x) = a_{ji}(x)$, the functions $a_{ij}(x)$, $c(x)$ are smooth (e.g., C^1) functions in $\bar{\mathscr{D}}$, and L is uniformly elliptic in $\bar{\mathscr{D}}$. Consider the partial differential equation

$$Lu = f(x, u), \quad x \in \mathscr{D}; \qquad u = 0, \quad x \in \partial\mathscr{D}. \tag{2.1}$$

We assume

(f_1) $f \in C(\bar{\Omega} \times \mathbb{R}, \mathbb{R})$ and there is a constant $M > 0$ such that $|f(x, z)| \le M$ for all $x \in \bar{\mathscr{D}}$, $z \in \mathbb{R}$.

Let $E = W_0^{1,2}(\mathscr{D})$. We are interested in weak solutions of (2.1), i.e., $u \in E$ such that

$$(Lu, \varphi) \equiv \int_{\mathscr{D}} \left(\sum_{i,j=1}^{n} a_{ij}(x) u_{x_i} \varphi_{x_j} + c(x) u \varphi \right) dx = \int_{\mathscr{D}} f(x, u(x)) \varphi \, dx \tag{2.2}$$

for all $\varphi \in C_0^{\infty}(\mathscr{D})$. Condition ($f_1$) implies that the right hand side of (2.2) makes sense if $u \in E$. Standard regularity theorems for (2.1) in conjunction with (f_1) imply that a weak solution u in fact has Hölder continuous first derivatives and if f is, e.g., Hölder continuous, u is a classical solution of (2.1).

We further assume that

(L) L has a null space, $N(L)$, spanned by $\varphi_1(x), \ldots, \varphi_p(x)$ and f satisfies one of the following two conditions distinguished by the superscript $\pm$:

($f_2^{\pm}$) If $F(x, z) = \int_0^z f(x, t) \, dt$, then $\int_{\mathscr{D}} F(x, \sum_{i=1}^{p} \alpha_i \varphi_i(x)) \, dx \to \pm\infty$ as $|\alpha| = (\sum_{i=1}^{p} \alpha_i^2)^{1/2} \to \infty$.

The interesting feature here is that L has a null space, for otherwise, existence would be trivial without ($f_2^{\pm}$) via the Schauder fixed-point theorem.

Ahmed *et al.* [1] proved that

THEOREM 2.3 If conditions (L), (f_1), and either of (f_2^{+}), (f_2^{-}) are satisfied, (2.1) possesses at least one weak solution.

The proof of Theorem 2.3 given in [1] uses a Galerkin argument which involves finding a critical point for an approximate finite dimensional problem in an interesting although indirect fashion. We will show how the result can be obtained directly from Theorem 1.2. A few preliminaries are required. For $u \in E$, define

$$I(u) = \tfrac{1}{2}(Lu, u) - \int_{\mathscr{D}} F(x, u(x)) \, dx. \tag{2.4}$$

Condition (f_1) implies that $I \in C^1(E, \mathbb{R})$ (see, e.g., Rabinowitz [3]). Let N^+ (resp. N^-) denote the subspace of E on which L is negative (resp. positive) definite. Elliptic theory implies N^- is finite dimensional, N^+ is infinite dimensional, and $E = N(L) \oplus N^+ \oplus N^-$ where these subspaces are mutually orthogonal. Let P, P^+, P^- denote the orthogonal projectors of E onto $N(L)$, N^+, N^-, respectively.

LEMMA 2.5 If f satisfies (f_1)–(f_2^-), given any $K > 0$, there is an $r > 0$ such that $I(u) \geq K$ for $u \in N(L) \oplus N^+$ and $\|u\| \geq r$.

Proof By (f_1) for $u \in N(L) \oplus N^+$,

$$I(u) = \tfrac{1}{2}(LP^+u, P^+u) - \int_{\mathscr{D}} F(x, u)\, dx \geq \tfrac{1}{2}\alpha\|P^+u\|^2 - \int_{\mathscr{D}} F(x, Pu)\, dx - \int_{\mathscr{D}} (F(x, u) - F(x, Pu))\, dx \geq \tfrac{1}{2}\alpha\|P^+u\|^2 - \int_{\mathscr{D}} F(x, Pu)\, dx - M\int_{\mathscr{D}} |P^+u|\, dx. \tag{2.6}$$

The result now follows from (f_2^-) and the Schwarz and Poincaré inequalities.

Remark 2.7 Similarly $I(u) \to \infty$ in a uniform fashion as $\|u\| \to \infty$, $u \in N^-$. Likewise if f satisfies (f_1)–(f_2^+), $I(u) \to \infty$ (resp. $-\infty$) in a uniform fashion as $\|u\| \to \infty$, $u \in N^+$ (resp. $N(L) \oplus N^-$).

LEMMA 2.8 If f satisfies (f_1) and (f_2^+) or (f_2^-), I satisfies the (PS) condition.

Proof Let (u_m) be a sequence in E such that $I(u_m)$ is bounded and $I'(u_m) \to 0$. Then

$$I'(u_m)\varphi = \int_{\mathscr{D}} \left[\sum_{i,j=1}^{n} a_{ij}(x) u_{mx_j}\varphi_{x_j} + (cu_m - f(x, u_m))\varphi\right] dx \tag{2.9}$$

and $|I'(u_m)\varphi| \leq \|\varphi\|$ for all $\varphi \in E$ and m sufficiently large. By successively choosing, e.g., $\varphi = P^+u_m$ and arguing as in Lemma 2.5, we obtain bounds for $((I - P)u_m)$ in E. Since $I(u_m)$ is bounded, Lemma 2.5 and Remark 2.7 then furnish bounds for (Pu_m). Thus (u_m) is bounded in E and therefore possesses a subsequence converging weakly in E and strongly in L^2 to $\tilde{u} \in E$. Since $N(L) \oplus N^-$ is finite dimensional, $(P + P^-)u_m \to (P + P^-)\tilde{u}$ along this subsequence. The Fréchet derivative of $\int_{\mathscr{D}} F(x, u)\, dx$ is compact and maps weakly convergent to strongly convergent sequences. (See, e.g., Lemma 2.19 of Rabinowitz [3]). Thus we can pass to a limit in (2.9) and get $I'(\tilde{u}) = 0$. Hence along $u_{m_i} \to \tilde{u}$, we have

$$I'(u_{m_i})\varphi - I'(\tilde{u})\varphi = \tfrac{1}{2}(L(u_{m_i} - \tilde{u}), \varphi) - \int_{\mathscr{D}} (f(x, u_{m_i}) - f(x, \tilde{u})\varphi\, dx \to 0, \qquad m_i \to \infty. \tag{2.10}$$

Choosing $\varphi = P^+(u_{m_i} - \tilde{u})$, (2.10) shows $P^+u_{m_i} \to P^+\tilde{u}$. Hence the (PS) condition is satisfied.

Proof of Theorem 2.3 Suppose first that (f_2^-) is satisfied. Choosing $E_1 = N^-$ and $E_2 = N(L) \oplus E^+$, Lemma 2.5 and Remark 2.2 show that $I|_{E_2}$ is bounded from below and $I|_{E_1} \to -\infty$ as $\|u\| \to \infty$. Thus the hypotheses of Theorem 1.2 are satisfied and I has a critical point. If (f_2^+) is satisfied, we take $E_1 = N(L) \oplus N^-$, $E_2 = E^+$ and argue similarly.

Remark 2.11 Theorem 2.3 generalizes earlier work of Landesman and Lazer [6] as well as other results. See Ahmad *et al.* [1] for a bibliography of related work.

Remark 2.12 A sufficient condition for $(f_2^\pm)$ to occur is that $F(x, z) \to \pm\infty$ as $|z| \to \infty$ uniformly for $x \in \bar{\mathscr{D}}$. To see this, e.g., for (f_2^+), let a_0 be such that $F(x, z) \geq 0$ for $|z| \geq a_0$ and $x \in \mathscr{D}$, and let $a_1 = a_1(K)$ be such that $F(x, z) \geq K$ if $|z| \geq a_1$ and $x \in \bar{\mathscr{D}}$. Choose $v \in N(L)$ and set $v(x) = t\varphi(x)$, where $\|\varphi\| = 1$ and $t \in \mathbb{R}^+$. Then for some constant K_1 (independent of v and K)

$$(2.13) \qquad \int_{\mathscr{D}} F(x, v(x))\, dx = \sum_{i=1}^{3} \int_{\mathscr{D}_i} F(x, v(x))\, dx \geq -K_1 + K \operatorname{meas} \mathscr{D}_3,$$

where $\mathscr{D}_1 = \{x \in \mathscr{D} \mid |v(x)| \leq a_0\}$, $\mathscr{D}_2 = \{x \in \mathscr{D} \mid a_0 < |v(x)| < a_1\}$, and $\mathscr{D}_3 = \{x \in \mathscr{D} \mid |v(x)| \geq a_1\}$. For $t = t(K, \varphi)$ sufficiently large, meas $\mathscr{D}_3 \geq \frac{1}{2}$ meas $\mathscr{D}$. Since $B_1 \cap N(L)$ is compact, t can be chosen independently of $\varphi \in N(L)$. Thus (f_2^+) now follows since K is arbitrary.

Remark 2.14 The restriction that f be uniformly bounded is fairly severe. If $f(x, z)$ is permitted to grow at a more rapid rate than linear, the F term in I will dominate at ∞ and the absence or presence of a nontrivial $N(L)$ may have no effect on the problem. See, e.g., Ambrosetti and Rabinowitz [2] for results in this direction.

As our next application, we will show (2.1) possesses a solution if f is perturbed slightly.

THEOREM 2.15 Let (L), (f_1), and (f_2^+) or (f_2^-) be satisfied. For any $g \in C(\bar{\mathscr{D}} \times \mathbb{R}, \mathbb{R})$ there is a constant $T > 0$ such that if $|t| < T$, the problem

$$(2.16) \qquad Lu = f(x, u) + tg(x, u), \quad x \in \mathscr{D}; \qquad u = 0, \quad x \in \partial\mathscr{D}$$

possesses a weak solution.

Proof It suffices to show that the perturbed functional

$$\tilde{I}(u) = \tfrac{1}{2}(Lu, u) - \int_{\mathscr{D}} (F(x, u) + tG(x, u))\, dx$$

has a critical point in E (where G is defined in the obvious fashion). Since no growth restrictions have been placed on g, $\tilde{I}$ may not be defined for all $u \in E$. We will now get around this difficulty in the following fashion. By Theorem 1.2 we have a critical point I of u with $I(u) = c$. Lemma 2.5 and Remark 2.7 then furnish us with a bound for $\|u\|$ for all $u \in K_c$. It is easily seen that this bound depends on c in a monotonic fashion. For $p > n$, the L^p theory of elliptic equations and some simple interpolation inequalities (see Agmon *et al.* [7]) show

$$\max_{x \in \mathscr{D}} |u(x)| \leq K_1[M + \|u\|] \leq K = K(c), \tag{2.17}$$

where K_1 is a constant independent of u.

Define

$$\begin{aligned} \tilde{g}(x, z) &= g(x, z) \qquad \text{if} \quad x \in \bar{\mathscr{D}}, \\ |z| &\leq K(c+1) + 1 \\ &= g(x, (((K(c+1)+1)z/|z|)) \qquad \text{if} \quad x \in \mathscr{D}, \\ |z| &> K(c+1) + 1. \end{aligned}$$

Thus $\tilde{g}$ satisfies (f_1) and these arguments show if t is sufficiently small, any weak solution $\tilde{u}$ of

$$Lu = f(x, u) + t\tilde{g}(x, u), \quad x \in \mathscr{D}; \qquad u = 0, \quad x \in \partial\mathscr{D} \tag{2.18}$$

satisfying $J(\tilde{u}) \leq c + 1$ will have $\max_{x \in \Omega} |\tilde{u}(x)| \leq K(c+1) + 1$. Here

$$J(u) = \tfrac{1}{2}(Lu, u) - \int_{\mathscr{D}} (F(x, u) + t\tilde{G}(x, u))\, dx.$$

Therefore, $\tilde{u}$ is a weak solution of (2.16). Consequently to solve (2.16), it suffices to show that for t sufficiently small, J has a critical point $\tilde{u}$ with $J(\tilde{u}) \leq c + 1$.

Since $\tilde{g}$ satisfies (f_1), $J \in C^1(E, \mathbb{R})$. Unfortunately $f + t\tilde{g}$ may not satisfy $(f_2{}^{\pm})$ and therefore J may not satisfy the (PS) condition. However we can take advantage of Remark 1.5. Observe that

$$\begin{aligned} I'(u)(P^+u) &= \tfrac{1}{2}(LP^+u, P^+u) - \int_{\mathscr{D}} f(x, u)P^+u\, dx \\ &\geq \alpha\|P^+u\|^2 - K_2\|P^+u\| \end{aligned} \tag{2.19}$$

for some constant K_2. Hence there is an $R^+ > 0$ such that

$$I'(u)(P^+u) > 0, \tag{2.20}$$

provided that $\|P^+u\| \geq R^+$ independently of $(I - P^+)u$. Similarly there is an

$R^- > 0$ such that

$$I'(u)(P^-u) < 0, \tag{2.21}$$

provided that $\|P^-u\| \geq R^-$ independently of $(I - P^-)u$.

Suppose that (f_2^+) is satisfied. Then in the context of Theorem 1.2, $E_1 = N(L) \oplus N^-$ and $E_2 = N^+$. Let $\mathcal{O}^+ = \{u \in N^+ \mid \|P^+u\| \leq R^+\}$. By Remark 2.7, there is an $R_1 > 0$ such that $I(u) < b_2 - 1$ for all $u \in E$ such that $\|(I - P^+)u\| \geq R_1$ and $\|P^+u\| \leq R^+$. Let $\mathcal{O}_1 = \{u \in N(L) \oplus N^- \mid \|u\| \leq R_1\}$ Consider $\mathcal{O} = \mathcal{O}_1 \oplus \mathcal{O}^+$. An examination of the proof of Lemma 1.1 (for E a Hilbert space—see, e.g., Rabinowitz [3]) shows that since (2.20) is satisfied, we can choose η such that if $u \in \partial\mathcal{O}$ and $\|P^+u\| = R^+$, then $\|P^+\eta(t, u)\| \leq R^+$. (In brief this is so since η is determined by solving an ordinary differential equation, $(d\eta/dt) = -V(\eta)$, where by (2.20) we can assume $(V(x), P^+x) \geq 0$ if $x \in \partial\mathcal{O}$ and $\|P^+x\| = R^+$. Thus $\|P^+\eta(t, u)\|$ cannot exceed R^+.) Moreover since $I(u) \leq b_2 - 1 < c$ for $u \in \partial\mathcal{O}$ and $\|(I - P^+)u\| = R_1$, an appropriate choice of $\bar{\varepsilon}$ makes $\eta(t, u) = u$ for such u. It therefore follows that $\eta(t, \cdot)\colon \mathcal{O} \to \mathcal{O}$, so by Remark 1.5 with $\Omega = B_{R_1} \cap E_1$ we can characterize c as

$$c = \inf_{\chi \in \mathcal{S}_{\mathcal{O}}} \max_{z \in \bar{\Omega}} I(\chi(u)).$$

Now consider $J(u)$ for small $|t|$. Since $\mathcal{O}$ is bounded, the proof of Lemma 2.8 shows $J|_{\mathcal{O}}$ satisfies the (PS) condition. Moreover

$$c_t = \inf_{\chi \in \mathcal{S}_{\mathcal{O}}} \max_{z \in \bar{\Omega}} J(\chi(u))$$

is, e.g., within $\frac{1}{3}$ of c. We can also assume $J(u) \geq b_2 - \frac{1}{3}$ for $u \in \mathcal{O} \cap E_2$ and $J'(u)(P^+u) > 0$ for $u \in \partial\mathcal{O}$ and $\|P^+u\| = R^+$. Hence an analog of Lemma 1.1 holds for $J|_{\mathcal{O}}$ so c_t is a critical value of J. By our above remarks, this provides us with a weak solution of (2.16).

Finally, suppose that (f_2^-) is satisfied. Then the argument parallels that just given except that we use Corollary 1.8 instead of Theorem 1.2.

For the final result in this section, we shall give an application of Theorem 1.9.

THEOREM 2.22 Suppose (L), (f_1), (f_2^-) are satisfied and in addition $f(x, 0) = 0$, $f(x, z)$ is odd in z, and

(f_3) there is a constant $r > 0$ so that $F(x, z) > 0$ if $0 < |z| < r$ and $x \in \bar{\mathcal{D}}$.

Then (2.1) possesses at least $p = \dim N(L)$ distinct pairs of nontrivial solutions.

Proof Recall $E_1 = N^-$ for this case. Let

$$A = \{u \in N^- \oplus N(L) \mid \|u\| = s\}.$$

Then for s sufficiently small, (f_3) implies that

$$I(u) = \tfrac{1}{2}(Lu, u) - \int_{\mathscr{D}} F(x, u)\, dx < 0$$

for all $u \in A$. Since $\gamma(A) = \dim N(L) + \dim N^-$, the result is immediate from Theorem 1.9.

3. An Application to a Hyperbolic Partial Differential Equation

In this section an existence theorem for a hyperbolic partial differential equation will be proved with the aid of Theorem 1.2. To avoid unduly complicating matters we will prove the result under more restrictive hypotheses than are necessary and remark later on how they can be weakened.

Consider the nonlinear wave equation

$$\square u \equiv u_{tt} - u_{xx} = cu + g(x, t, u); \quad 0 < x < \pi, \quad t \in [0, \pi] \tag{3.1}$$

together with the boundary and periodicity conditions

$$u(0, t) = 0 = u(\pi, t), \qquad u(x, t + 2\pi) = u(x, t). \tag{3.2}$$

In (3.1), c is a constant and g satisfies

(g_1) $g \in C^1([0, \pi] \times \mathbb{R}^2, \mathbb{R})$, is 2π periodic in t, and there is an $M \geq 0$ such that $|g(x, t, z)| \leq M$ if $x \in [0, \pi]$ and $(t, z) \in \mathbb{R}^2$.

Such equations have been considered by several authors. See, e.g., Rabinowitz [8] for some references. Let H^j denote the closure with respect to

$$\|u\|_j = \left(\sum_{|\sigma| \leq j} \int_0^\pi \int_0^{2\pi} |D^\sigma u(x, t)|^2\, dx\, dt \right)^{1/2} \tag{3.3}$$

of infinitely differentiable functions in (x, t) on $[0, \pi] \times \mathbb{R}$ which satisfy (3.2). The standard multi-index notation is being used in (3.3). Letting $E = H^1$, we seek a weak solution of (3.1)–(3.2) in E.

Using Fourier series, it is easy to see that the spectrum of $\square$ consists of $\{k^2 - j^2 \mid j, k \in \mathbb{Z}, j \neq 0\}$ with corresponding eigenvectors $\sin jx(\alpha \cos kt + \beta \sin kt)$. In particular $N(\square)$ is infinite dimensional. Let $Lu = \square u - cu$. It is easy to see that $N(L) = \{0\}$ or is finite dimensional. The following conditions on g will be used below:

(g_2) $|c^{-1} g_z(x, t, z)| < 1$ and $g_t(x, t, z)$ is uniformly bounded on $[0, \pi] \times \mathbb{R}^2$;

$(g_3^{\pm})$ if $G(x, t, z) = \int_0^z g(x, t, s)\, ds$, $G(x, t, z) \to \pm\infty$ as $|z| \to \infty$.

Then we have

THEOREM 3.4 Let $c \neq 0$ and let g satisfy (g_1)–(g_2). If $\dim N(L) > 0$, further suppose that (g_3^+) or (g_3^-) is satisfied. Then (3.1)–(3.2) possesses a weak solution $u \in H^1$.

Proof Let N^+, N^- again denote the orthogonal subspaces of $E = H^1$ on which $\Box$ is positive and negative definite, respectively. Thus $E = N(L) \oplus N^+ \oplus N^-$. For this problem $N(\Box)$ and N^- are infinite dimensional so we are not in the framework of Theorem 1.2. However, we bypass this difficulty by using a Galerkin argument. There is considerable flexibility in how to do this. To minimize technicalities, let

$$X_m = \left\{ \sum_{0 \leq j,\, |k| \leq m} a_{jk} \sin jx e^{ikt} \right\},$$

where the a_{jk} are complex but the sums are real. Thus $X_m \subset E$. For $u \in E$, set

$$I(u) = \int_0^\pi \int_0^{2\pi} [\tfrac{1}{2}(u_x^2 - u_t^2 - cu^2) - G(x, t, u)]\, dx\, dt. \tag{3.5}$$

As in Section 2, using (g_1), it follows that $I \in C^1(E, \mathbb{R})$ and a fortiori $I \in C^1(X_m, \mathbb{R})$. The notation of Section 2 will be employed in what follows and we define $P, P^\pm, P_m, P_m^\pm$ in the obvious fashion. Let $Q_m = P_m + P_m^+ + P_m^-$.

At this point we have to distinguish between the cases where $\dim N(L) = 0$ and $\dim N(L) > 0$. Suppose first that $\dim N(L) = 0$. Also for convenience we take $c > 0$. Choose $R \geq 0$ so that if $u \in N(L) \oplus N$ and $\|u\| > R$,

$$I(u) < \inf_{w \in N^+} I(w) - 1 \equiv \underline{b}.$$

That such an R can be found follows since the u^2 term dominates the $G(x, t, u)$ term in (3.5). It is easy to verify that the hypotheses of Theorem 1.2 are satisfied on X_m with $E_1 = (N(L) \oplus N^-) \cap X_m$, $E_2 = N^+ \cap X_m$, and $\Omega = B_R \cap E_1$. From Theorem 1.2 we obtain a critical point u_m of $I|_{X_m}$ with corresponding critical value $b_m > \underline{b}$.

We will show that a subsequence of u_m converges to a weak solution of (3.1)–(3.2). To do so requires a priori bounds for $(\|u_m\|_1)$. Indeed if (u_m) is bounded in E, a subsequence converges weakly to $\tilde{u} \in E$ and along this subsequence we have for all $\varphi \in X_k$ and $m \geq k$,

$$0 = \int_0^\pi \int_0^{2\pi} [-u_{mt}\varphi_t + u_{mx}\varphi_x - (cu_m + g(x, t, u_m))\varphi]\, dx\, dt$$

$$\to \int_0^\pi \int_0^{2\pi} [-\tilde{u}_t\varphi_t + \tilde{u}_x\varphi_x - (c\tilde{u} + g(x, t, \tilde{u}))\varphi]\, dx\, dt.$$

It follows that $\tilde{u}$ is a weak solution of (3.1)–(3.2).

To obtain the bounds for $\|u_m\|_1$, observe first that

$$Lu_m = Q_m g(x, t, u_m). \tag{3.6}$$

For the equation

$$Lw = h(x, t),$$

if $w, h \in (I - P)H^0$, expanding w and h in Fourier series,

$$w = \sum_{j, |k| = 0}^{\infty} w_{jk} \sin jx e^{ikt},$$

etc., we find

$$w_{jk} = (j^2 - k^2 - c)^{-1} h_{jk}.$$

Since $N(L) = \{0\}$, there is a $\beta > 0$ so that $|j^2 - k^2 - c| \geq \beta > 0$ for all $j, |k| \in \mathbb{N}$. Therefore, $|w_{jk}| \leq \beta^{-1} |h_{jk}|$ and

$$\|w\|_0 \leq K_1 \|h\|_0. \tag{3.7}$$

In (3.7) and the sequel, K_1, K_2, etc. denote positive constants. Taking in particular $h = (P_m{}^+ + P_m{}^-)g(x, t, u_m)$ and $w_m = (P_m{}^+ + P_m{}^-)u_m$, (3.6) and (3.7) yield

$$\|w_m\|_0 \leq K_1 \|g(x, t, u_m)\|_0 \leq K_2 \tag{3.8}$$

via (g_1). Rewriting (3.6) as

$$\Box u_m = cu_m + Q_m g(x, t, u_m), \tag{3.9}$$

and observing that if

$$\Box w = h \tag{3.10}$$

for h and w in $(I - P)H^0$, then $w_{jk} = (k^2 - j^2)^{-1} h_{jk}$ and

$$\|w\|_1 \leq K_3 \|h\|_0. \tag{3.11}$$

Equations (3.8), (3.9), and (3.11) imply

$$\|w_m\|_1 \leq K_3(c\|w_m\|_0 + \|(P_m{}^+ + P_m{}^-)g(x, t, u_m)\|_0) \leq K_4. \tag{3.12}$$

It remains to bound $v_m = P_m u_m$. From (3.9) we have

$$\int_0^{\pi} \int_0^{2\pi} (cu_m + Q_m g(x, t, u_m))\varphi \, dx \, dt = 0 \tag{3.13}$$

for all $\varphi \in P_m X_m$. Choosing in particular $\varphi = -v_{mtt}$ and substituting in (3.13) gives

$$\int_0^{\pi} \int_0^{2\pi} cv_{mt}^2 \, dx \, dt = -\int_0^{\pi} \int_0^{2\pi} (g_t(x, t, u_m)v_{mt} + Q_m g_u(x, t, u_m)u_{mt} v_{mt}) \, dx \, dt. \tag{3.14}$$

Using (g_2) together with (3.12) and (3.14) yields a bound for $\|v_{mt}\|_0 = \|v_{mx}\|_0$ and therefore for $\|v_m\|_1$.

Thus we have the theorem when $N(L) = \{0\}$. Next suppose $\dim N(L) = p > 0$ and $N(L)$ is spanned by $\varphi_1(x, t), \ldots, \varphi_p(x, t)$. The argument of Remark 2.12 shows that if g satisfies $(g_3{}^{\pm})$,

$$\int_0^\pi \int_0^{2\pi} G(x, t, \varphi(x, t))\, dx\, dt \to \pm\infty \tag{3.15}$$

as $\|\varphi\|_1 \to \infty$ for $\varphi \in N(L)$. With the aid of this observation, it again follows that $I|_{X_m}$ satisfies the hypotheses of Theorem 1.2 (e.g., if $(g_3{}^+)$ is satisfied, we take $E_1 = (N(L) \oplus N^-) \cap X_m$, $E_2 = N^+ \cap X_m$, and Ω as in the case treated above with R being determined using (3.15) as in Section 2).

Let u_m denote a critical point of $I|_{X_m}$ with b_m its corresponding critical value. The proof reduces as earlier to obtaining bounds for (u_m). The difficulty in treating this case is that we no longer have $|j^2 - k^2 - c| \geq \beta > 0$ for some β and all j, $|k| \in \mathbb{N}$ so (3.7) is no longer valid. However, if we write $w_m = Tw_m + (I - T)w_m$, where T is the orthogonal projector of E onto $N(L)$, our earlier estimates show

$$\|(I - T)w_m\|_1 \leq K_5. \tag{3.16}$$

We specialize to the case in which $(g_3{}^+)$ is satisfied. The remaining case is treated similarly. Theorem 1.2 implies that

$$b_m = I(u_m) \geq \inf_{u \in N^+} I(u) \geq \underline{b} > -\infty. \tag{3.17}$$

But

$$\begin{aligned} b_m &= \frac{1}{2}(\Box w_m, w_m) - \int_0^\pi \int_0^{2\pi} \left(\frac{c}{2} u_m{}^2 + G(x, t, u_m)\right) dx\, dt \\ &= \frac{1}{2}(L(I - T)w_m, w_m) \\ &\quad - \int_0^\pi \int_0^{2\pi} \left(\frac{c}{2} |Pu_m|^2 + G(x, t, u_m)\right) dx\, dt. \end{aligned} \tag{3.18}$$

Hence (3.16)–(3.18) and the argument of Remark 2.7 provide a bound for $\|Tw_m\|_0$ and therefore for $\|Tw_m\|_1$ since $N(L)$ is finite dimensional as well as for $\|v_m\|_0$. Finally (3.13) and (3.14) give a bound on $\|v_m\|_1$ as earlier. Hence the proof is complete.

Remark 3.19 Using more delicate estimates, e.g., as in Rabinowitz [8], (g_2) can be weakened to eliminate the condition on g_t and only require the condition on $g_u(x, t, z)$ in some ball in z.

References

1. S. Ahmad, A. C. Lazer, and J. L. Paul, Elementary critical point theory and perturbations of elliptic boundary value problems at resonance (preprint), Indiana Univ. Math. J., **25** (1976), 933–944.
2. A. Ambrosetti and P. H. Rabinowitz, Dual variational methods in critical point theory and applications. *J. Functional Analysis* **14** (1973), 349–381.
3. P. H. Rabinowitz, Variational methods for nonlinear eigenvalue problems. *Proc. Symp. Eigenvalues Nonlinear Problems, Varenna, Italy*, pp. 141–195. Edizioni Cremonese, Rome, 1974.
4. J. T. Schwartz, "Nonlinear Functional Analysis." New York Univ. Lecture Notes, 1965.
5. D. C. Clark, A variant of the Ljusternik–Schnirelmann theory. *Indiana Univ. Math. J.* **22** (1972), 65–74.
6. E. Landesman and A. Lazer, Nonlinear perturbations of linear elliptic boundary value problems at resonance. *J. Math. Mech.* **19** (1970), 609–623.
7. S. Agmon, A. Douglis, and L. Nirenberg, Estimates near the boundary for solutions of elliptic partial differential equations satisfying general boundary conditions, I. *Comm. Pure Appl. Math.* **12** (1959), 623–727.
8. P. H. Rabinowitz, Time periodic solutions of nonlinear wave equations, *Manuscripta Math.* **5** (1971), 165–194.

This research was sponsored in part by the United States Army under Contract No. DAAG29-75-C-0024 and by the Office of Naval Research under Contract No. N00014-76-C-0300. Reproduction in whole or in part is permitted for any purpose of the United States Government.

AMS (MOS) 1970 Subject Classification: 35A15, 35B10, 35D05, 35J60, 35L60, 47H15, 49G99, 58E05.

Branching and Stability for Nonlinear Gradient Operators

D. Sather

University of Colorado

1. Introduction

The problem of determining the equilibrium states of a physical system leads in a number of cases to the problem of finding the nontrivial solutions near $w = 0$ of a nonlinear equation of the form

$$Lw + N(w) = \lambda w, \qquad w \in \mathscr{B}, \qquad \lambda \in \mathbb{R}^1, \tag{1.1}$$

where $\mathscr{B}$ is a real Banach space, $L\colon \mathscr{D}(L) \to \mathscr{B}$ is a linear Fredholm operator with index zero, and $N\colon \mathscr{D}(N) \to \mathscr{B}$ is a nonlinear operator with $N(0) = 0$. The basic problem of determining the number of nontrivial solutions near $w = 0$ of an equation of the form (1.1) has led to an entire theory of branching of solutions of nonlinear equations, and although the general problem remains unsolved, there has been a considerable amount of progress in recent years (e.g., see references [2,18,22,24] and the references therein). In particular, it is well known that the Lyapunov–Schmidt method provides a general constructive approach to establishing the existence of *continuous* solution branches of (1.1).

On the other hand, compared to the voluminous amount of recent work on the existence of nontrivial solution branches of (1.1), there have been relatively few mathematical papers on the equally important physical problem of the "stability" (or "instability") in infinite-dimensional spaces of solution branches, and a general theory of stability for such problems does not exist at the present time. However, there have been some noteworthy papers in recent years on special topics of stability (e.g., see references [6,9,11,21,22] and the references therein), and in some problems they provide useful information on both the existence and stability of continuous solution branches. In particular, McLeod and Sattinger [11] have shown that the method of Lyapunov–Schmidt also contains many of the important elements required for a stability analysis of various solution branches of (1.1).

In the present paper we study both the branching and the stability of solutions near $w = 0$ of a special case of Eq. (1.1), namely, an equation of the form

$$w - \mu A w + T(w) + R(w) = 0, \qquad w \in \mathscr{H}, \quad \mu \in \mathbb{R}^1, \tag{$*$}$$

where $\mathscr{H}$ is a real Hilbert space with inner product $(\cdot, \cdot)$ and norm $\|\cdot\|$, $A: \mathscr{H} \to \mathscr{H}$ is a positive compact self-adjoint operator, $T: \mathscr{H} \to \mathscr{H}$ is a continuous homogeneous operator of degree k $(k \geq 2)$, and R: $\mathscr{H} \to \mathscr{H}$ is an analytic "higher order" operator with $R(0) = 0$. In addition, we assume for the branching results in Section 3 that T (but not R) is a gradient operator, and we assume for the stability results in Section 4 that both T and R are gradient operators. Under these assumptions, we obtain in Section 3 a new branching result for equation $(*)$, which establishes in a general situation the existence of Hölder continuous solution branches of $(*)$ (see Theorem 3). In addition, by sharpening some of the stability results of McLeod and Sattinger [11], we show that the Lyapunov–Schmidt method, together with some elementary ideas of nonlinear functional analysis, provides in many cases not only the existence of Hölder continuous solution branches but also a relatively complete stability analysis of such solution branches.

The concept of stability used throughout the paper is essentially "linearized stability," namely, the stability (or instability) of a solution $w^*(\mu)$ of $(*)$ at μ is determined by the spectrum of the derived operator

$$D(w^*, \mu) = I - \mu A + T'(w^*) + R'(w^*), \tag{1.2}$$

where, here and in the remainder of the paper, the Fréchet derivative of an operator F at w is denoted by $F'(w)$.

The methods used throughout the paper depend heavily on some of the basic properties of homogeneous gradient operators developed in Rothe's early papers [14,15]; such properties are particularly useful in the development of constructive methods for finding continuous solution branches of Eq. ($*$) and in carrying out a stability analysis of such solution branches. The proof of the main branching result (Theorem 3) employs, in addition, finite-dimensional (topological) degree theory together with a known "curve selection lemma" from the theory of analytic sets (see Böhme [1] and Milnor [12]).

Although we do not consider any specific applications in the present paper, the results obtained here can be used, for example, to carry out a complete branching and stability analysis of some buckling problems in nonlinear elasticity (see Knightly and Sather [10] and Sather [20]).

2. Preliminaries

In this section we set the hypotheses on the nonlinear operators T and R, and outline the Lyapunov–Schmidt method for Eq. ($*$).

Our assumptions throughout the paper on the operators T and R are as follows (see Vainberg [23] and Vainberg and Trenogin [24, p.17 ff]) for the various definitions):

(H1) T is a continuous homogeneous polynomial of degree k ($k \geq 2$) which is the (strong) gradient of the function $\tau(w) = (1/(k+1))(T(w), w)$, i.e., there exists a functional $r\colon \mathscr{H} \times \mathscr{H} \to \mathbb{R}^1$ such that for $w, h \in \mathscr{H}$

$$\tau(w+h) - \tau(w) = (T(w), h) + r(w, h)$$

and $r(w, h)/\|h\| \to 0$ as $\|h\| \to 0$.

(H2) R is analytic in a neighborhood of the origin, $R(0) = 0$, and $\|R(w)\|/\|w\|^k \to 0$ as $\|w\| \to 0$.

In addition, we assume for the stability results in Section 4 that R is a gradient operator.

The assumption that T is the gradient of the particular functional $\tau(w) = (1/(k+1))(T(w), w)$ is a consequence of the homogeneity of T (e.g., see Rothe [14, p.272]). The assumption of analyticity on R is imposed mainly for the sake of convenience. In fact, all of the results of the paper except our main branching result in Theorem 3, in which the analyticity of R plays a crucial role, continue to hold under much less stringent conditions on R (e.g., see McLeod and Sattinger [11, Section 6]), however the various solution branches obtained under such conditions are, of course, no longer analytic.

The Lyapunov–Schmidt method for equation ($*$) is well-known so that we only outline here the main points (see also Sather [18] and Vainberg and Trenogin [24]). Let μ_0 be a (positive) characteristic value of A and let $\mathcal{N} = \mathcal{N}(I - \mu_0 A)$ denote the finite-dimensional null space of $I - \mu_0 A$. We assume throughout the paper that $\dim \mathcal{N} = n \geq 2$. Let $\{w_1, \ldots, w_n\}$ be an orthonormal basis for $\mathcal{N}$. Then the "branching equations" for Eq. $(*)$ are given by

$$-\eta\xi_i + \left(T\left(\sum_{j=1}^{n} \xi_j w_j\right), w_i\right) + r^i(\xi, \eta) = 0 \qquad (i = 1, \ldots, n), \tag{2.1}$$

where $\eta = (\mu/\mu_0) - 1$, $\xi = (\xi_1, \ldots, \xi_n)$ belongs to $\mathbb{R}^n$, and

$$r^i(\xi, \eta) = \left(T\left(\sum_{j=1}^{n} \xi_j w_j + v(\xi, \eta)\right) - T\left(\sum_{j=1}^{n} \xi_j w_j\right), w_i\right) + \left(R\left(\sum_{j=1}^{n} \xi_j w_j + v(\xi, \eta)\right), w_i\right) \qquad (i = 1, \ldots, n). \tag{2.2}$$

Here $v = v(\xi, \eta)$ is an element in $\mathcal{N}^\perp$ which is analytic in ξ and η and satisfies $\|v(\xi, \eta)\| \leq c|\xi|^k$ for $|\xi| < \rho_0$ and $|\eta| < \eta_0$, where the constant c depends only on ρ_0 and η_0, and the analytic functions r^i are "higher order" in the sense that if $|\eta| < \eta_0$, then $r^i(\xi, \eta)/|\xi|^k \to 0$ as $|\xi| \to 0$ (e.g., see Sather [18, p.231]). If $\xi = \xi(\eta) \in \mathbb{R}^n$ is a solution of the "branching equations" in (2.1), then $w = \sum_{j=1}^n \xi_j(\eta) w_j + v(\xi(\eta), \eta)$ is clearly a solution of $(*)$.

If we now make the substitution

$$\xi = \eta^\alpha \beta, \qquad \beta \in \mathbb{R}^n, \quad \alpha = 1/(k-1) \tag{2.3}$$

(for k odd we consider only the case $\eta \geq 0$) and divide by $\eta^{-k\alpha}$, the system (2.1) becomes

$$F^i(\beta, \eta) \equiv -\beta_i + \left(T\left(\sum_{j=1}^{n} \beta_j w_j\right), w_i\right) + \eta^{-k\alpha} r^i(\eta^\alpha \beta, \eta) = 0 \qquad (i = 1, \ldots, n). \tag{2.4}$$

Formally setting $\eta = 0$ in (2.4) we obtain the closely related system

$$f^i(\beta) \equiv -\beta_i + \left(T\left(\sum_{j=1}^{n} \beta_j w_j\right), w_i\right) = 0 \qquad (i = 1, \ldots, n). \tag{2.5}$$

Clearly, if β^* is a nontrivial solution of (2.5) from which we can obtain a solution of (2.4) by an argument involving some sort of an implicit function theorem, then the substitution (2.3) yields a nontrivial solution of the branching equations (2.1), which in turn generates a nontrivial solution of $(*)$. Thus, the problem of finding nontrivial solutions of $(*)$ near $w = 0$ reduces to the problem of finding nontrivial solutions of (2.5) together with the development of suitable implicit function theorems for the system described in (2.4). Such a program will be carried out in the next section by exploiting the fact that T is a homogeneous gradient operator.

3. The Branching Theorems for Equation (*)

The first branching theorem is similar to results of McLeod and Sattinger [11] and Sather [16] and is stated for the convenience of the reader; the theorem is a consequence of the ordinary finite-dimensional implicit function theorem and holds without the assumption that T is a gradient operator.

THEOREM 1 Suppose that $\mathcal{N} \equiv \mathcal{N}(I - \mu_0 A)$ is n-dimensional and that the degree k of homogeneity of T in (H_1) is even [resp., odd]. Suppose that $\tilde{\beta}$ is a nontrivial solution of (2.5) and that $j(f, \tilde{\beta}) \neq 0$, where $j(f, \cdot)$ denotes the Jacobian $\partial(f^1, \ldots, f^n)/\partial(\beta_1, \ldots, \beta_n)$. Then there exists a positive constant δ such that for $0 < |\mu - \mu_0| < \delta$ [resp., $\mu_0 < \mu < \mu_0 + \delta$] the equation (*) has a branch of nontrivial solutions of the form

$$\tilde{w}(\eta) = \eta^\alpha \tilde{u} + U(\eta^\alpha), \qquad \eta = (\mu/\mu_0) - 1, \quad \alpha = 1/(k-1), \tag{3.1}$$

where $\tilde{u} = \sum \tilde{\beta}_j w_j$ belongs to $\mathcal{N}$, U is an analytic function of η^α, and $\lim_{|\eta| \to 0} \eta^{-\alpha} U = 0$ [resp., $\lim_{\eta \to 0^+} \eta^{-\alpha} U = 0$].

Since we are assuming in this chapter that T is also a gradient operator, for certain solutions of (2.5) the condition $j(f, \tilde{\beta}) \neq 0$ can be replaced by conditions which may be somewhat easier to verify. For example, we have the following result.

THEOREM 2 Suppose that $\mathcal{N} \equiv \mathcal{N}(I - \mu_0 A)$ is n-dimensional and that the degree of homogeneity k in (H_1) is even [resp., odd]. Suppose that the functional $(T(u), u)$ restricted to the unit sphere $\mathcal{S}$ in $\mathcal{N}$ has a *positive* relative extremum at u^*, and suppose that $\theta = (T(u^*), u^*)$ is not an eigenvalue of $QT'(u^*)$, where Q denotes the orthogonal projection of $\mathcal{H}$ onto $\mathcal{N}$. Then there exists a positive constant δ such that for $0 < |\mu - \mu_0| < \delta$ [resp., $\mu_0 < \mu < \mu_0 + \delta$] the equation (*) has a branch of nontrivial solutions of the form

$$w^*(\eta) = (\eta/\theta)^\alpha u^* + U^*(\eta^\alpha), \qquad \eta = (\mu/\mu_0) - 1, \quad \alpha = 1/(k-1), \tag{3.2}$$

where U^* is an analytic function of η^α and $\lim_{|\eta| \to 0} \eta^{-\alpha} U^* = 0$ [resp., $\lim_{\eta \to 0^+} \eta^{-\alpha} U^* = 0$].

Remark 3.1 In the important special case where $k = 2$ and u^* corresponds to a positive relative minimum, the additional restriction in Theorem

2 that θ is not an eigenvalue of $QT'(u^*)$ can be omitted (see the proof of a related result by Sather [19, Theorem 2] and also Grundmann [3]).

In order to set the notation for Section 4 we outline a proof of Theorem 2 in the case when $n \geq 0$; a complete proof may be found in reference [16]. Because of the extremum property of u^*, there exists a special orthonormal basis $\{v_1, \ldots, v_n\}$ for $\mathcal{N}$ such that $v_1 = u^*$, $(T(v_1), v_j) = 0$ $(j = 2, \ldots, n)$, $(T'(v_1)v_i, v_j) = 0$ for $i \neq j$, and

$$vd_n \leq vd_{n-1} \leq \cdots \leq vd_2 \leq v\theta, \tag{3.3}$$

where $d_j = (T'(v_1)v_j, v_j)$ and the parameter v is such that $v = 1$ when u^* corresponds to a maximum and $v = -1$ when u^* corresponds to a minimum; the existence of the special basis $\{v_1, \ldots, v_n\}$ is an easy consequence of some the basic properties of homogeneous gradient operators established by Rothe [14] (see Sather [16, Lemma 3]). Using the basis $\{v_1, \ldots, v_n\}$ in (2.1), (2.4), and (2.5), it follows that $\beta^* = (\theta^{-\alpha}, 0, \ldots, 0)$ is a nontrivial solution of (2.5) and that the Jacobian $j(f, \cdot) \equiv \partial(f^1, \ldots, f^n)/\partial(\beta_1, \ldots, \beta_n)$ at β^* is given by (see reference [16, p.54])

$$j(f, \beta^*) = (k - 1)\left(\prod_{m=2}^{n} [-1 + (d_m/\theta)]\right). \tag{3.4}$$

Since θ is not an eigenvalue of $QT'(u^*)$, it follows that $vd_2 < v\theta$ (see reference [16, p.59]) which together with the inequalities (3.3) implies $j(f, \beta^*) \neq 0$. Hence, the ordinary implicit function theorem implies the existence of a solution of (2.4) of the form $\beta = \beta^* + b(\eta^\alpha)$, where b is an analytic function of η^α and $\lim_{\eta \to 0^+} b(\eta^\alpha) = 0$. Such a solution generates a solution $\xi^*(\eta) = \eta^\alpha(\beta^* + b(\eta^\alpha))$ of (2.1), which in turn generates the desired solution w^* of $(*)$.

Clearly, if there are M points on $\mathcal{S}$ that are extreme points for $(T(u), u)|_{\mathcal{S}}$ satisfying the conditions of Theorem 2, then for the appropriate choices of θ Eq. $(*)$ has at least M solution branches of the form in (3.2).

The condition in Theorem 2 that the extremum value be positive is essentially a normalization. If k is even and if $(T(u, u)$ restricted to $\mathcal{S}$ has a positive relative extremum at u^*, then it also has a negative relative extremum at $(-u^*)$. Therefore, since $(T(u^*), u^*)u^* = -(T(-u^*), -u^*)u^*$, the solutions constructed in Theorem 2 for u^* and $-u^*$ may not be distinct. If k is odd and if $(T(u), u)$ restricted to $\mathcal{S}$ is always negative then a result like Theorem 2 holds with the interval $\mu_0 < \mu < \mu_0 + \delta$ replaced by $\mu_0 - \delta < \mu < \mu_0$; in this case one considers only $\eta \leq 0$ and the substitution (2.3) becomes $\xi = |\eta|^\alpha \beta$.

The situation described in Theorem 2 represents the nondegenerate case in which the existence of "analytic" solution branches can be inferred from the nonvanishing of certain Jacobians of Eqs. (2.5). The next result shows that, even in the degenerate case when Theorem 2 does not apply, there are at least as many nontrivial solution branches of $(*)$ as there are points on $\mathscr{S}$ at which $(T(u), u)|_{\mathscr{S}}$ has an isolated positive relative extremum. The result is a consequence of a known "curve selection lemma" in the theory of real analytic sets together with the following lemma on calculating the topological index of certain types of fixed points; by the index of an isolated fixed point ξ_0 of a homogeneous mapping $H: \mathbb{R}^n \to \mathbb{R}^n$ we mean

$$\operatorname{ind}(H, \xi_0) = d(H - \mathbb{1}, D, 0),$$

where $\mathbb{1}$ denotes the identity mapping in $\mathbb{R}^n$, D is an open ball about ξ_0 with radius so small that $\bar{D}$ contains no fixed point of H except ξ_0, and $d(G, \Omega, z)$ denotes, in general, the topological degree of a mapping G at z relative to a bounded open set Ω in $\mathbb{R}^n$ (e.g., see Heinz [4]).

LEMMA 1 Let $h: \mathbb{R}^n \to \mathbb{R}^1$ be a homogeneous mapping of degree $(k + 1)$, and let $H = \operatorname{grad} h$. If $\beta^* \neq 0$ is a fixed point of H and if $h(\beta^*) < h(\beta)$ [resp., $h(\beta^*) > h(\beta)$] for all β such that $|\beta| = |\beta^*|$ and $0 < |\beta - \beta^*| < r$, then $\operatorname{ind}(H, \beta^*) = 1$ [resp., $\operatorname{ind}(H, \beta^*) = (-1)^{n-1}$].

A proof of the lemma in the case $n = 2$ may be found in reference [17]. Much more elegant proofs which also hold in the general case $n \geq 2$ may be found in Grundmann [3] and Ize [7].

Our main branching result for equation $(*)$ is contained in the following theorem (compare the results of Böhme [1], where R is also assumed to be a gradient operator, and see the closely related results for $n = 2$ in Sather [16,17]).

THEOREM 3 Suppose that $\mathcal{N} \equiv \mathcal{N}(I - \mu_0 A)$ is n-dimensional and that the degree k of homogeneity of T in (H_1) is even [resp., odd]. Suppose that the functional $(T(u), u)$ restricted to the unit sphere $\mathscr{S}$ in $\mathcal{N}$ has isolated positive relative extrema at the m points $u_1^*, u_2^*, \ldots, u_m^*$. Then there exists a positive constant δ such that, for $0 < |\mu - \mu_0| < \delta$ [resp., $\mu_0 < \mu < \mu_0 + \delta$], the equation $(*)$ has at least m nontrivial solution branches of the form $(i = 1, \ldots, m)$

$$w_i(\eta) = (\eta/\theta_i)^\alpha u_i^* + U_i^*(|\eta|^{\alpha/s}), \qquad \eta = (\mu/\mu_0) - 1, \quad \alpha = 1/(k - 1), \tag{3.5}$$

where $\theta_i = (T(u_i^*), u_i^*)$ and U_i^* may be expressed, on each of the intervals $\mu_0 \leq \mu < \mu_0 + \delta$ and $\mu_0 - \delta < \mu \leq \mu_0$, as a power series in some fractional power $|\eta|^{\alpha/s}$ such that $\lim_{|\eta| \to 0} \eta^{-\alpha} U_i^* = 0$ [resp., $\lim_{\eta \to 0^+} \eta^{-\alpha} U_i^* = 0$].

Proof The first part of the proof makes use of an implicit function theorem that is based on topological degree theory.

It will be sufficient to consider the case when $\eta \geq 0$.

Suppose that $(T(u), u)|_{\mathscr{S}}$ has an isolated positive relative extremum at u^*. Let $\{v_1, \ldots, v_n\}$ be the special orthonormal basis for $\mathscr{N}$ with $v_1 = u^*$ that was introduced in the proof of Theorem 2. Then the functional $h: \mathbb{R}^n \to \mathbb{R}^1$ defined by

$$h(\beta) = \frac{1}{k+1}\left(T\left(\sum_{j=1}^{n} \beta_j v_j\right), \sum_{j=1}^{n} \beta_j v_j\right), \qquad \beta \in \mathbb{R}^n,$$

is homogeneous of degree $(k + 1)$ and has an isolated positive relative extremum on $|\beta| = 1$ at $(1, 0, \ldots, 0)$. Moreover, if $H = \operatorname{grad} h$ and $\theta = (T(u^*), u^*)$, then $\beta^* = (\theta^{-\alpha}, 0, \ldots, 0)$ is a nontrivial (isolated) fixed point of H which satisfies the conditions of Lemma 1. Hence, if $f = \{f^1, \ldots, f^n\}$ is the vector field defined by (2.5), then $f = H - \mathbb{1}$ does not vanish on the boundary of a sufficiently small ball $B^* = \{\beta: |\beta - \beta^*| < \rho_1\}$ so that $d(f, B^*, 0)$ is defined and, by Lemma 1,

$$d(f, B^*, 0) = \operatorname{ind}(H, \beta^*) \neq 0.$$

Since the vector field $F = \{F^1, \ldots, F^n\}$ defined by (2.4) is continuous near $|\beta| = \eta = 0$ and since there exists a positive number η_1 such that F does not vanish on the boundary of B^* for $0 < \eta < \eta_1$, it follows that for each fixed η satisfying $0 < \eta < \eta_1$, $F(\cdot, \eta)$ is homotopic to f so that

$$d(F(\cdot, \eta), B^*, 0) = d(f, B^*, 0) \neq 0$$

(e.g., see Heinz [4, p.237]). Therefore, by the basic existence theorem of degree theory (e.g., see Heinz [4, p.237]), for each fixed η in $0 < \eta < \eta_1$, there exists a point β_η in B^* such that $\beta_\eta \neq 0$ and $F(\beta_\eta, \eta) = 0$. In particular then Eqs. (2.4) have (possibly an infinite number of) solution pairs (β_η, η) in $B^* \times (0, \eta_1)$.

It remains to show that, in fact, there exists a branch of solutions of (2.4) of the form $\beta(\eta) = \beta^* + b(\eta^{\alpha/s})$, where b is an analytic function of $\eta^{\alpha/s}$ and $b(\eta^{\alpha/s}) \to 0$ as $\eta \to 0^+$. If so, then the substitution (2.3) generates a solution branch $\xi(\eta) = \eta^\alpha(\beta^* + b(\eta^{\alpha/s}))$ of the branching equations (2.1) which, in turn, generates a solution $w^* = w^*(\eta)$ of the form (3.5).

In order to establish the existence of a branch of solutions of (2.4) of the desired form, let us show first of all that $(\beta^*, 0)$ is a limit point of the set $\mathscr{C} = \{(\beta, \eta) \in B^* \times (0, \eta_1): F(\beta, \eta) = 0\}$. If $\eta_k = 1/k$ $(k = k_0, k_0 + 1, \ldots)$, then for k_0 sufficiently large, the corresponding sequence $\{\beta_{\eta_k}\} \subset B^*$ constructed as above using degree theory, has a limit point in $\bar{B}^*$, say $\tilde{\beta}$, and a subsequence, call it $\{\tilde{\beta}_k\}$, such that $\tilde{\beta}_k \to \tilde{\beta}$ as $k \to \infty$. Since $F(\tilde{\beta}_k, \tilde{\eta}_k) = 0$ $(k = 1, 2, \ldots)$, where $\{\tilde{\eta}_k\}$ is the subsequence of $\{\eta_k\}$ corresponding to $\{\tilde{\beta}_k\}$ and since $\eta^{-k\alpha} r(\eta^\alpha \beta, \eta) \to 0$ as $\eta \to 0^+$, it follows that $f(\tilde{\beta}) = 0$. But β^* is an isolated zero of the vector field f so that $\tilde{\beta} = \beta^*$. Thus, $(\beta^*, 0)$ is a limit point of the sequence $\{(\tilde{\beta}_{\eta_k}, \tilde{\eta}_k)\}$.

We are now in a position to apply a known "curve selection lemma" from the theory of real analytic sets (see references [1,12]). Since F is analytic in β and $\sigma = \eta^\alpha$ near $(\beta, \sigma) = (\beta^*, 0)$, Eqs. (2.4) together with the inequalities $|\beta - \beta^*| < \rho$ and $|\sigma| < \sigma_1$ define a (real) analytic set $\mathscr{A}$ in $\mathbb{R}^{n+1}$. In addition, the inequalities $|\beta - \beta^*| < \rho_1$ and $0 < \sigma < \eta_1{}^\alpha$ define a semianalytic set $\mathscr{U}$ in $\mathbb{R}^{n+1}$. Therefore, since $(\beta^*, 0)$ belongs to the closure of $\mathscr{A} \cap \mathscr{U}$, there exists an analytic curve $p^*: [0, \gamma) \to \mathbb{R}^n \times \mathbb{R}_+{}^1$ such that $p^*(0) = (\beta^*, 0)$ and $p^*(t) \equiv (b^*(t), \sigma^*(t)) \in \mathscr{A} \cap \mathscr{U}$ for $0 < t < \gamma$. Moreover, since the analytic set $\mathscr{A}$ is not contained in the coordinate plane $\sigma = 0$, the parametrization of p^* can be chosen so that $\sigma = \sigma^*(t) = t^s$, $0 < t < \gamma$, where s is a positive integer. The curve $b(\eta^{\alpha/s}) = b^*(\eta^{\alpha/s}) - \beta^*$, $0 \le \eta < \gamma^{s/\alpha}$, then provides the desired solution branch of (2.4).

Since these arguments can be carried out for each point $u_i \in \mathscr{S}$ $(i = 1, \ldots, m)$ which provides an isolated relative extremum value of $(T(u), u)|_{\mathscr{S}}$ and since the corresponding balls B_i $(i = 1, \ldots, m)$ can be chosen to be disjoint, the proof of the theorem now follows easily.

Since R is not necessarily a gradient operator in Theorem 3, the following example by Dancer [2, p.722] shows that the condition of having isolated extrema on $\mathscr{S}$ in Theorem 3 is in some sense "best possible." Let $\mathscr{H} = \mathbb{R}^2$ and consider the system

$$\begin{pmatrix} \xi_1 \\ \xi_2 \end{pmatrix} - \mu \begin{pmatrix} \xi_1 \\ \xi_2 \end{pmatrix} + \begin{pmatrix} \xi_1(\xi_1{}^2 + \xi_2{}^2) \\ \xi_2(\xi_1{}^2 + \xi_2{}^2) \end{pmatrix} + \begin{pmatrix} \xi_2{}^5 \\ -\xi_1{}^5 \end{pmatrix} = 0, \qquad \xi \in \mathbb{R}^2, \quad \mu \in \mathbb{R}^1. \tag{3.6}$$

Multiplying the first component by ξ_2, the second component by ξ_1, and subtracting the resultant equations, one sees that $\xi_1{}^6 + \xi_2{}^6 = 0$ so that there are no nontrivial solutions of (3.6). On the other hand, $(T(\xi), \xi) = (\xi_1{}^2 + \xi_2{}^2)^2$ so that Theorem 3 does not apply.

4. The Stability Results

In this section we establish the stability properties of some of the nontrivial solution branches of $(*)$ which were constructed in Section 3.

For the sake of simplicity we assume throughout the section that, in addition to (H_2), R is a gradient operator; in particular then the Fréchet derivative of R is symmetric (e.g., see Vainberg [23, p.56]) so that the "derived" operators introduced in the sequel are bounded symmetric perturbations of self-adjoint operators and are themselves self-adjoint (e.g., see Kato [8, p.287]).

We begin with a discussion of what we mean by "stability" of solutions of Eq. (*) (see also the closely related discussion by McLeod and Sattinger [11, Section 4]). Let μ_1 denote the smallest (positive) characteristic value of A and let $\tilde{w}$ be either the trivial solution $\tilde{w} = 0$ or a nontrivial solution branch of (*) constructed as in Theorem 1 with $\mu_0 = \mu_1$. Then $\tilde{w}$ satisfies (*) for all $\mu \in \mathscr{I}$ (here, and in the sequel, $\mathscr{I}$ denotes a sufficiently small interval of the form $0 < |\mu - \mu_1| < \delta$ or $\mu_1 < \mu < \mu_1 + \delta$). The "derived" operator associated with $\tilde{w}$ is given by

$$D(\tilde{w}, \mu) = I - \mu A + T'(\tilde{w}) + R'(\tilde{w}), \qquad \mu \in \mathscr{I}.$$

Since $\tilde{w}$ is of the form $\tilde{w} = \eta^\alpha \tilde{u} + U$, where $\alpha = 1/(k-1)$, $\eta = (\mu - \mu_1)/\mu_1$, and $\|U(\eta)\| = o(\eta^\alpha)$ as $|\eta| \to 0$, the operator $D(\tilde{w}, \mu)$ can be written as

$$D(\tilde{w}, \mu) = A_1 + \eta\tilde{L} + \tilde{G}(\eta), \qquad \mu \in \mathscr{I}, \tag{4.1}$$

where $A_1 = I - \mu_1 A$,

$$\tilde{L} = -\mu_1 A + T'(\tilde{u}), \tag{4.2}$$

and $\tilde{G}(\eta)$ is a bounded operator which satisfies $\|\tilde{G}(\eta)\| = o(\eta)$ as $|\eta| \to 0$; here we have also used that R is "higher order" than T and that the Fréchet derivative at w of a homogeneous polynomial of degree k is itself a homogeneous polynomial in w which has degree $(k-1)$ (e.g., see Hille and Phillips [5, p.761]).

Let $\{w_1, \ldots, w_n\}$ be the orthonormal basis for $\mathscr{N} \equiv \mathscr{N}(I - \mu_1 A)$ used in the formulation of Theorem 1. Then, as in Lemma 4.1 of McLeod and Sattinger [11], for η sufficiently small there exists a unique n-dimensional invariant subspace of $D(\tilde{w}, \mu)$ spanned by vectors of the form $\tilde{\phi}_i = w_i + 0(1)$ $(i = 1, \ldots, n)$, and a well-defined $n \times n$ matrix $\tilde{B}(\eta) = [\tilde{b}_{ij}(\eta)]$ with

$$\tilde{b}_{ij}(0) = (\tilde{L}w_i, w_j) \qquad (i, j = 1, \ldots, n), \tag{4.3}$$

such that

$$D\tilde{\phi}_i = \sum_{k=1}^{n} \eta \tilde{b}_{ik} \tilde{\phi}_k \qquad (i = 1, \ldots, n). \tag{4.4}$$

Moreover, since $\tilde{L}$ is self-adjoint, the matrix $\tilde{B}(0)$ is symmetric.

We are now in a position to formulate the definition of "stability" used throughout the paper. If for fixed η sufficiently small the eigenvalues of $\eta\tilde{B}(0)$ are all positive, the solution $\tilde{w}$ is said to be stable at η. If for fixed η sufficiently small at least one of the eigenvalues of $\eta\tilde{B}(0)$ is negative, the solution $\tilde{w}$ is said to be unstable at η. (If neither of these cases holds, the stability is said to be indeterminate at η and an analysis of "higher order" terms must be carried out.) Note that our definition of stability for solutions of (*) differs from that of McLeod and Sattinger [11, p.71] because of the form of $(*)$ and because $D(\tilde{w}, \mu)$ is self-adjoint for $\mu \in \mathscr{I}$. Let us also note that, since $\tilde{L} = -\mu_1 A$ for $\tilde{w} = 0$, the trivial solution $\tilde{w} = 0$ is stable for $\mu < \mu_1$ and unstable for $\mu > \mu_1$ so that the trivial solution loses stability as μ passes through μ_1.

Remark 4.1 Since it follows easily from the definition of μ_1 that $(A_1 z, z) \geq [1 - (\mu_1/\mu_2)]\|z\|^2$ for all $z \in \mathscr{N}^{\perp}$, the isolation distance $2d$ of the eigenvalue $v = 0$ of A_1 is $2d = [1 - (\mu_1/\mu_2)] > 0$. Hence for η sufficiently small, it follows that the spectrum of the derived operator $D(\tilde{w}, \mu)$ consists of a part containing n eigenvalues $v_i(\eta)$ which lie in the interval $(-d, d)$, and a second part which lies in the interval (d, ∞) (e.g., see Kato [8, p.290 ff]). In particular then, since the eigenvalues of $\eta\tilde{B}(0)$ determine the signs of the n eigenvalues $v_j(\eta)$ (see (4.4)), the above definition of stability is consistent with the usual notions of "linearized stability."

Remark 4.2 Since, for each $\mu \in \mathscr{I}$, $D(\tilde{w}, \mu)$ is a bounded self-adjoint operator, one is tempted to use instead of (4.4) a more standard perturbation theorem such as that by Rellich [13, p.57]. Namely, since $D(\tilde{w}, \mu)$ is a power series in μ, there exist analytic functions $v_i(\eta)$ (in $\mathbb{R}^1$) and $\psi_i(\eta)$ (in $\mathscr{H}$) such that $\{\psi_1(\eta), \ldots, \psi_n(\eta)\}$ is an orthonormal set for $\mu \in \mathscr{I}$,

$$D(\tilde{w}, \mu)\psi_i(\eta) = v_i(\eta)\psi_i(\eta) \qquad (i = 1, \ldots, n), \tag{4.5}$$

and the spectrum of $D(\tilde{w}, \mu)$ near $\eta = 0$ consists of exactly the n eigenvalues $v_1(\eta), \ldots, v_n(\eta)$. However, in such a perturbation result the vectors $\{\psi_1(0), \ldots, \psi_n(0)\}$ cannot, in general, be prescribed in advance. For example (see Rellich [13, p.36]), if $\mathscr{H} = \mathbb{R}^2$ and $A(\eta) = \left[\begin{smallmatrix} \eta & 0 \\ 0 & -\eta \end{smallmatrix}\right]$ then the eigenvalues of $A(\eta)$ are $\pm\eta$ and the corresponding eigenvectors are determined up to a factor of unit magnitude by $\psi_1(\eta) = \left[\begin{smallmatrix} 0 \\ 1 \end{smallmatrix}\right]$ and $\psi_2(\eta) = \left[\begin{smallmatrix} 1 \\ 0 \end{smallmatrix}\right]$; thus, the basis $\psi_1{}^0 = (1/\sqrt{2})\left[\begin{smallmatrix} 1 \\ 1 \end{smallmatrix}\right]$ and $\psi_2{}^0 = (1/\sqrt{2})\left[\begin{smallmatrix} -1 \\ 1 \end{smallmatrix}\right]$ cannot be chosen in advance as the unperturbed basis at $\eta = 0$. In connection with this example, let us note that if $w_1 = \psi_1{}^0$ and $w_2 = \psi_2{}^0$, then the matrix $\tilde{B}(\eta)$ is given by $\tilde{B}(\eta) = \left[\begin{smallmatrix} 0 & -1 \\ -1 & 0 \end{smallmatrix}\right]$. Thus, the eigenvalues of $\eta\tilde{B}(\eta)$ are also $\pm\eta$ but the relationship (4.4) implies $A(\eta)\tilde{\phi}_1(\eta) = -\eta\tilde{\phi}_2(\eta)$ and $A(\eta)\tilde{\phi}_2(\eta) = -\eta\tilde{\phi}_1(\eta)$ so that $\tilde{\phi}_1(\eta)$ and $\tilde{\phi}_2(\eta)$ are not eigenvectors for the eigenvalues $\pm\eta$ of $A(\eta)$.

By relating the matrix $[(\tilde{L}w_i, w_j)]$ to the Jacobian matrix $J(f, \cdot)$ associated with the system (2.5); the following interesting theorem is established by McLeod and Sattinger [11, p.75].

THEOREM 4 Suppose that the hypotheses of Theorem 1 are satisfied for $\mu_0 = \mu_1$, where μ_1 denotes the smallest (positive) characteristic value of A, and suppose that T and R are gradient operators. Let $J(f, \tilde{\beta})$ be the nonsingular Jacobian matrix of (2.5) evaluated at $\tilde{\beta}$, and let $\tilde{w}$ be the nontrivial solution branch of $(*)$ constructed as in Theorem 1. If the eigenvalues of

$J(f, \tilde{\beta})$ are all positive, then for $\mu \in \mathscr{I}$, $\tilde{w}$ is stable for $\mu > \mu_1$ and, if it exists, unstable for $\mu < \mu_1$. If the eigenvalues of $J(f, \tilde{\beta})$ are all negative then the stability properties of $\tilde{w}$ are reversed. If the eigenvalues of $J(f, \tilde{\beta})$ are both positive and negative, then $\tilde{w}$ is unstable for all $\mu \in \mathscr{I}$.

The proof of Theorem 4 consists of showing that $J(f, \tilde{\beta}) = \tilde{B}(0)$. This last theorem holds under much less stringent conditions on T and R and, in particular, it holds without the assumptions that T and R are gradient operators, provided that one takes instead the "real parts" of the eigenvalues of $J(f, \tilde{\beta})$. Not surprisingly then, by exploiting the fact that T is a gradient operator, a stronger stability theorem (see Theorem 5) can be established for the solution branches of $(*)$ constructed in Theorem 2.

Let w^* be the nontrivial solution branch of $(*)$ constructed as in Theorem 2 with $\mu_0 = \mu_1$. Since w^* is of the form $w^* = (\eta/\theta)^\alpha u^* + U^*$, where $\theta = (T(u^*), u^*)$ and $\|U^*(\eta)\| = o(\eta^\alpha)$ as $|\eta| \to 0$, the "derived" operator associated with w^* can be written as

$$D(w^*, \mu) = A_1 + \eta L[u^*] + H(\eta), \qquad \mu \in \mathscr{I}, \tag{4.6}$$

where $A_1 = I - \mu_1 A$ as before,

$$L[u^*] = -\mu_1 A + \theta^{-1} T'(u^*), \tag{4.7}$$

and $H(\eta)$ is a bounded operator which satisfies $\|H(\eta)\| = o(\eta)$ as $|\eta| \to 0$.

Let $\{v_1, \ldots, v_n\}$ be the special orthonormal basis for $\mathscr{N}$ introduced in the proof of Theorem 2. Then, as in the above there exists a well-defined $n \times n$ matrix $B(\eta) = [b_{ij}(\eta)]$ with

$$b_{ij}(0) = (L[u^*]v_i, v_j) \qquad (i, j = 1, \ldots, n) \tag{4.8}$$

such that $B(\eta)$ determines the stability properties of w^* by means of the eigenvalues of $B(0)$. Hence, Theorem 4 implies that the stability properties of w^* are determined by the eigenvalues of the nonsingular Jacobian matrix $J(f, \beta^*)$, where $\beta^* = (\theta^{-\alpha}, 0, \ldots, 0)$ and the functions f^i in (2.5) and $J(f, \cdot)$ are now calculated with respect to the special basis $\{v_1, \ldots, v_n\}$. On the other hand, by making proper use of the extremum property of the element u^*, the following type of stability theorem can be established.

THEOREM 5 Suppose that the hypotheses of Theorem 2 are satisfied for $\mu_0 = \mu_1$, where μ_1 denotes the smallest (positive) characteristic value of A, and suppose that R is a gradient operator. Let w^* be the nontrivial solution branch of $(*)$ constructed as in Theorem 2 with $\mu_0 = \mu_1$. If the degree of homogeneity k of T is *even* and *if* $(T(u), u)|_{\mathscr{S}}$ has a positive relative minimum at $u^* \in \mathscr{S}$, then w^* is stable for $\mu_1 < \mu < \mu_1 + \delta$ and unstable for $\mu_1 - \delta < \mu < \mu_1$; whereas if $(T(u), u)|_{\mathscr{S}}$ has a positive relative maximum at $u^* \in \mathscr{S}$, then w^* is unstable for $0 < |\mu - \mu_1| < \delta$. If k is *odd* and if

$(T(u), u)|_{\mathscr{S}}$ has a positive relative minimum [resp., maximum] at $u^* \in \mathscr{S}$, then w^* is stable [resp., unstable] for $\mu_1 < \mu < \mu_1 + \delta$.

Proof In view of Theorem 4 and the fact that $J(f, \beta^*) = B(0)$, it is sufficient to determine the elements of $B(0) = [b_{ij}(0)]$. By making use of the properties of the special basis $\{v_1, \ldots, v_n\}$ for $\mathscr{N}$ and the definition of $L[u^*]$ in (4.7), some simple calculations show that $b_{ij}(0) = 0$ for $i \neq j$ and

$$b_{jj}(0) = ((-\mu_1 A + \theta^{-1} T'(u^*))v_j, v_j) = -1 + (d_j/\theta) \qquad (j = 1, \ldots, n). \tag{4.9}$$

Hence, using the standard Euler identity $(T'(u^*)u^*, u^*) = k\theta$ (e.g., see the proof of a related Euler identity by Rothe [14, p.272]), one easily sees that the eigenvalues of $J(f, \beta^*)$ are $\lambda_1 = (k-1)$ and $\lambda_j = -1 + (d_j/\theta)$ for $j = 2, \ldots, n$. Since $j(f, \beta^*) = \det J(f, \beta^*) \neq 0$ by hypothesis (see also the proof of Theorem 2), it then follows from (3.3) that $\nu d_2 < \nu\theta$ so that all of the eigenvalues of $J(f, \beta^*)$ are positive when u^* corresponds to a minimum ($\nu = -1$), and $(n-1)$ of the eigenvalues of $J(f, \beta^*)$ are negative when u^* corresponds to a maximum ($\nu = 1$). The conclusions of the theorem then follow from Theorem 4.

The stability (or instability) of the solution branches constructed in Theorem 3 is not considered here. Since the case $\det J(f, \cdot) = 0$ cannot, in general, be excluded for such solutions, the stability is indeterminate and higher order terms must also be considered. This more complicated stability problem will be treated in a second paper by means of topological degree theory.

References

1. R. Böhme, Die Lösung der Verzweigungsgleichungen für Nichtlineare Eigenwertprobleme. *Math. Z.* **127** (1972), 105–126.
2. E. N. Dancer, Bifurcation theory in real Banach spaces. *Proc. London Math. Soc.* **23**, No. 3 (1971), 699–734.
3. A. Grundmann, Der topologische Abbildungsgrad homogener Polynomoperatoren. Dissertation, Universität Stuttgart, 1974.
4. E. Heinz, An elementary analytic theory of the degree of mapping in n-dimensional space. *J. Math. Mech.* **8** (1959), 231–247.
5. E. Hille and R. S. Phillips, "Functional Analysis and Semi-groups." Amer. Math. Soc., Providence, Rhode Island, 1957.
6. G. Iooss, Théorie non linéaire de la stabilité des écoulements laminaires dans le cas de le'échange des stabilitiés. *Arch. Rational Mech. Anal.* **40** (1971), 166–208.
7. G. Ize, Bifurcation theory for Fredholm operators. Ph.D. Thesis, New York University, 1974.
8. T. Kato, "Perturbation Theory for Linear Operators." Springer-Verlag, Berlin and New York, 1966.
9. K. Kirchgässner and H. Kielhöfer, Stability and bifurcation in fluid dynamics. *Proc.*

Seminar Nonlinear Eigenvalue Problems, Santa Fe, 1971. Rocky Mountain J. Math. **3** (1973), 275–318.

10. G. H. Knightly and D. Sather, Nonlinear buckled states of rectangular plates, *Arch. Rational Mech. Anal.* **54** (1974), 356–372.
11. J. B. McLeod and D. H. Sattinger, Loss of stability and bifurcation at a double eigenvalue. *J. Functional Analysis* **14** (1973), 62–84.
12. J. Milnor, "Singular Points of Complex Hypersurfaces." Annals of Mathematical Studies, Vol. 61. Princeton Univ. Press, Princeton, 1968.
13. F. Rellich, Perturbation Theory of Eigenvalue Problems. Gordon and Breach, New York, 1968.
14. E. H. Rothe, Completely continuous scalars and variational methods. *Ann. of Math.* **49** (1948), 265–278.
15. E. H. Rothe, Gradient mappings. *Bull. Amer. Math. Soc.* **59** (1953), 5–19.
16. D. Sather, Branching of solutions of an equation in Hilbert space. *Arch. Rational Mech. Anal.* **36** (1970), 47–64.
17. D. Sather, Nonlinear gradient operators and the method of Lyapunov–Schmidt. *Arch. Rational Mech. Anal.* **43** (1971), 222–244.
18. D. Sather, Branching of solutions of nonlinear equations. *Proc. Seminar Nonlinear Eigenvalue Problems, Santa Fe, 1971. Rocky Mountain J. Math.* **3** (1973), 203–250.
19. D. Sather, "Bifurcation and stability for a class of shells." *Arch. Rational Mech. Anal.* **63** (1977), 295–304.
20. D. Sather, Branching and stability for nonlinear shells. *Proc. IUTAM/IMU Symp. Applications Methods Funct. Anal. Problems Mechanics, Marseille, 1975.* Lecture Notes in Mathematics, Vol. 503, pp. 462–473. Springer-Verlag, Berlin and New York, 1976.
21. D. H. Sattinger, Stability of bifurcating solutions by Leray–Schauder degree. *Arch. Rational Mech. Anal.* **43** (1971), 154–166.
22. D. H. Sattinger, "Topics in Stability and Bifurcation Theory." Lecture Notes in Mathematics Vol. 309. Springer-Verlag, Berlin and New York, 1973.
23. M. M. Vainberg, "Variational Methods for the Study of Nonlinear Operators." Holden-Day, San Francisco, 1964.
24. M. M. Vainberg and V. A. Trenogin, The methods of Lyapunov and Schmidt in the theory of nonlinear equations and their further development. *Russ. Math. Surveys* **17** No. 2 (1962), 1–60.

This research was supported in part by NSF Grant No. MPS 73-08948 and in part by the Council on Research and Creative Work of the University of Colorado.

The author wishes to thank Professor George H. Knightly for several helpful discussions related to the present paper.

AMS (MOS) 1970 Subject Classification: 47H15.

Recent Progress in Bifurcation Theory

D. H. Sattinger
University of Minnesota

To Erich H. Rothe

1. Bifurcation Theory

Bifurcation theory deals with the analysis of branch points of nonlinear functional equations in a vector space, usually a Banach space. Thus, suppose $G(\lambda, u)$ is a smooth (Fréchet differentiable) mapping from $K \times \mathscr{E}$ to $\mathscr{F}$, where K denotes the field of real or complex numbers, and $\mathscr{E}$ and $\mathscr{F}$ are Banach spaces; and that $G(0, 0) = 0$. One is then interested in constructing all solutions of the functional equation

$$G(\lambda, u) = 0 \tag{1.1}$$

in a neighborhood of $\lambda = 0$, $u = 0$. If $G_u(0, 0)$ is invertible, that is, if it is an isomorphism from $\mathscr{E}$ to $\mathscr{F}$, then by the implicit function theorem there is a smooth $\mathscr{E}$-valued function $u(\lambda)$, defined for small λ, such that

$$G(\lambda, u(\lambda)) = 0.$$

Furthermore, all solutions of (1.1) in a sufficiently small neighborhood of (0, 0) in $K \times \mathscr{E}$ lie on the curve $(\lambda, u(\lambda))$. If $\mathscr{E}$ and $\mathscr{F}$ are complex Banach spaces and K is the field of complex numbers, then u is an analytic function of λ.

(In applications it is often the case that $\mathscr{E}$ and $\mathscr{F}$ are complex and $K = \mathbb{C}$, but that only real solutions of (1.1) are of physical significance. In such cases, the operator G is real analytic in the sense that $\overline{G(\lambda, u)} = G(\bar{\lambda}, \bar{u})$. We assume these conditions to be true throughout this chapter.)

Bifurcation theory concerns itself with the situation where $G_u(0, 0)$ is not invertible. The simplest case is that in which $L_0 = G_u(0, 0)$ is a Fredholm operator: L_0 has a finite dimensional kernel $\mathscr{N}$ and a closed range $\mathscr{R}$, with codim $\mathscr{R}$ = dim $\mathscr{N}$. In that case the bifurcation problem may be reduced, via an alternative method to a system of n algebraic equations in n unknowns, viz.

$$F_i(\lambda, v) = 0, \qquad i = 1, \ldots, n, \quad v = (v_1, \ldots, v_n). \tag{1.2}$$

For such problems the main task is the analysis of Eqs. (1.2). In practice this problem can be quite complex,since it is difficult to compute any but the terms of lowest order in (1.2).

The subject of bifurcation is an important topic for applied mathematics inasmuch as it arises naturally in any physical system described by a nonlinear set of equations depending on a set of parameters. Equation (1.1) may typically represent a nonlinear system of partial differential equations, ordinary differential equations, etc. When discussing applied problems stability considerations are unavoidable. In fact, the phenomenon of bifurcation is intimately associated with "loss of stability." A complete resolution of the branching problem therefore requires an analysis of the stability of the bifurcating solutions.

Let us suppose (as is often the case) that (1.1) is the time independent equation for a dynamical system

$$\partial u/\partial t = G(\lambda, u). \tag{1.3}$$

We state loosely the following:

Principle of Linearized Stability An equilibrium point u_0 of (1.3) is stable if the spectrum of the linear operator $G_u(\lambda, u_0)$ lies strictly in the left half plane and unstable if points in the spectrum of $G_u(\lambda, u_0)$ lie in the right half plane.

In the case of finite dimensions (where (1.3) is a system of ordinary differential equations) the principle of linearized stability is valid and known as Lyapunov's first theorem. The principle is also valid for a number of important problems in partial differential equations, for example the Navier–

Stokes equations and other nonlinear parabolic systems. There is more latitude in the infinite dimensional case, since stability may be measured in a variety of nonequivalent norms. The validity of the linearized stability principle has been firmly established for the Navier–Stokes equations by a number of workers, including Prodi [35], Judovic [18], Kirchgässner and Sorger [25], Sattinger [44], Iooss [13], and Kielhöfer [22]. The question has been treated for other parabolic problems by Kielhöfer, Chafee [6], and Auchmuty [1], among others. A good exposition of these and other matters may be found in the forthcoming lecture notes by Henry [11]. Therefore, throughout the remainder of this article we shall assume the principle of linearized stability.

The zero solution therefore loses stability if one or several eigenvalues of $G_u(\lambda, 0)$ cross the imaginary axis as λ crosses some critical value λ_c. For convenience we may as well normalize the parameter so that $\lambda_c = 0$. If a simple eigenvalue crosses the origin or a pair of simple complex conjugate eigenvalues cross the imaginary axis the situation is understood relatively well. The following theorems (again stated somewhat loosely) are well-known.

THEOREM 1 Let $\sigma(\lambda)$ be a simple real eigenvalue of $G_u(\lambda, 0)$ such that $\sigma(0) = 0$ and $\sigma'(0) > 0$. Then the solution set of (1.1) in a neighborhood of the origin consists of the curve $(\lambda, 0)$ and a second nontrivial branch $(\lambda(\varepsilon), u(\varepsilon))$. Solutions on the nontrivial branch are stable above criticality $(\lambda(\varepsilon) > 0)$ and unstable below criticality $(\lambda(\varepsilon) < 0)$.

If the Banach spaces are complex, $\lambda(\varepsilon)$ and $u(\varepsilon)$ are analytic functions of the parameter ε. Theorem 1 was stated by Hopf [12] under the assumption $\lambda'(0) \neq 0$ or $\lambda'(0) = 0$, $\lambda''(0) \neq 0$. It was proved in general, without these assumptions, by Sattinger [45] using a topological degree argument, and subsequently, by Crandall and Rabinowitz [8] using a perturbation argument.

THEOREM 2 Let $\sigma(\lambda)$, $\bar{\sigma}(\lambda)$ be a pair of simple complex conjugate eigenvalues of $G_u(\lambda, 0)$ such that $\sigma(0) = i\omega_0$, $R_e \sigma'(0) > 0$. Then (1.3) has an analytic one-parameter family of solutions

$$\lambda(\varepsilon), \qquad u(s, e), \qquad s = \omega(\varepsilon)t,$$

where $u(s, \varepsilon)$ is 2π-periodic in s. Supercritical solutions are stable and subcritical solutions are unstable.

Theorem 2 was first proved by Hopf [12] for general systems of ordinary differential equations. An extension of Hopf's results to parabolic systems of partial differential equations has been given by a number of workers, including Sattinger [46], Iooss [14], Judovic [19], Joseph and Sattinger [17], and

Marsden [29]. The computational aspects of the problem, even in the finite dimensional case, are nontrivial. Some of these questions have recently been pursued by McCracken [31].

There are basically three approaches to the Hopf bifurcation theorem: a Poincaré map argument (Iooss [14], Hopf [12]), a Lyapunov–Schmidt argument (Joseph and Sattinger [17]), and the center manifold approach (Marsden [29]). Iooss [14] and Judovic [21] give rigorous proofs of the stability of the bifurcating time periodic solutions. Stability is immediate in the center manifold approach. The forthcoming notes by Marsden and McCracken [30] discuss Hopf's theorem in depth using the center manifold approach.

The question of loss of stability of a periodic solution, resulting in the bifurcation of an invariant torus, has been discussed by G. Iooss [15] and by Ruelle and Takens [39]. O. E. Lanford has written an excellent account of the work of Ruelle and Takens (see Lanford [27a]. Lanford's article also contains a proof of the center manifold theorem in a Banach space. Iooss has investigated the bifurcation of invariant tori in a number of situations, including the case where two pair of complex conjugate eigenvalues cross the imaginary axis.

2. Bifurcation at Multiple Eigenvalues

In this section we address ourselves to the multiple dimensional case where the critical eigenvalue is no longer simple. In physical applications the presence of a multiple eigenvalue generally seems to be due to the invariance of the problem under a transformation group. (Physicists even go so far as to assert as a matter of faith that this is *always* the case. Ruelle [40] asserts that reducibility is nongeneric.) The phenomenon is well known in quantum mechanics, where invariance of the Hamiltonian under a symmetry group leads to a degeneracy of the energy levels. When the nonlinear equations (1.1) are invariant under a transformation group $\mathscr{G}$, the invariance is inherited by the bifurcation equations, and this fact may aid considerably in their analysis.

We begin with a quick review of alternative methods for bifurcation problems. (See Cesari [4], Bancroft *et al.* [2], and Sather [41,42]. We shall assume that $\mathscr{E} \subseteq \mathscr{F}$, with the injection $\mathscr{E} \to \mathscr{F}$ continuous in the $\mathscr{F}$-topology. This setup is particularly convenient when dealing with nonlinear systems of partial differential equations. Let $L_0 = G_u(0, 0)$ have a finite dimensional kernel $\mathscr{N}_0 = \mathscr{E}$. The projection onto $\mathscr{N}_0$ is given by

$$P_0 = \frac{1}{2\pi i} \int_C (z - L_0)^{-1} \, dz,$$

where C encloses the origin and no other points in the spectrum of L_0. Clearly $P_0: \mathscr{F} \to \mathscr{E}$, and so we may regard P_0 as a bounded mapping from $\mathscr{E}$ to $\mathscr{E}$ or from $\mathscr{F}$ ro $\mathscr{F}$. In the same way, $Q_0 = I - P_0$ is then a bounded mapping from $\mathscr{F}$ to $\mathscr{F}$ or from $\mathscr{E}$ to $\mathscr{E}$. Equation (1.1) can be decomposed into a coupled system

$$Q_0 G(\lambda, u) = 0, \qquad P_0 G(\lambda, u) = 0,$$

and we may write $u = P_0 u + Q_0 u = v + \Psi$. We choose Ψ to satisfy the functional equation

$$Q_0 G(\lambda, v + \Psi(\lambda, v)) = 0. \tag{2.1}$$

This equation is easily solved by virtue of the implicit function theorem in a Banach space. In fact, the Fréchet derivative at $\lambda = 0$, $v = 0$ is simply $Q_0 L_0$, which by our assumption that L_0 is a Fredholm operator, is an isomorphism between $Q_0\mathscr{E}$ and $Q_0\mathscr{F}$. Therefore there exists a unique function $\Psi(\lambda, v)$, analytic in λ and v, with values in $Q_0\mathscr{E}$. Furthermore, $\Psi(\lambda, v)$ is uniquely defined for small $|\lambda| + |v|$.

The bifurcation equations are

$$F(\lambda, v) = P_0 G(\lambda, v + \Psi(\lambda, v)) = 0. \tag{2.2}$$

These reduce to a system of n equations in n unknowns if we introduce a basis $\{\varphi_1, \ldots, \varphi_n\}$ for $\mathscr{N} = P_0\mathscr{E}$. (We assume here that the Riesz index v of $\mathscr{N}$ is one, i.e., that $\mathscr{N} = \{\varphi : L_0 \varphi = 0\} = \mathscr{N}_2 = \{\varphi : L_0{}^2\varphi = 0\}$. The result is still true if $v > 1$, that is, we still get n equations in n unknowns; however a little more care is required in their derivation.) Then the projection P_0 has the form

$$P_0 u = \sum_{j=1}^{n} c_j(u)\varphi_j .$$

From the linearity of P_0 we may infer that the $c_j(u)$ are linear functionals on $\mathscr{F}$. They therefore have the form $c_j(u) = \langle u, \varphi_j{}^* \rangle$ where $\varphi_j{}^* \in \mathscr{F}^* \subseteq \mathscr{E}^*$. From the relationship $P_0 L_0 = L_0 P_0$ it is easily proved that $L_0{}^*\varphi_j{}^* = 0$; and from the relationship $P_0{}^2 = P_0$ one sees that $\langle \varphi_i, \varphi_j^* \rangle = \delta_{ij}$. Putting $v = z_1 \varphi_1 + \cdots + z_n \varphi_n$ we find that (2.2) is equivalent to the system of equations

$$F_j(\lambda, z_1, \ldots, z_n) = \langle G(\lambda, v + \Psi(\lambda, v)), \varphi_j{}^* \rangle = 0, \qquad j = 1, \ldots, n.$$

Now suppose the original problem is invariant under a transformation group. Let T_g be a representation of a group $\mathscr{G}$ onto the Banach space $\mathscr{F}$:

$$T_{g_1 g_2} = T_{g_1} T_{g_2}, \qquad T_e = I$$

(e the identity in $\mathscr{G}$). Then we have

THEOREM 3 The kernel $\mathcal{N}$ of L_0 and the bifurcation equations (1.2) are invariant under $T_g : T_g F(\lambda, v) = F(\lambda, T_g v)$.

Proof The following is immediate:

$$T_g G_u(\lambda, v) = G_u(\lambda, T_g v) T_g.$$

It follows that $T_g L_0 = L_0 T_g$ for all $g \in G$ and consequently the kernel $\mathcal{N}$ is invariant under T_g. The solution $\Psi(\lambda, v)$ of (2.1) is also invariant under T_g. In fact,

$$T_g Q_0 G(\lambda, v + \Psi(\lambda, v)) = Q_0 G(\lambda, T_g v + T_g \Psi(\lambda, v)) = 0,$$

while

$$Q_0 G(\lambda, T_g v + \Psi(\lambda, T_g v)) = 0.$$

By the uniqueness of the solution obtained by the implicit function theorem,

$$T_g \Psi(\lambda, v) = \Psi(\lambda, T_g v).$$

We therefore have

$$T_g F(\lambda, v) = T_g P_0 G(\lambda, v + \Psi(\lambda, v)) = P_0 G(\lambda, T_g v + \Psi(\lambda, T_g v)) = F(\lambda, T_g v),$$

and the invariance of the bifurcation equations is established.

By the way, if the original equations are real the same argument shows that $\overline{F(\lambda, v)} = F(\bar{\lambda}, \bar{v})$.

Let us recall that in physical applications a complete resolution of the branching problem requires a stability analysis of the bifurcating solutions. The stability question reduces to perturbation theory for a multiple isolated eigenvalue of an analytic family of operators. Let $(\lambda(\varepsilon), u(\varepsilon))$ be a one-parameter branch of solutions of (1.1):

$$G(\lambda(\varepsilon), u(\varepsilon)) = 0.$$

The stability of $u(\varepsilon)$ is determined by the eigenvalues of the linear operator

$$L(\varepsilon) = G_u(\lambda(\varepsilon), u(\varepsilon)).$$

At $\varepsilon = 0$, we have, by hypothesis, an isolated n-fold eigenvalue at the origin; and the rest of the spectrum of L_0 lies in the left half plane since $\lambda = 0$ is supposed to be the critical value of the parameter. The stability of the branch $(\lambda(\varepsilon), u(\varepsilon))$ thus depends on the behavior of this n-fold eigenvalue along the one parameter branch curve. Under our assumption that L_0 is a Fredholm operator, classical perturbation theory applies. The basic result is (Dunford and Schwartz [8a]).

LEMMA 4 Let L_0 have a finite dimensional kernel $\mathcal{N}$ and closed range $\mathcal{R} \subseteq \mathcal{F}$, with codim $\mathcal{R} = \dim \mathcal{N}$. Let $L(\varepsilon)$ be an analytic family of bounded operators mapping $\mathcal{E}$ to $\mathcal{F}$. Then for small $|\varepsilon|$ there is an analytic projection-valued operator $E(\varepsilon)$ and an analytic operator $B(\varepsilon) : E(\varepsilon)\mathcal{N} \to E(\varepsilon)\mathcal{N}$ such that

$$L(\varepsilon)E(\varepsilon) = B(\varepsilon)E(\varepsilon). \tag{2.3}$$

The subspace $E(\varepsilon)\mathcal{N}$ is invariant under $L(\varepsilon)$. The projection $E(\varepsilon)$ is given by

$$E(\varepsilon) = \frac{1}{2\pi i} \int_C (z - L(\varepsilon))^{-1} \, dz,$$

where C is a small circle enclosing the origin and no other points in $\sigma(L_0)$ (the spectrum of L_0).

An immediate consequence of (2.3) is that the splitting of the n-fold eigenvalue at the origin is given by the eigenvalues of the finite dimensional operator $B(\varepsilon)$. The following theorem is proved by Sattinger [47].

THEOREM 5 Let the bifurcation problem (1.1) have a one-parameter family of solutions $(\lambda(\varepsilon), v(\varepsilon) + \Psi(\lambda(\varepsilon), v(\varepsilon))$, where

$$\lambda(\varepsilon) = \varepsilon^m \gamma_0 + \varepsilon^{m+1}\gamma_1 + \cdots, \qquad v(\varepsilon) = \varepsilon^n v_0 + \varepsilon^{n+1} v_1 + \cdots.$$

(Here $v(\varepsilon) \in \mathcal{N}$.) Suppose that

$$F(\varepsilon^m \gamma, \varepsilon^n v) = \varepsilon^k Q(\gamma_0, v_0) + O(\varepsilon^{k+1}), \tag{2.4}$$

where $k > \max\{m, n\}$. Then the operator $B(\varepsilon)$ given in (2.3) has the form

$$B(\varepsilon) = \varepsilon^{k-n} Q_w(\gamma_0, v_0) + O(\varepsilon^{k-n+1}).$$

The equations

$$Q(\gamma_0, v_0) = 0 \tag{2.5}$$

are called the *reduced bifurcation equations*. They are the equations obtained after a scaling of variables as in (2.4) (see Graves [10], Sather [41], and Sattinger [47]. An appropriate scaling can often be determined by a Newton diagram argument. If (γ_0, v_0) is a solution of the reduced bifurcation equations (2.5) and $Q_v(\gamma_0, v_0)$ is nonsingular, the existence of a solution of the full bifurcation equations can be established by the application of the implicit function theorem. The assertion of Theorem 5 is that, to lowest order, the stability of the corresponding bifurcating family is determined by the eigenvalues of the operator $Q_v(\gamma_0, v_0)$.

Theorem 5 implies, or contains as a special case, a number of previously known results (see Sattinger [45], McLeod and Sattinger [32], and Crandall and Rabinowitz [8]). For example, let us see how Theorem 5 applies to the case of bifurcation at a simple eigenvalue. We first show that the bifurcation equations have the form

$$f(\lambda, v) = \sigma'(0)\lambda v - bv^n + \cdots = 0, \tag{2.6}$$

where v is a scalar, and $\sigma(\lambda)$ is the critical eigenvalue of $G_u(\lambda, 0)$.
Differentiating

$$G_u(\lambda, 0)\varphi(\lambda) = \sigma(\lambda)\varphi(\lambda)$$

with respect to λ, setting $\lambda = 0$, and taking the scalar product with the adjoint eigenfunction $\varphi_0{}^*$, we get

$$\langle G_{u\lambda}(0, 0)\varphi_0, \varphi_0{}^*\rangle = \sigma'(0)\langle \varphi_0, \varphi_0{}^*\rangle.$$

Since σ is a simple eigenvalue we may normalize $\varphi_0{}^*$ so that $\langle \varphi_0, \varphi_0{}^*\rangle = 1$. Then

$$\sigma'(0) = \langle G_{u\lambda}(0, 0)\varphi_0, \varphi_0{}^*\rangle.$$

Remark If $\langle \varphi_0, \varphi_0{}^*\rangle = 0$, then φ_0 is in the range of L_0 and there is a vector φ_1 such that $L_0\varphi_1 = \varphi_0$. It follows that $L_0{}^2\varphi_1 = 0$, and therefore the index of the eigenvalue is greater than 1. We assume σ is simple in the strict sense that its index is 1.

We now turn to the bifurcation equation, which is

$$f(\lambda, v) = \langle G(\lambda, v\varphi_0 + \Psi(\lambda, v)), \varphi_0{}^*\rangle = 0,$$

where v is a scalar. Differentiating, we have

$$f_{v\lambda}(0, 0) = \langle G_{\lambda u}(0, 0)\varphi_0, \varphi_0{}^*\rangle + \langle G_{\lambda u}(0, 0)\Psi_v(0, 0), \varphi_0{}^*\rangle + \langle G_u(0, 0)\Psi_{v\lambda}, \varphi_0{}^*\rangle.$$

Now $\langle G_u(0, 0)\Psi_{v\lambda}, \varphi_0{}^*\rangle = \langle L_0\Psi_{v\lambda}, \varphi_0{}^*\rangle = \langle \Psi_{v\lambda}, L_0{}^*\varphi_0{}^*\rangle = 0$; and furthermore, differentiating $Q_0 G(\lambda, v\varphi_0 + \Psi(\lambda, v)) = 0$ with respect to v and setting $\lambda = v = 0$, we get

$$Q_0 G_u(0, 0)[\varphi_0 + \Psi_v(0, 0)] = Q_0 G_u(0, 0)\Psi_v(0, 0) = 0,$$

which implies that $\Psi_v(0, 0) = 0$. Therefore

$$f_{v\lambda}(0, 0) = \sigma'(0).$$

Finally, if Eq. (1.1) has no solutions of the form $\lambda = 0, 0 < \|u\| < \delta$, then the branching function $f(0, v)$ cannot vanish identically. It therefore has the form $f(0, v) = -bv^n + \cdots$, from which Eq. (2.6) follows.

The reduced bifurcation equations, obtained by scaling $\lambda = \varepsilon^{n-1}\gamma$, $v = \varepsilon w$, are

$$Q(\gamma, w) = \sigma'(0)\gamma w - bw^n = 0.$$

This equation has the real solution $\gamma = b$ and

$$w = \begin{cases} \pm (\sigma'(0))^{1/n-1}, & n \text{ odd}, \\ (\sigma'(0))^{1/n-1}, & n \text{ even}. \end{cases}$$

In either case,

$$Q_w(\gamma, w) = b\sigma'(0)(1-n).$$

According to Theorem 5, the critical eigenvalue (call it $\xi(\varepsilon)$) has the form

$$\xi(\varepsilon) = b\sigma'(0)(1-n)\varepsilon^{n-1} + \cdots,$$

while

$$\lambda(\varepsilon) = \varepsilon^{n-1}b + \cdots.$$

Therefore

$$\xi(\varepsilon)/\lambda(\varepsilon) = \sigma'(0)(1-n) + 0(\varepsilon)$$

for small ε. Since $\sigma'(0) > 0$, we see that $\xi(\varepsilon)$ and $\lambda(\varepsilon)$ have opposite signs near the branch point; so solutions are stable above criticality ($\lambda > 0$) and unstable below criticality ($\lambda < 0$).

3. Bifurcation in the Presence of a Symmetry Group

The branching problem in the case of higher multiplicities is considerably more complicated. The assumption of some kind of group invariance is, as we have said, a natural one; and the importance of this aspect of bifurcation theory has attracted the attention of a number of workers in recent years. (See references [9,33,34,40,47–49], and also the article by S. Schecter in Chapter 7 of Marsden and McCracken [30].) We discuss here some examples of problems which are invariant under a transformation group.

(1) The nonlinear equation

$$\Delta u + \lambda u + f(\lambda, u) = 0 \tag{3.1}$$

is invariant under the rotation group $O(n)$ (when it is cast in $\mathbb{R}^n$). If O is an orthogonal matrix, define the representation of $O(n)$ by

$$T_o U(\mathbf{x}) = U(O^{-1}\mathbf{x}).$$

Then T_g commutes with the nonlinear operator in (3.1). The problem will be invariant under the rotation group or one of its subgroups if the equation is set in a domain possessing such a symmetry. For example, we may consider (3.1) on the disk $x^2 + y^2 \leq 1$, the sphere in three dimensions $x^2 + y^2 + z^2 = 1$, or on a regular polygon in the plane or polyhedron in space. More generally, the equations of continuum mechanics are invariant under the orthogonal group. Some natural physical problems then would be buckling of square or rectangular plates, or buckling of a sphere.

(2) The autonomous equation

$$\partial u/\partial t = G(\lambda, u) \tag{3.2}$$

is invariant under time translations $t \to t + \gamma$ but not under reflections $t \to -t$. In order to construct time periodic solutions of (3.2) of period $2\pi/\omega$ we put $s = \omega t$ and look for 2π-periodic solutions of

$$\omega\, \partial u/\partial s = G(\lambda, u). \tag{3.3}$$

Since u is 2π-periodic, eq. (3.3) is invariant under the subgroup $\mathbb{R}/\mathbb{Z} \cong SO(2)$.

(3) The *Taylor Problem* is the problem of a viscous fluid contained between two concentric cylinders which rotate at constant speed. (cf. Kirchgässner and Sorger [25] and Chandrasekhar [7]). At low speeds the fluid circulates in a rotational manner. That is $v_r = v_z = 0$, $v_\theta = v_\theta(r)$; but at critical speeds of the cylinders this flow loses stability and another, more complicated flow may appear. Let us write the equations of motion symbolically as in (3.2). The Navier–Stokes equations are invariant under a representation of the rotation group and $t \to t + \gamma$. If we take the common axis of the cylinders to be the z-axis, and choose cylindrical coordinates (r, θ, z), the geometry is invariant under rotations $\theta \to \theta + \theta_0$; reflections $\theta \to -\theta$ and $z \to -z$; and translations $z \to z + z_0$, $t \to t + t_0$. When the problem is restricted to solutions periodic in z and t, the symmetric group is $O(2) \times O(2) \times SO(2)$. For a further discussion of the symmetry of the Taylor problem see the article by S. Schecter in the work by Marsden and McCracken [30].

(4) In the Bénard problem in fluid mechanics one considers a layer of fluid, infinite in horizontal extent, heated from below. The equations used to describe convection are the Boussinesq equations (cf. Chandrasekhar [7], and Busse [3]). These are a system of five nonlinear partial differential equations in the unknowns (u_1, u_2, u_3, T, p), where (u_1, u_2, u_3) are the Cartesian components of the fluid velocity, T is the temperature, and p is the hydrodynamic pressure. These equations are invariant under the following represen-

tation of the group of rigid motions:

$$T_{\{O,\,\mathbf{a}\}} = \begin{bmatrix} u_1 \\ u_2 \\ u_3 \\ \\ T \\ p \end{bmatrix} = \left[\begin{array}{ccc|cc} & & & 0 & 0 \\ & \bigcirc & & 0 & 0 \\ & & & 0 & 0 \\ \hline 0 & 0 & 0 & 1 & 0 \\ 0 & 0 & 0 & 0 & 1 \end{array}\right] \begin{bmatrix} u_1 \\ u_2 \\ u_3 \\ \\ T \\ p \end{bmatrix} (O^{-1}\mathbf{x} + \mathbf{a}), \quad (3.4)$$

where O is a 3×3 orthogonal matrix and $\mathbf{a}$ is a vector in $\mathbb{R}^3$.

The geometry of the Bénard problem is invariant under the subgroup of rigid motions in the plane. In experiments, "cellular solutions" are observed at the onset of convection, usually in the form of hexagons or rolls. (See Segal [50, the plate on p. 173].) The bifurcation of cellular solutions in the Bénard problem has been investigated by many authors. (See references [3,19,24,28,36,37,48].) An investigation of the problem based on group theoretic considerations has been given by Sattinger [48]. In the next section we give a brief account of this work.

4. Bifurcation of Doubly Periodic Solutions

The appearance of cellular solutions can be described mathematically as the bifurcation of doubly periodic solutions. Let $\mathscr{E}$ be a Banach space, and consider the vector space $\mathscr{F}$ of functions $u(\mathbf{x})$ mapping $\mathbf{x} \in \mathbb{R}^2$ into $\mathscr{E}$. Suppose $\mathscr{F}$ is a Banach space under a suitable norm. Let $\mathscr{E}(2)$ be the group of rigid motions on $\mathbb{R}^2$ and let $S(\sigma)$ be a representation of $\mathscr{E}(2)$ onto $\mathscr{E}$; then

$$(T(\sigma)u)(\mathbf{x}) = (S(\sigma)u)(\sigma^{-1}\mathbf{x})$$

is a representation of $\mathscr{E}(2)$ onto $\mathscr{F}$. (In (3.4) $S(\sigma)$ is the matrix on the right-hand side; the Banach space $\mathscr{F}$ consists of the five-tuples (u_1, u_2, u_3, T, p) defined on the interval $0 \leq z \leq 1$, the layer of fluid.)

Let the mapping $G(\lambda, u)$ take functions u in $\mathscr{F}$ into another such Banach space $\mathscr{F}'$ of vector valued functions defined on the plane. We assume G is invariant under our representation: $T(\sigma)G(\lambda, u) = G(\lambda, T(\sigma)u)$.

Let Λ be a lattice of vectors in the plane: if $\boldsymbol{\omega}_1, \boldsymbol{\omega}_2$ belong to Λ, then $n\boldsymbol{\omega}_1 + m\boldsymbol{\omega}_2 \in \Lambda$ for any integers n and m. Any such lattice is generated by two basis vectors, so we can assume $\Lambda = \{n\boldsymbol{\omega}_1 + m\boldsymbol{\omega}_2 \mid n, m \in \mathbb{Z}\}$. The parallelogram generated by $\boldsymbol{\omega}_1$ and $\boldsymbol{\omega}_2$ is called a *basic period cell.* We denote by $\mathscr{T}(2)$ the group of translations in the plane, and by $\mathscr{T}(\Lambda)$ the lattice subgroup

$$\mathscr{T}(\Lambda) = \{T_{\mathbf{a}} : \mathbf{a} \in \Lambda\}.$$

DEFINITION The function $u(\mathbf{x})$ is doubly periodic if $T_{\mathbf{a}}u = u$ for all $T_{\mathbf{a}} \in \mathcal{T}(\Lambda)$. We denote the subspaces of doubly periodic solutions by $\mathcal{F}(\Lambda)$ and $\mathcal{F}'(\Lambda)$. Since G is invariant under translation $G\colon \mathcal{F}(\Lambda) \to \mathcal{F}'(\Lambda)$.

Using group theoretic considerations alone one may derive the general structure of the finite dimensional equations which arise when one considers the bifurcation of doubly periodic solutions. Let $\mathcal{N}(\Lambda)$ be the kernel of $L_0 = G_u(0, 0)$ in the Banach space $\mathcal{F}'(\Lambda)$. We must find the action of the symmetry group of $\mathcal{F}(\Lambda)$ on $\mathcal{N}(\Lambda)$.

When restricted to the subspace $\mathcal{F}(\Lambda)$ the action of the group $\mathcal{T}(2)$ is $\mathcal{T}(2)/\mathcal{T}(\Lambda)$. We denote this subgroup by $H(\Lambda)$; it is a compact two parameter Lie group and can be identified with the basic period cell. The subspace $\mathcal{F}(\Lambda)$ is not invariant under arbitrary rotations and reflections but only under the subgroup which leaves Λ invariant. We denote the maximal such subgroup, called the *holohedry* of Λ, by $\mathcal{D}(\Lambda)$. The only holohedries in the plane are C_2, D_2, D_4, and D_6.

The irreducible representations of $H(\Lambda)$ are $\chi_{\boldsymbol{\omega}}(\mathbf{a}) = e^{2\pi i\langle \boldsymbol{\omega}, \mathbf{a}\rangle}$, where $\boldsymbol{\omega} \in \Lambda'$, the dual lattice. (The dual lattice Λ' is the lattice generated by $\boldsymbol{\omega}_1'$, $\boldsymbol{\omega}_2'$, the vectors forming a dual basis to $\boldsymbol{\omega}_1$, and $\boldsymbol{\omega}_2 : \langle \boldsymbol{\omega}_i, \boldsymbol{\omega}_j' \rangle = \delta_{ij}$.) We may certainly decompose $\mathcal{N}(\Lambda)$ into irreducible invariant subspaces of $H(\Lambda)$, so suppose $\Psi \in \mathcal{N}(\Lambda)$ and $T_a \Psi = e^{i\langle \boldsymbol{\omega}', \mathbf{a}\rangle}\Psi$. Thus

$$\Psi(\mathbf{x} + \mathbf{a}) = e^{i\langle \boldsymbol{\omega}', \mathbf{a}\rangle}\Psi(\mathbf{x}).$$

From this relationship it follows immediately that

$$\Psi(\mathbf{x}) = e^{i\langle \boldsymbol{\omega}', \mathbf{x}\rangle}\Psi(0).$$

Now let r be an element of $\mathcal{D}(\Lambda)$. Then

$$T_r \Psi = e^{i\langle \boldsymbol{\omega}', r^{-1}\mathbf{x}\rangle} S_r \Psi(0) = e^{i\langle r\boldsymbol{\omega}', \mathbf{x}\rangle} S_r \Psi(0).$$

The kernel $\mathcal{N}(\Lambda)$ thus contains the vectors

$$\left\{ S_r \Psi(0) e^{i\langle r\boldsymbol{\omega}', \mathbf{x}\rangle} \right\}\Big|_{r \in \mathcal{D}(\Lambda)} \tag{4.1}$$

Let us consider the case of the hexagonal lattice generated by two vectors making an angle of 60° with each other. Then the dual lattice is also generated by a pair of vectors making an angle of 60° with each other. We drop the primes and denote these two vectors by $\boldsymbol{\omega}_1$ and $\boldsymbol{\omega}_2$. As r ranges over D_6, $r\boldsymbol{\omega}_1$ ranges over six vectors (although there are twelve elements in D_6). If

$$S_r \Psi(0) = S_{r'} \Psi(0) \qquad \text{whenever} \quad r\boldsymbol{\omega}_1 = r'\boldsymbol{\omega}_1,$$

there are only six vectors in (4.1). Thus $\dim \mathcal{N}(\Lambda) = 6$, where Λ is the hexagonal lattice. The situation is shown in Fig. 1. We denote the six wave

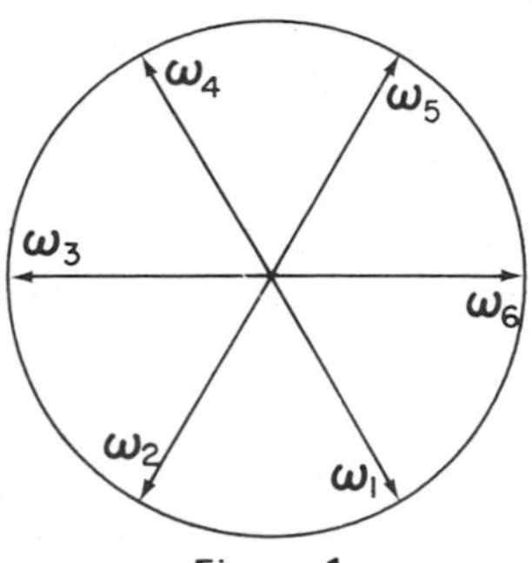

Figure 1

functions in $\mathcal{V}(\Lambda)$ by

$$\Psi_j(\mathbf{x}) = v_j e^{i\langle \boldsymbol{\omega}_j, \mathbf{x}\rangle},$$

where the vectors $\{v_j\}$ are appropriate $\{S_r \Psi(0)\}$ and the $\boldsymbol{\omega}_j$ are as shown in Fig. 1. The following is immediate.

THEOREM 6 Let α be the counterclockwise rotation through 60° and θ the reflection in the plane across the $\boldsymbol{\omega}_1$–$\boldsymbol{\omega}_4$ axis. Then D_6 is generated by α and θ, and the action of the symmetry group on the basis $[\Psi_1, \ldots, \Psi_6]$ is as follows:

(i) $T(\alpha)$ acts on $\{\Psi_1, \ldots, \Psi_6\}$ as the cyclic permutation $\alpha = (654321)$;
(ii) $T(\theta)$ acts on $\{\Psi_1, \ldots, \Psi_6\}$ as the permutation $\theta = (26)(35)$;
(iii) $T_{\mathbf{a}} \Psi_j = e^{i\langle \boldsymbol{\omega}_j, \mathbf{a}\rangle} \Psi_j$.

There is an additional property shared by the wave vectors Ψ_j which is true for the Bénard problem (cf. Sattinger [48, Appendix]), and that is $T_{\alpha^3} \Psi_j = \overline{\Psi_j}$, i.e.,

$$\overline{\Psi_1} = \Psi_4, \qquad \overline{\Psi_2} = \Psi_5, \qquad \overline{\Psi_3} = \Psi_6.$$

If $v \in \mathcal{V}(\Lambda)$, we expand $v = \sum_j z_j \Psi_j$; for real vectors we require $\overline{z_1} = z_4$, $\overline{z_2} = z_5$, $\overline{z_3} = z_6$.

Let us compute the general mapping $F(v)$ which is invariant under the group actions. From $\alpha F(v) = F(\alpha v)$ we get

$$F_2(v_1, \ldots, v_6) = F_1(v_2, v_3, \ldots, v_6, v_1) \tag{4.2}$$

Thus once we determine F_1 the other components of the mapping are generated by cyclic permutation of $v_1, \ldots, v_6$. From $\theta F = F\theta$ we see that

$$F_1(v_1, \ldots, v_6) = F_1(v_1, v_6, v_5, v_4, v_3, v_2). \tag{4.3}$$

Finally, from $T_{\mathbf{a}} F = F T_{\mathbf{a}}$ we get

$$e^{i\langle \boldsymbol{\omega}_1, \mathbf{a}\rangle} F_1(v_1, \ldots, v_6) = F_1(e^{i\langle \boldsymbol{\omega}_1, \mathbf{a}\rangle} v_1, \ldots, e^{i\langle \boldsymbol{\omega}_6, \mathbf{a}\rangle} v_6). \tag{4.4}$$

Let us compute the general mapping homogeneous of degree 2. Suppose F_1 contains a term $v_j v_k$. From (4.4) we see that

$$e^{i\langle \boldsymbol{\omega}_1, \mathbf{a}\rangle} v_j v_k = e^{i\langle \boldsymbol{\omega}_j, \mathbf{a}\rangle} v_j e^{i\langle \boldsymbol{\omega}_k, \mathbf{a}\rangle} v_k$$

for all $\mathbf{a}$, v_j and v_k. Hence $\boldsymbol{\omega}_1 = \boldsymbol{\omega}_j + \boldsymbol{\omega}_k$. The only possibility is $\boldsymbol{\omega}_2 + \boldsymbol{\omega}_6 = \boldsymbol{\omega}_1$, so the general quadratic term is $v_2 v_6$. From (4.2) we get $F_2 = v_3 v_1$, $F_3 = v_4 v_2, \ldots,$ etc.

Now let us compute the general mapping homogeneous of degree 3. The translational invariance (4.4) leads to the condition $\boldsymbol{\omega}_1 = \boldsymbol{\omega}_j + \boldsymbol{\omega}_k + \boldsymbol{\omega}_l$. It is easy to see that the only possible terms are $v_1 v_4 v_1$, $v_2 v_5 v_1$, $v_3 v_6 v_1$. The invariance (4.3) implies that F_1 must have the form

$$a(z_2 z_5 + z_3 z_6)z_1 + b z_1{}^2 z_4.$$

From the reality condition $(\overline{F(\lambda, z)} = F(\lambda, \bar{z}))$ we see that a and b must be real. (They depend on λ and on the original parameters of the problem.)

We leave it to the reader to show that the linear terms are necessarily of the form

$$F_j(z_1, \ldots, z_6) = c z_j \tag{4.5}$$

(c real). Incidentally, this shows that the kernel $\mathcal{N}(\Lambda)$ is irreducible under the transformation group. (The irreducibility is an immediate consequence of Schur's lemma. The significance of (4.5) is that the only linear transformation which commutes with all elements of the group is a scalar multiple of the identity.)

One can now write down the reduced bifurcation equations. However, from the invariance $\overline{F(z)} = F(\bar{z})$ and $\alpha^3 F = F\alpha^3$ it follows that

$$F_4(z_1, z_2, z_3, \bar{z}_4, \bar{z}_5, \bar{z}_6) = \overline{F_1(z_1, z_2, z_3, \bar{z}_4, \bar{z}_5, \bar{z}_6)},$$

etc., so the six equations reduce to three. These equations simplify further when one introduces "action-angle" variables: $z_j = x_j e^{i\theta_j}$, $j = 1, 2, 3$. The bifurcation equations then reduce to

$$x_1 = a x_2 x_3, \qquad x_2 = a x_3 x_1, \qquad x_3 = a x_1 x_2 \tag{4.6}$$

if the leading nonlinear terms are quadratic, and, if they are cubic, to

$$\begin{aligned} x_1(1 + a(x_2{}^2 + x_3{}^2) + b x_1{}^2) &= 0 \\ x_2(1 + a(x_3{}^2 + x_1{}^2) + b x_2{}^2) &= 0 \\ x_3(1 + a(x_1{}^2 + x_2{}^2) + b x_3{}^2) &= 0. \end{aligned} \tag{4.7}$$

A similar analysis can be carried out for the other lattices in the plane. A complete analysis of the branching problem can then be given. The details are carried out by Sattinger [48]. The point we wish to make here is that the structure of the branching equations (4.6), (4.7) which is quite simple, can be derived from group-theoretic considerations alone.

Other problems similar in nature to the Bénard problem are pattern formation phenomena in reaction–diffusion processes and phase transitions (see Kirkwood and Monroe [26], and Raveche and Stuart [38]).

References

1. J. F. G. Auchmuty, Lyapunov methods and equations of parabolic type, "Nonlinear Problems in the Physical Sciences and Biology," Springer Lecture Notes in Mathematics, Vol. 322. Springer-Verlag, Berlin and New York, 1973.
2. S. Bancroft, J. K. Hale, and D. Sweet, Alternative problems for nonlinear functional equations. *J. Differential Equations* **4** (1968), 40–56.
3. F. H. Busse, The Stability of finite amplitude cellular convection and its relation to an extremum principle. *J. Fluid Mech.* **30** (1967), 625–650.
4. L. Cesari, Functional analysis and Galerkin's method. *Michigan Math. J.* **11** (1964), 385–418.
5. N. Chafee, The bifurcation of one or more closed orbits from an equilibrium point. *J. Differential Equations* **4** (1968), 661–679.
6. N. Chafee, A stability analysis for a semilinear parabolic equation. *J. Differential Equations* **15** (1974), 522–540.
7. S. Chandrasekhar, "Hydrodynamic and Hydromagnetic Stability," Oxford Univ. Press, London and New York, 1961.
8. M. Crandall and P. Rabinowitz, Bifurcation, perturbation of simple eigenvalues and linearized stability. *Arch. Rational Mech. Anal.* **52** (1973), 161–180.

8a. N. Dunford and J. T. Schwartz, "Linear Operators," Part I. Wiley (Interscience), New York, 1966.

9. J. I. Gimitro, Concentration patterns generated by reaction and diffusion. Ph.D. Thesis, University of Minnesota, Minneapolis, 1969.
10. L. Graves, Remarks on singular points of functional equations. *Trans. Amer. Math. Soc.* **79** (1955), 150–157.
11. D. Henry, "Geometric theory of Parabolic Equations" (to appear in Lecture Notes in Mathematics Series, Springer-Verlag, Berlin and New York).
12. E. Hopf, Abzweigung einer periodischer Lösung eines Differential Systems. *Ber. Verb. Saechs. Akad. Wiss. Leipzig, Math.-Phys. Kl.* **64** (1942), 1–22.
13. G. Iooss, Théorie non linéaire de la stabilité des écoulements laminaires dans le cas de l'echange des stabilités. *Arch. Rational Mech. Anal.* **40** (1971), 166–208.
14. G. Iooss, Existence et stabilité de la solution périodique secondaire intervenant dans les problèmes d'évolution du type Navier–Stokes. *Arch. Rational Mech. Anal.* **47** (1972), 301–329.
15. G. Iooss, Bifurcation of a periodic solution of the Navier–Stokes equations into an invariant torus. *Arch. Rational Mech. Anal.* **58** (1975), 35–56.
16. D. D. Joseph, "Nonlinear Theory of Stability of Fluid Motions," Springer Tracts in Natural Philosophy. Springer Publ., New York, 1976.
17. D. Joseph and D. H. Sattinger: Bifurcating time periodic solutions and their stability. *Arch. Rational Mech. Anal.* **45** (1972), 75–109.
18. V. I. Judovic, Stability of stress flows of viscous incompressible fluids. *Dokl. Akad. Nauk. SSSR* **16** (1965), 1037–1040.

19. V. Judovic, On the origin of convection. *Prikl. Mat. Meh.* (transl. *J. Appl. Math Mech.*) **30** (1966), 1193–1199.
20. V. Judovic, Appearance of auto-oscillations in a fluid. *Prikl. Mat. Meh.* (transl. *J. Appl. Math. Mech.*) **35** (1971), 638–655.
21. V. Judovic, On the stability of self oscillations of a fluid. *Soviet Math. Dokl.* **11**, (1970), 1543–1546.
22. H. Kielhöfer, Stability and semilinear evolution equations in Hilbert space. *Arch. Rational Mech. Anal.* **57** (1974), 150–165.
23. H. Kielhöfer, Verzweigung Periodischer Lösungen in Mehnfachen Eigenverten. "Habilitationschrift," Stuttgart, 1976.
24. H. Kielhöfer and K. Kirchgässner, Stability and bifurcation in fluid mechanics. *Rocky Mountain J. Math.* **3** (1973), 275–318.
25. K. Kirchgässner and P. Sorger, Branching analysis for the Taylor problem, *Quart. J. Mech. Appl. Math.* **22** (1969), 183–210.
26. J. G. Kirkwood and E. Monroe, Statistical mechanics of fusion. *J. Chem. Phys.* **9** (1941), 514–526.
27. O. A. Ladyzhenskaya, Mathematical Analysis of the Navier–Stokes equations for incompressible liquids. *Annu. Rev. Fluid Mech.* **7** (1975), 249.
27a. O. E. Lanford, Bifurcation of periodic solutions into invariant tori: The work of Ruelle and Takens, "Nonlinear Problems in the Physical Sciences and Biology." Lecture Notes in Mathematics, Vol. 322. Springer-Verlag, Berlin and New York, 1973.
28. W. V. R. Malkus and G. Veronis, Finite amplitude cellular convection. *J. Fluid Mech.* **4** (1958), 225–260.
29. J. Marsden, The Hopf bifurcation for nonlinear semi-groups. *Bull. Amer. Math. Soc.* **79** (1973), 537–541.
30. J. Marsden and M. McCracken, *The Hopf Bifurcation Theorem and its Applications*, Applied Mathematical Sciences **19**, Springer-Verlag, New York, Heidelberg, Berlin.
31. M. McCracken, Computation of stability for the Hopf bifurcation theorem and the Lorenz equations (to appear in *SIAM J. Appl. Math.*).
32. J. B. McLeod and D. H. Sattinger, Loss of stability and bifurcation at double eigenvalues. *J. Functional Analysis* **14** (1973), 62–84.
33. H. G. Othmer and L. E. Scriven, Nonlinear aspects of dynamic pattern in cellular networks. *J. Theoretical Biol.* **43** (1974), 83–112.
34. H. G. Othmer and L. E. Scriven, Instability and dynamic pattern in cellular networks. *J. Theor. Biol.* **32** (1971), 507–537.
35. G. Prodi, Theoremi di tipo locale per il sistema di Navier–Stokes e stabilita delle soluzioni stazionare. *Rend. Sem. Mat. Univ. Padova* **32** (1962), 374–397.
36. E. Palm, T. Ellingsen, and B. Gjevck, On the occurrence of cellular motion in Bénard convection. *J. Fluid Mech.* **30** (1967), 651.
37. P. H. Rabinowitz, Existence and nonuniqueness of rectangular solutions of the Bénard problem. *Arch. Rational Mech. Anal.* **24** (1968), 32–57.
38. H. J. Raveche and C. A. Stuart, Towards a molecular theory of freezing. *J. Chem. Phys.* **63** (1975), 1099–1111.
39. D. Ruelle and F. Takens, On the nature of turbulence. *Comm. Math. Phys.* **20** (1971), 167–192.
40. D. Ruelle, Bifurcations in the presence of a symmetry group. *Arch. Rational Mech. Anal.* **51** (1973), 136–152.
41. D. Sather, Branching of solutions of an equation in Hilbert space. *Arch. Rational Mech. Anal.* **36** (1970), 47–64.

42. D. Sather, Branching of solutions of nonlinear equations. *Rocky Mountain J. Math.* **3** (1973), 203–250.
43. D. H. Sattinger, *Topics in stability and bifurcation theory*, Springer Lecture Notes in Mathematics, Vol. 309. Springer-Verlag, Berlin and New York, 1973.
44. D. H. Sattinger, The mathematical problem of hydrodynamic stability. *J. Math. Mech.* **19** (1970), 797–817.
45. D. H. Sattinger, Stability of bifurcating solutions by Leray–Schauder degree. *Arch. Rational Mech. Anal.* **43** (1971), 154–166.
46. D. H. Sattinger, Bifurcation of periodic solutions of the Navier–Stokes equations. *Arch. Rational Mech. Anal.* **41** (1971), 66–80.
47. D. H. Sattinger, Group representation theory and branch points of nonlinear functional equations. *SIAM J. Math. Anal.* **8** (1977), 179–201.
48. D. H. Sattinger, Group representation theory, bifurcation theory, and pattern formation *J. Functional Analysis* (to appear).
49. T. H. Schwab, Dynamic concentration patterns generated by reaction and diffusion in open planar systems. Ph.D. Dissertation, University of Minnesota, 1975.
50. L. A. Segel, Nonlinear problems in hydrodynamic stability. *In* " Nonequilibrium Thermodynamics, Variational Techniques, and Stability." Univ. of Chicago Press, Chicago, Illinois, 1965. (The plate showing hexagonal cellular motion is due to H. L. Koschmieder.)

On the Subgradient of Convex Functionals

M. M. Vainberg
Ped. Institute

This paper is closely related to the investigations of E. Rothe on the minimum of nonlinear functionals.

Let E be a real space with a norm, f a real functional defined on an open convex set $\omega \subset E$, and

$$V_+ f(x, h) = \lim_{t \to +0} \frac{f(x + th) - f(x)}{t}$$

the variation of f at the fixed point $x \in \omega$ for any chosen direction $h \in \omega$. It is well known that if φ is a sublinear functional, that is,

$$\varphi(\alpha x) = \alpha\varphi(x) \qquad \text{for any} \quad \alpha > 0$$

and

$$\varphi(x + y) \leq \varphi(x) + \varphi(y) \qquad \text{for} \quad \forall\, x, \ y \in \omega,$$

then $V_+ \varphi(x, h)$ exists for any point $x \in \omega$. Further, if $V_+ f(x, h)$ exists for any point of ω, then it is a sublinear functional on ω.

The linear continuous functional y_0 is called a subgradient of the functional f at the point $x_0 \in \omega$, if for any $x \in \omega$

$$f(x) - f(x_0) \geq \langle y_0, x - x_0 \rangle,$$

where $\langle y_0, h \rangle$ is the value of y_0 on the vector $h \in \omega$. If f has a subgradient at every point of ω, then the functional f is convex and weakly semicontinuous from below on ω.

The following main result of this paper may be of some interest because the two mentioned properties are connected with the problem of the minimum of nonlinear functionals.

THEOREM 1 If the finite convex functional f, defined on ω, has a variation $V_+ f(x, h)$, continuous in h at $h = 0$, at each point $x \in \omega$, then it has a subgradient at every point of ω.

The paper also contains other theorems.

1. Введение

Данная работа примыкает к исследованиям Э. Роте и, в частности, к работе [1]. В ней рассматриваются вопросы, связанные с проблемой о минимуме нелинейных вещественных функционалов.

Хорошо известно, что если на ограниченном секвенциально слабо замкнутом множестве рефлексивного банахова пространства задан слабо полунепрерывный снизу функционал, то на этом множестве он ограничен снизу и достигает своей точной нижней границы. В связи с этим представляют интерес предложения о слабой полунепрерывности снизу функционалов. Далее, известно, что если конечный вещественный функционал f, заданный на открытом выпуклом множестве ω нормированного пространства имеет субградиент в каждой точке ω, то f — выпуклый и слабо полунепрерывный снизу на ω функционал. По этой причине представляет интерес следующая доказываемая здесь .

ТЕОРЕМА 1 Если конечный выпуклый функционал f, заданный на открытом выпуклом множестве ω нормированного пространства, имеет непрерывную по h вариацию $V_+ f(x, h)$ для любого $x \in \omega$, то он имеет субградиент в каждой точке $x \in \omega$.

2. Основные Понятия и Вспомогательные Предложения

Здесь мы приведем известные понятия и некоторые вспомогательные предложения.

Пусть E — вещественное нормированное пространство, E^* — сопряженное пространство, то есть пространство всех линейных непрерывных функционалов, заданных на E и f — вещественный функционал, заданный на E.

ОПРЕДЕЛЕНИЕ 1 Если существует линейный функционал $y_0 \in E^*$ такой, что для любого $x \in E$

$$f(x) - f(x_0) \geq \langle y_0, x - x_0 \rangle, \tag{1}$$

где $\langle y, x \rangle$ это значение линейного непрерывного функционала y на векторе $x \in E$, то y_0 называется субградиентом функционала f в точке x_0. Для субградиента y_0 мы воспользуемся обозначением

$$\nabla f(x_0) = y_0 .$$

Разумеется, субградиент может оказаться многозначным отображением из E в E^*. Такие примеры известны.

Сходимость последовательности $\{x_n\}$ к вектору x_0 в смысле нормы мы будем обозначать так: $x_n \to x_0$, а слабую сходимость этой последовательности к x_0 обозначим $x_n \rightharpoonup x_0$. Слабая сходимость означает, что для любого непрерывного функционала $y \in E^*$

$$\lim_{n \to \infty} \langle y, x_n \rangle = \langle y, x_0 \rangle.$$

ОПРЕДЕЛЕНИЕ 2 Функционал f, линейность которого не предполагается, называется полунепрерывным (слабо полунепрерывным) снизу на $U \subset E$, если для любого вектора $x \in U$ и любой последовательности $\{x_n\} \subset U$, сходящейся (слабо сходящейся) к x выполняется неравенство

$$f(x) \leq \underline{\lim}_{n} f(x_n).$$

ОПРЕДЕЛЕНИЕ 3 Функционал f называется выпуклым на выпуклом множестве $\omega \subset E$, если для любых x_1, $x_2 \in \omega$ и любого $\lambda (0 < \lambda < 1)$ выполняется неравенство

$$f(\lambda x_1 + (1 - \lambda) x_2) \leq \lambda f(x_1) + (1 - \lambda) f(x_2)$$

и строго выпуклым, если равенство справедливо лишь при $x_1 = x_2$.

ОПРЕДЕЛЕНИЕ 4 Функционал f называется сублинейным, если он является положительно однородным и полуаддитивным, то есть $f(\alpha x) = \alpha f(x)$ при $\alpha > 0$ и для любых x_1, x_2 справедливо неравенство

$$f(x_1 + x_2) \leq f(x_1) + f(x_2).$$

Для дальнейшего мы приведем следующее предложение.

ТЕОРЕМА 2 Если на открытом выпуклом множестве $\omega \subset E$ вещественный функционал f имеет субградиент в каждой точке ω, то f — выпуклый и слабо полунепрерывный снизу функционал на ω.

Доказательство Применим неравенство (1) к двум произвольным векторам $x_1, x_2 \in \omega$, где $x_1 \neq x_2$ и напишем

$$f(x_1) \geq f(x_0) + \langle y_0, x_1 - x_0 \rangle$$

$$f(x_2) \geq f(x_0) + \langle y_0, x_2 - x_0 \rangle.$$

Умножая эти неравенства соответственно на λ и $1 - \lambda$, где $0 < \lambda < 1$, и полагая $x_0 = \lambda x_1 + (1 - \lambda)x_2$, получим

$$\lambda f(x_1 + (1 - \lambda)f(x_2) \geq f(x_0)$$
$$+ \langle y_0, \lambda x_1 + (1 -)x_2 - x_0 \rangle = f(x_0) + \langle y_0, 0 \rangle = f(x_0).$$

Данное неравенство доказывает выпуклость функционала f (или строгую выпуклость, если в (1) равенство возможно лишь при $x_1 = x_2$).

Для доказательства второго утверждения теоремы рассмотрим произвольную последовательность $\{x_n\} \subset \omega$, слабо сходящуюся к x_0, то есть $x_n \rightharpoonup x_0$. Так как y_0 — линейный непрерывный функционал и $(x_n - x_0) \rightharpoonup 0$, то из (1) имеем

$$\underline{\lim_n}[f(x_n) - f(x_0)] \geq \underline{\lim_n}\langle y_0, x_n - x_0 \rangle = 0,$$

то есть,

$$f(x_0) \leq \underline{\lim_n}\, f(x_n).$$

Данное неравенство означает слабую полунепрерывность снизу функционала f в точке x_0. Теорема 1 доказана.

Приведем еще следующее предложение (см. [2, Теорема 8.1]).

ТЕОРЕМА 3 Для того, чтобы функционал f, заданный в нормированном пространстве, был слабо полунепрерывен снизу, необходимо и достаточно выполнения следующего условия:

каково бы не было вещественное число c, множество

$$E_c = \{x \in E : f(x) \leq c\}$$

секвенциально слабо замкнуто.

ТЕОРЕМА 4 Для того, чтобы выпуклый функционал f, заданный в нормированном пространстве E, был слабо полунепрерывен снизу, необходимо и достаточно, чтобы он был полунепрерывен снизу.

Доказательство Необходимость вытекает из определения.

Достаточность Рассмотрим при любом фиксированном $c \in R^1$ множество

$$E_c = \{x \in E : f(x) \leq c\}$$

и произвольную последовательность $\{x_n\} \subset E_c$, сходящуюся к x_0, и покажем, что $x_0 \in E$. Действительно, если допустить, что $x_0 \notin E_c$, то

$$f(x_0) > c,$$

а потому найдется $\varepsilon > 0$ такое, что $f(x_0) > c + \varepsilon$. Но из условия, в силу полунепрерывности снизу f в точке x_0, вытекает существование $n_0 = n_0(\varepsilon)$ такого, что для $n \geq n_0$ будет

$$f(x_n) > f(x_0) - \tfrac{1}{2}\varepsilon > c + \tfrac{1}{2}\varepsilon.$$

Данное неравенство противоречит тому, что $\{x_n\} \subset E_c$. Полученное противоречие показывает, что $x_0 \in E_c$, а следовательно E_c замкнуто. Далее, из выпуклости f следует выпуклость множества E_c. Раз E_c выпукло и замкнуто, то по теореме Мазура E_c секвенциально слабо замкнуто и следовательно, по Теореме 3 функционал f слабо полунепрерывен снизу. Теорема доказана.

Пусть f — вещественный функционал, заданный в нормированном пространстве E.

ОПРЕДЕЛЕНИЕ 5 Если существует предел

$$V_+ f(x, h) = \lim_{t \to +0} \frac{f(x + th) - f(x)}{t}$$

для любого $h \in E$, то $V_+ f(x, h)$ называется вариацией f в точке x по направлению вектора h.

Отметим, что если выпуклый функционал f имеет вариацию по направлению, то эта вариация представляет собою сублинейный функционал. Действительно, из определения $V_+ f(x, h)$ следует ее положительная однородность по h, то есть

$$V_+ f(x, \alpha h) = \alpha V_+ f(x, h) \text{ при } \alpha > 0.$$

Далее,

$$V_+ f(x, h_1 + h_2) = \lim_{t \to +0} \frac{f(x + (t/2)(h_1 + h_2)) - f(x)}{t/2}$$

$$= \lim_{t \to 0} \frac{2f[\frac{1}{2}(x + th_1) + \frac{1}{2}(x + th_2)] - 2f(x)}{t}$$

так что в силу выпуклости f

$$V_+ f(x, h_1 + h_2) \le \lim_{t \to +0} \frac{f(x + th_1) - f(x)}{t} + \lim_{t \to +0} \frac{f(x + th_2) - f(x)}{t}$$
$$= V_+ f(x, h_1) + V_+ f(x, h_2),$$

то есть $V_+ f(x, h)$ полуаддитивна и положительно однородна по h.

Отметим еще, что если выпуклый функционал f является сублинейным, то есть положительно однородным и полуаддитивным, то при любом x он имеет вариацию по направлению $V_+ f(x, h)$ (см., например, [2, Лемма 2.7]).

3. Доказательство Теоремы 4

Пусть ω — открытое выпуклое множество вещественного нормированного пространства E, f — конечный выпуклый функционал, заданный на ω и $x \in E$ — произвольно фиксированный вектор. Зафиксируем ненулевой вектор $h_0 \in E$ и рассмотрим в E прямую

$$z = \lambda h_0 \qquad (-\infty < \lambda < +\infty).$$

Подберем теперь линейный функционал y_0 так, чтобы

$$\zeta y_0(h_0) = V_+ f(x, h_0). \tag{2}$$

Так как $\zeta V_+ f(x, h)$ — сублинейный по h функционал, то из (2) имеем

$$\zeta y_0(\lambda h_0) = V_+ f(x, \lambda h_0), \qquad \lambda > 0.$$

Далее, в силу полуаддитивности $V_+ f(x, h)$ по h

$$0 = V_+ f(x, h - h) \le V_+ f(x, h) + V_+ f(x, -h)$$

или

$$-V_+ f(x, -h) \le V_+ f(x, h) \tag{3}$$

и, полагая $V_- f(x, h) = -V_+ f(x, -h)$, получим

$$V_- f(x, h) \le V_+ f(x, h). \tag{4}$$

Напишем еще, что в силу (3) при $\lambda < 0$:

$$y_0(\lambda h_0) = \lambda V_+ f(\alpha, h_0) = -|\lambda| V_+ f(x, h_0)$$
$$\le |\lambda| V_+ f(x, -h_0) = V_+ f(x, -|\lambda| h_0) = V_+ f(x, \lambda h_0).$$

Таким образом для любого значения $z = \lambda h_0$

$$y_0(z) \le V_+ f(x, z). \tag{5}$$

Отсюда по теореме Хана–Банаха линейный функционал y_0, заданный на прямой и удовлетворяющий неравенству (5) можно, распространить на все E с сохранением условия (5), то есть

$$y(h) \leq V_+ f(x, h), \tag{6}$$

причем

$$y(h_0) = V_+ f(x, h_0). \tag{7}$$

Но из неравенства (6) имеем

$$y(-h) \leq V_+ f(x, -h),$$

или в силу неравенства (4),

$$y(h) = -y(-h) \geq -V_+ f(x, -h) = V_- f(x, h).$$

Отсюда и из неравенства (6) следует

$$V_- f(x, h) \leq y(h) \leq V_+ f(x, h). \tag{8}$$

Так как по условию $V_+ f(x, h)$ непрерывна по h и при $h = 0$

$$V_+ f(x, 0) = V_- f(x, 0),$$

то из неравенства (8) следует, что линейный функционал $y(h)$ непрерывен в нуле, а значит непрерывен всюду, то есть

$$y(h) = \langle y, h \rangle,$$

Неравенство (8) принимает поэтому вид

$$V_- f(x, h) \leq \langle y, h \rangle \leq V_+ f(x, h). \tag{9}$$

Рассмотрим, наконец, выпуклую функцию

$$\varphi(t) = f(x + th).$$

Как известно для выпуклой функции φ справедливо неравенство

$$\varphi'(+0) \leq (\varphi(1) - \varphi(0))/(1 - 0) \leq \varphi'(1 - 0).$$

Отсюда

$$\varphi(1) - \varphi(0) \geq \varphi'(+0),$$

или

$$f(x + h) - f(x) \geq V_+ f(x, h).$$

Из данного неравенства и неравенства (9) мы находим

$$f(x + h) - f(x) \geq \langle y, h \rangle.$$

Теорема доказана.

Замечание Согласно Теоремам 2 и 3, если функционал имеет в каждой точке открытого выпуклого множества субградиент, то на этом множестве он полунепрерывен снизу. Поэтому, если линейный функционал имеет в каждой точке некоторой окрестности нуля субградиент, то он всюду непрерывен. Отсюда вытекают утверждения:

(1) Линейный неограниченный функционал нигде не имеет субградиента.

(2) Существуют строго выпуклые функционалы, которые нигде не имеют субградиента.

Для этого достаточно к непрерывному строго выпуклому функционалу прибавить неограниченный линейный функционал.

Литература

1. E. Rothe, A note on the Banach spaces of Calkin and Morrey. *Pacific J. Math.* **3** (1953), 493–499.
2. М. М. Вайнберг. Вариационный метод и метод монотонных операторов в теории нелинейных уравнений. Наука (1972).

On the Stability of Bifurcating Solutions

H. F. Weinberger

University of Minnesota

To Erich H. Rothe on his eightieth birthday

1. Introduction

The equilibrium solutions of the real scalar differential equation

$$\frac{d\xi}{dt} = f(\xi, \lambda), \tag{1.1}$$

where f is a real analytic function of ξ and λ, are simply the points (ξ, λ) where $f(\xi, \lambda) = 0$. If $f(0, 0) = 0$ and $f_\xi(0, 0) = 0$, there may be several branches of such solutions emanating from the origin. We then say that the origin is a bifurcation point. If ξ_0 is an isolated zero of $f(\xi, \lambda_0)$, the stability of the solution $\xi \equiv \xi_0$ as a solution of (1.1) with $\lambda = \lambda_0$ is determined by the behavior of $f(\xi, \lambda_0)$ near $\xi = \xi_0$. If $f(\xi, \lambda_0)$ decreases through zero as ξ increases through ξ_0, then ξ_0 is stable, while otherwise ξ_0 is unstable. In particular, it follows that at most one of two successive zeros of $f(\xi, \lambda_0)$ can be stable.

It is the purpose of this paper to give a partial generalization of these results to nonlinear equations of the form

$$K\frac{dx}{dt} = F(x, \lambda), \tag{1.2}$$

where $x(t)$ is a function of the real variable t with values in a Banach space X_1, the parameter λ lies in a Banach space Λ, K is a continuous linear transformation from X_1 to a Banach space X_2, and F is a smooth, usually nonlinear, mapping from $X_1 \times \Lambda$ to X_2. This formulation subsumes systems of partial as well as ordinary differential, difference, and integrodifferential equations.

For the case where $\Lambda = R$, X_1 and X_2 are finite dimensional, and K is the identity, it was shown by Lyapounov [9] and Schmidt [12] that if $F(0, \lambda) \equiv 0$ and an eigenvalue ρ of the problem

$$F_x(0, \lambda)q = \rho(\lambda)q \tag{1.3}$$

increases through zero at $\lambda = 0$, one can compute a bifurcating branch near (0, 0).

Hopf [5] showed that if for some complex conjugate pair of eigenvalues ρ and $\bar{\rho}$, Re $\rho(0) = 0$ and Re $\rho'(\lambda) > 0$ while no other eigenvalue has the real part zero at $\lambda = 0$, then a branch of periodic solutions bifurcates from (0, 0).

Under the additional hypotheses that the eigenvalues $\rho(0)$, $\bar{\rho}(0)$ are algebraically simple and that all other eigenvalues near $\lambda = 0$ have negative real parts, Hopf obtained the following result: If the periodic solutions lie in the halfspace $\lambda > 0$ where Re $\rho > 0$ (supercritical case), then they are stable, while if they lie in $\lambda < 0$ where Re $\rho < 0$ (subcritical case), they are unstable.

This result can be interpreted in the following way: In the supercritical case for a fixed $\lambda > 0$ the equilibrium solution is unstable while the periodic solution is stable. In the subcritical case with $\lambda < 0$ the equilibrium solution is stable and the periodic solution is unstable. In both cases, two solutions with the same λ have opposite stabilities. Moreover, if we consider the solutions of largest amplitude on the two sets $\lambda < 0$ and $\lambda > 0$, they have the same stability. That is, in the supercritical case both the solution $x \equiv 0$ for $\lambda < 0$ and the periodic solution for $\lambda > 0$ are stable, while in the subcritical case both the periodic solution for $\lambda < 0$ and the solution $x \equiv 0$ for $\lambda > 0$ are unstable.

Hopf indicated without proof that similar results are true in the case of subcritical and supercritical steady-state bifurcations when $\rho(0) = 0$, $\rho'(0) > 0$, the zero eigenvalue has algebraic multiplicity one, and all other eigenvalues have negative real parts near $\lambda = 0$.

In all these cases X_1 and X_2 are finite dimensional, and the stability results follow from the principle of linearized stability (Lyapounov's theorem): The asymptotic stability or instability of the linearized equation implies that of the original equation.

For infinite dimensional spaces this principle is a conjecture which has been proved only for some special, though important, classes of problems. (See references [4, Section 5.1;8] and the references given in these papers.) Infinite dimensional results, including those which will be presented here, only deal with linearized stability.

The (linear) stability of supercritical and instability of subcritical equilibrium solutions when X_1 and X_2 are Banach spaces was proved by Sattinger [10] under the hypotheses that $K = I$, the eigenvalue $\rho(0) = 0$ is algebraically simple, $\rho'(0) > 0$, all other eigenvalues are in the left-half plane near $\lambda = 0$, as well as some technical hypotheses needed to introduce the Leray–Schauder degree of an equivalent problem. Some of the technical hypotheses were removed by Crandall and Rabinowitz [1], who also replaced the assumptions that $K = I$ and that $\rho(0) = 0$ is algebraically simple by the assumption that $\rho(0) = 0$ is K-simple.

Sattinger [11] also showed that if, under some assumptions, a branch of equilibrium solutions doubles back on itself, one of the half branches is stable and the other is unstable.

The extension of the Hopf result on the stability of supercritical or subcritical bifurcating periodic solutions has been carried out by Joseph and Sattinger [7], by Joseph and Nield [6], and by Crandall and Rabinowitz [2].

In this paper we shall drop the condition that Re $\rho'(0) \neq 0$ and, in the steady-state case, the assumption that $F(0, \lambda) \equiv 0$. In Section 2 we simply assume that $F(0, 0) = 0$ and that $F_x(0, 0)$ is a Fredholm operator with null space spanned by a, that the null space of its adjoint is spanned by b^*, and that $(b^*, Ka) \neq 0$ so that the eigenvalue zero is K-simple. Under these conditions there may be no branches or many branches (or, for that matter a rather complicated set) of solutions near $(0, 0)$. In a neighborhood of $(0, 0)$ we order the solutions by the values of $\alpha = d^*u$ where $d^* \in X_1{}^*$ and $d^*a \neq 0$.

We let $\sigma(u, \lambda)$ be the eigenvalue which is the perturbation of the eigenvalue 0 of the problem

$$F_x(u, \lambda)q = \sigma Kq.$$

We find that the sign of $\sigma(u, \lambda)$ is equal to that of the partial derivative $\varphi_\alpha(\alpha, \lambda)$ of the bifurcation function φ, which vanishes if and only if $F(u, \lambda) = 0$. It follows that if there are no solutions between the solutions (u_1, λ) and (u_2, λ) in the above ordering, then $\sigma(u_1, \lambda)$ and $\sigma(u_2, \lambda)$ cannot be both positive or both negative.

Under some hypotheses on the other eigenvalues which will not be discussed here, the eigenvalue $\sigma(u, \lambda)$ determines the linear stability of the solution. Our results then show that adjacent solutions have opposite stability properties, and that maximal (or minimal) solutions have the same stability properties.

Two examples at the end of Section 2 show that the K-simplicity of the eigenvalue 0 is needed for these results.

In Section 3 we prove the analogous results for the bifurcation of periodic solutions from a steady-state solution.

Some generalizations of the above results will be presented elsewhere (see Weinberger [13]).

2. The Steady-State Bifurcation

We are concerned with an equation of motion of the form

$$F(x, \lambda) = K\frac{dx}{dt}, \tag{2.1}$$

where $x(t)$ is a C^1 function of t with values in a real Banach space X_1. F is a mapping from a neighborhood N_1 of (0, 0) in the product space $X_1 \times \Lambda$ to the real Banach space X_2. K is a continuous linear transformation from X_1 to X_2. We assume that

$$F(0, 0) = 0, \tag{2.2}$$

that F has a continuous Fréchet derivative in N_1, and that $F_x(0, 0)$ is a Fredholm operator whose null space is spanned by the element a of X_1 and whose range is the null space of the element $b^* \in X_2{}^*$.

An equilibrium solution of (2.1) is an element (u, λ) of $X_1 \times \Lambda$ such that

$$F(u, \lambda) = 0. \tag{2.3}$$

The equilibrium solutions near (0, 0) can be found by the method of Lyapounov and Schmidt. One chooses an element w of X_2 with $b^*w \neq 0$ and any linear functional d^* in $X_1{}^*$ such that $d^*a \neq 0$. One then considers the pair of equations

$$F(u, \lambda) - \gamma w = 0, \qquad d^*u = \alpha \tag{2.4}$$

for $(u, \gamma) \in X_1 \times R$ for given $(\alpha, \lambda) \in R \times \Lambda$.

For $\alpha = \lambda = 0$ this system has the solution (0, 0). Moreover, the Fréchet derivative of the transformation on the left at $\alpha = \lambda = u = \gamma = 0$ is

$$\begin{pmatrix} F_x(0, 0) & -w \\ d^* & 0 \end{pmatrix},$$

which is easily seen to have a continuous inverse. By the implicit function theorem there is a neighborhood $\hat{N}_1$ of (0, 0) in $R \times \Lambda$ such that for each $(\alpha, \lambda) \in \hat{N}_1$ the equation (2.4) has the unique solution $u = y(\alpha, \lambda)$, $\gamma = \varphi(\alpha, \lambda)$, and y and φ are C^1 functions of α and λ. Thus

$$F(y(\alpha, \lambda), \lambda) - \varphi(\alpha, \lambda)w = 0, \qquad d^*y(\alpha, \lambda) = \alpha, \qquad \forall(\alpha, \lambda) \in \hat{N}_1. \quad (2.5)$$

Clearly, $(y(\alpha, \lambda), \lambda)$ is a solution of (2.3) if and only if

$$\varphi(\alpha, \lambda) = 0. \quad (2.6)$$

Moreover, any solution of (2.3) in the neighborhood

$$N_2 = \{(u, \lambda) \in N_1 : (d^*u, \lambda) \in \hat{N}_1\}$$

of (0, 0) satisfies (2.4) with $\alpha = d^*u$ and is therefore of the form $(y(\alpha, \lambda), \lambda)$. Thus the bifurcation equation (2.6) describes the set of equilibrium solutions in N_2.

We now suppose that the eigenvalue 0 of $F_x(0, 0)$ is K-simple in the sense of Crandall and Rabinowitz [1]. That is, $b^*Ka \neq 0$. This is equivalent to assuming that the only solutions of the linearized equation

$$F_x(0, 0)q = K\frac{dq}{dt}$$

that are polynomials in t are of the form βa with β constant.

The linear stability of an equilibrium solution (u, λ) is determined by the spectrum of the problem

$$F_x(u, \lambda)p - \sigma Kp = 0, \qquad p \in X_1$$

or equivalently, the spectrum of the adjoint problem

$$c^*F_x(u, \lambda) - \sigma c^*K = 0, \qquad c^* \in X_2^*.$$

We normalize c^* so that

$$c^*F_x(u, \lambda) - \sigma c^*K = 0, \qquad c^*Ka = b^*Ka. \quad (2.7)$$

For $u = \lambda = 0$ this system clearly has the solution $c^* = b^*$, $\sigma = 0$. The Fréchet derivative of the left-hand side with respect to (c^*, σ) at $u = \lambda = 0$, $c^* = b^*$, $\sigma = 0$ is

$$\begin{pmatrix} F_x(0, 0) & -b^*K \\ Ka & 0 \end{pmatrix},$$

which is easily seen to have a bounded inverse from $X_1^* \times R$ to $X_2^* \times R$.

By the implicit function theorem there is a neighborhood N_3 of (0, 0) in $X_1 \times \Lambda$ on which the system (2.7) has a C^1 solution $(c^*(u, \lambda), \sigma(u, \lambda))$ with $c^*(0, 0) = b^*$ and $\sigma(0, 0) = 0$.

In stating our principal proposition we shall use the standard function

$$\operatorname{sgn} \xi = \begin{cases} 1 & \text{if } \xi > 0, \\ 0 & \text{if } \xi = 0, \\ -1 & \text{if } \xi < 0. \end{cases}$$

PROPOSITION 1 Let $F \in C^1(N_1, X_2)$ and $K \in \mathscr{L}(X_1, X_2)$. Suppose that $F(0, 0) = 0$, that $F_x(0, 0)$ has the K-simple eigenvalue 0 with the corresponding eigenvector a, and that the range of F_x is the orthogonal complement of $b^* \in X_2{}^*$. If w and d^* in the Lyapounov–Schmidt method (2.5) are chosen so that

$$(b^*w)(d^*a)(b^*Ka) > 0, \tag{2.8}$$

then there is a neighborhood $\hat{N}$ of (0, 0) in $R \times \Lambda$ such that $\hat{N} \subset \hat{N}_1$ and for every $(\alpha, \lambda) \in \hat{N}$

$$\operatorname{sgn} \sigma(y(\alpha, \lambda), \lambda) = \operatorname{sgn} \varphi_\alpha(\alpha, \lambda). \tag{2.9}$$

Proof We differentiate the first equation of (2.5) with respect to α and apply $c^*(y(\alpha, \lambda), \lambda)$. On using (2.7), we find that

$$\sigma(y(\alpha, \lambda), \lambda)c^*(y(\alpha, \lambda), \lambda)Ky_\alpha = \varphi_\alpha(\alpha, \lambda)c^*(y(\alpha, \lambda), \lambda)w. \tag{2.10}$$

We see from this equation that $\varphi_\alpha(0, 0) = 0$. By differentiating (2.5) with respect to α and setting $\alpha = \lambda = 0$, we see that

$$y_\alpha(0, 0) = (1/d^*a)a.$$

Thus,

$$c^*(y(\alpha, \lambda), \lambda)Ky_\alpha(\alpha, \lambda)\bigg|_{\alpha=\lambda=0} = b^*Ka/d^*a, \qquad c^*(y(\alpha, \lambda), \lambda)w\bigg|_{\alpha=\lambda=0} = b^*w.$$

By continuity there is a neighborhood $\hat{N}$ of (0, 0) in $R \times \Lambda$ in which

$$(b^*Ka/d^*a)c^*(y(\alpha, \lambda), \lambda)Ky_\alpha(\alpha, \lambda) > 0, \qquad b^*wc^*(y(\alpha, \lambda), \lambda)w > 0.$$

We then see from hypothesis (2.8) that

$$c^*(y(\alpha, \lambda), \lambda)Ky_\alpha(\alpha, \lambda)c^*(y(\alpha, \lambda), \lambda)w > 0 \qquad \text{in } \hat{N}.$$

Hence (2.9) follows from (2.10), and Proposition 1 is proved.

We shall now state some simple corollaries of Proposition 1. We define the neighborhood

$$N = \{(u, \lambda) \in N_2 : (d^*u, \lambda) \in \hat{N}\}$$

of (0, 0) in $X_1 \times \Lambda$.

COROLLARY 1 Suppose that the hypotheses of Proposition 1 are satisfied.

Let (u_1, λ_0) and (u_2, λ_0) be two solutions in N of (2.3), and suppose that the line segment $\{(u, \lambda_0) : u = \beta u_1 + (1 - \beta)u_2, \ \beta \in (0, 1)\}$ lies in N and that there is no solution (u_3, λ_0) with d^*u_3 between d^*u_1 and d^*u_2. Then

$$\sigma(u_1, \lambda_0)\sigma(u_2, \lambda_0) \leq 0.$$

Proof This follows immediately from Proposition 1 and the fact that the derivatives of the continuous differentiable function $\varphi(\alpha, \lambda_0)$ at two consecutive zeros cannot both be positive or negative.

COROLLARY 2 Let the hypotheses of Proposition 1 be satisfied. Suppose that two solutions (u_1, λ_1) and (u_2, λ_2) of (2.3) lie in N, and that the points (d^*u_1, λ_1) and (d^*u_2, λ_2) can be connected by a curve in $\hat{N}$ which contains no point (d^*u_3, λ_3) corresponding to a solution (u_3, λ_3) of (2.3) in N and such that the beginning of the curve is in the direction of constant λ and increasing α and the end is in the direction of constant λ and decreasing α. Then

$$\sigma(u_1, \lambda_1)\sigma(u_2, \lambda_2) \geq 0.$$

If the ends of the curve go in the same direction of constant λ, then

$$\sigma(u_1, \lambda_1)\sigma(u_2, \lambda_2) \leq 0.$$

Proof If the curve begins in the direction of increasing α at (u_i, λ_i) and $\varphi_\alpha(u_i, \alpha_i) \neq 0$, then $\operatorname{sgn} \varphi = \operatorname{sgn} \varphi_\alpha(u_i, \lambda_i)$ on the curve. Similar considerations when α decreases along the curve together with Proposition 1 give the result.

If the rest of the spectrum of $F_x(u, \lambda) - \sigma K$ remains in a half-plane $\operatorname{Re} \sigma \leq \eta < 0$ so that the linear stability is determined by $\sigma(u, \lambda)$ and if $\varphi(\alpha, 0) \not\equiv 0$, Corollary 2 says that if there are maximal or minimal solutions for $\lambda > 0$ and $\lambda < 0$, they must have the same stability properties. On the other hand, adjacent solutions for the same λ have opposite stability properties.

The following is an obvious consequence of Corollary 2.

COROLLARY 3 Let the hypotheses of Proposition 1 hold. If $(u(\lambda), \lambda)$ is a continuously differentiable family of equilibrium solutions with $u(0) = 0$ and if $\sigma(u(\lambda), \lambda)$ changes its sign at $\lambda = 0$, then in every neighborhood of $(0, 0)$ in $X_1 \times \Lambda$ there is a solution of (2.3) that is not contained in the family $(u(\lambda), \lambda)$.

The following examples show that Corollaries 1–3 may fail when the zero eigenvalue of $F_x(0, 0)$ is simple but not K-simple, so that the condition of K-simplicity cannot be removed.

EXAMPLE 1 The equilibrium states of the system

$$\frac{d\xi}{dt} = \lambda\xi + \eta, \qquad \frac{d\eta}{dt} = \lambda\eta + \xi^2$$

are easily seen to be $(0, 0, \lambda)$ and $(\lambda^2, -\lambda^3, \lambda)$. The operator $F_x(0, 0)$ is represented by the matrix

$$\begin{pmatrix} 0 & 1 \\ 0 & 0 \end{pmatrix},$$

which has the simple but not K-simple eigenvalue 0. An elementary computation shows that for $\lambda > 0$ problem (2.7) has a positive eigenvalue at each solution, so that both solutions are unstable. Thus Corollary 1 is not valid here.

EXAMPLE 2 At the equilibrium solution $(0, 0, \lambda)$ of the problem

$$\frac{d\xi}{dt} = \lambda\xi + \eta, \qquad \frac{d\eta}{dt} = \lambda\eta - \xi^3, \tag{2.11}$$

the eigenvalue problem (2.7) has the single simple, but not K-simple, eigenvalue $\sigma = \lambda$, which changes sign as λ passes through zero. Nevertheless it is easily verified that there are no other equilibrium solutions, so that Corollaries 2 and 3 are violated.

If in this example we replace the identity matrix K by the matrix

$$K_\varepsilon = \begin{pmatrix} 1 & \varepsilon \\ \varepsilon & 1 \end{pmatrix}$$

with $0 < \varepsilon < 1$, the zero eigenvalue at $(0, 0, 0)$ is K_ε-simple. In this case the equilibrium solution $(0, 0, \lambda)$ is stable for $\lambda < \frac{1}{2}\varepsilon$ and unstable for $\lambda > \frac{1}{2}\varepsilon$. Thus, Corollary 3 applies, but as ε approaches zero the boundary of N approaches the origin.

We remark that the origin $(0, 0)$ is a center for problem (2.11) with $\lambda = 0$. Thus a branch of periodic solutions rather than equilibrium solutions bifurcates from the zero solution at $\lambda = 0$.

For other examples, see Crandall and Rabinowitz [1].

3. The Hopf Bifurcation

We now consider the bifurcation of periodic solutions of the equation

$$F(x, \lambda) = K\frac{dx}{dt} \tag{3.1}$$

with $F \in C^2(N_1, X_2)$ and $K \in \mathscr{L}(X_1, X_2)$ from a family of equilibrium solutions. N_1 is a neighborhood of $(0, 0)$ in $X_1 \times \Lambda$. We again suppose that

$$F(0, 0) = 0,$$

but we shall adopt a hypothesis which implies that the eigenvalue problem

$$F_x(0, 0)p = \rho K p$$

has the simple and K-simple pure imaginary eigenvalues $\rho = \pm i\omega_0$ with $\omega_0 > 0$ and eigenvectors a and $\bar{a}$, and that no other eigenvalue with real part zero is an integral multiple of $i\omega_0$.

More precisely, we introduce real Banach spaces Π_1 and Π_2 of 2π-periodic functions $x(\tau)$ and $y(\tau)$ from R to X_1 and R to X_2, respectively. We assume that the injections $w \to w \sin n\tau$, $n = 1, 2, \ldots$ and $w \to w \cos n\tau$, $n = 0, 1, 2, \ldots$ from X_1 to Π_1 and from X_2 to Π_2 are bicontinuous, and that the τ-translations in Π_1 and Π_2 are continuous. Moreover, we assume that the induced map $(x(\tau), \lambda) \to F(x(\tau), \lambda)$ is in $C^2(N_1{}^0, \Pi_2)$, where $N_1{}^0 = \{(x(\tau), \lambda) \in \Pi_1 \times \Lambda : (x(\tau), \lambda) \in N_1 \ \forall \tau \in R\}$, and that $K(d/d\tau) \in \mathscr{L}(\Pi_1, \Pi_2)$. Our major hypothesis is that

$$F_x(0, 0) - \omega_0 K \, d/d\tau$$

is a Fredholm operator whose null space is spanned by the real and imaginary parts of $e^{i\tau}a$, and whose range is the intersection of the null spaces of the real and imaginary parts of $e^{-i\tau}b^*$, where b^* lies in the complexification of $X_2{}^*$ and $b^* F_x(0, 0) = i\omega_0 b^* K$. The action of $e^{-i\tau}b^*$ is defined by

$$(e^{-i\tau}b^*, y) = (1/2\pi) \int_0^{2\pi} e^{-i\tau} b^* y(\tau) \, d\tau.$$

We assume that $b^* K a \neq 0$, and that b^* and a are normalized so that

$$b^* K a = 1. \tag{3.2}$$

In order to find periodic solutions of (3.1), we introduce the new variable $\tau = \omega t$ and rewrite (3.1) as

$$F(x, \lambda) - \omega K \frac{dx}{d\tau} = 0. \tag{3.3}$$

If $x(\tau) \in \Pi_1$ solves this equation, then $x(\omega t)$ is a solution of period $2\pi/\omega$ of (3.1).

Under our hypotheses $F_x(0, 0)$ has a bounded inverse from X_2 to X_1. Since $F(0, 0) = 0$, the implicit function theorem shows that there is a neighborhood I of $\lambda = 0$ in which there is a unique equilibrium solution $(x_0(\lambda), \lambda)$. That is, $F(x_0(\lambda), \lambda) = 0$ for $\lambda \in I$. By introducing the new variable $x - x_0(\lambda)$ and shrinking N_1 if necessary, we shall assume without loss of generality that

$$F(0, \lambda) = 0 \qquad \forall (0, \lambda) \in N_1. \tag{3.4}$$

In order to generalize the results of Section 2 to this case we need to derive a single real bifurcation equation. For this purpose we introduce the following variant of the Hopf bifurcation method. (For another method which gives such a bifurcation equation see Hale [3, pp. 58–59, 92–93].)

Define the mapping

$$G(y, \lambda, \alpha) = \begin{cases} (1/\alpha)F(\alpha y, \lambda) & \text{for} \quad \alpha \neq 0 \\ F_x(0, \lambda)y & \text{for} \quad \alpha = 0. \end{cases}$$

Because $F \in C^2(N_1, \Pi_2)$ and because $F(0, \lambda) \equiv 0$, $G \in C^1(N_1, \Pi_2)$.

We choose any w in Π_2 such that

$$\operatorname{Re}(e^{-i\tau}b^*, w) = 1 \tag{3.5}$$

and any d^* in the complexification of $X_1{}^*$ with the property

$$d^*a = 1. \tag{3.6}$$

We consider the problem of finding $(y, \omega, \gamma) \in \Pi_1 \times R^2$ such that

$$G(y, \lambda, \alpha) - \omega K \frac{dy}{d\tau} - \gamma w = 0, \qquad \operatorname{Re}(e^{-i\tau}d^*, y) = 1, \qquad \operatorname{Im}(e^{-i\tau}d^*, y) = 0, \tag{3.7}$$

where $(\alpha, \lambda) \in R \times \Lambda$ is prescribed. It is easily seen that when $\alpha = \lambda = 0$, this problem has the solution

$$y = e^{i\tau}a + e^{-i\tau}\bar{a}, \qquad \omega = \omega_0, \qquad \gamma = 0.$$

Moreover, the Fréchet derivative of the left-hand side of (3.7) with respect to (y, ω, γ) at $y = e^{i\tau}a + e^{-i\tau}\bar{a}$, $\omega = \omega_0$, $\gamma = 0$, $\alpha = \lambda = 0$ is

$$\begin{pmatrix} F_x(0, 0) - \omega_0 K \dfrac{d}{d\tau} & -iK(e^{i\tau}a - e^{-i\tau}\bar{a}) & -w \\ \operatorname{Re} e^{-i\tau}d^* & 0 & 0 \\ \operatorname{Im} e^{-i\tau}d^* & 0 & 0 \end{pmatrix}$$

This is easily seen to have a bounded inverse from $\Pi_2 \times R^2$ to $\Pi_1 \times R^2$. Therefore, by the implicit function theorem, there is a neighborhood $\hat{N}_1$ of $(0, 0)$ in $R \times \Lambda$ such that problem (3.7) has a unique solution $(y(\alpha, \lambda), \omega(\alpha, \lambda), \gamma(\alpha, \lambda))$ for each $(\alpha, \lambda) \in \hat{N}_1$. Moreover, y, ω, and γ are C^1 functions of α and λ.

We now define

$$z(\alpha, \lambda) \equiv \alpha y(\alpha, \lambda), \qquad \varphi(\alpha, \lambda) \equiv \alpha\gamma(\alpha, \lambda).$$

Then (3.4), the definition of G, and (3.7) show that

$$F(z(\alpha, \lambda), \lambda) - \omega K \frac{dz(\alpha, \lambda)}{d\tau} = \varphi(\alpha, \lambda)w, \qquad (e^{-i\tau}d^*, z) = \alpha. \tag{3.8}$$

Clearly $(z(\alpha, \lambda), \omega(\alpha, \lambda), \lambda)$ is a solution of (3.3) if and only if

$$\varphi(\alpha, \lambda) = 0.$$

On the other hand, if $(x(\tau), \omega, \lambda) \in \Pi_1 \times R \times \Lambda$ is a solution of (3.3), then for any real θ

$$(e^{-i\tau}d^*, x(\tau + \theta)) = e^{i\theta}(e^{-i\tau}d^*, x(\tau)). \tag{3.9}$$

By a proper choice of θ we can make $z = x(\tau + \theta)$ satisfy (3.8) with

$$\alpha = |(e^{i\tau}d^*, x(\tau))|.$$

Therefore any solution (x, ω, λ) of (3.3) with (x, λ) in the neighborhood

$$N_2 = \{(x, \lambda) : (x, \lambda) \in N_1, (|(e^{-i\tau}d^*, x)|, \lambda) \in \hat{N}_1\}$$

of $(0, 0)$ in $\Pi_1 \times \Lambda$ is of the form $(z(\alpha, \lambda; \tau - \theta), \omega(\alpha, \lambda), \lambda)$, and $\varphi(\alpha, \lambda) = 0$. The real equation $\varphi(\alpha, \lambda) = 0$ is the bifurcation equation for our problem.

The linearized stability of a solution (x, ω, λ) in $(\Pi_1 \times R \times \Lambda)$ of (3.3) is defined in terms of the spectrum of the family of linear transformations

$$F_x(x, \lambda) - \omega K \frac{d}{d\tau} - \sigma K \tag{3.10}$$

from Π_1 to Π_2. For $x = 0$, $\omega = \omega_0$, $\lambda = 0$ the intersection of this spectrum with a neighborhood of $\sigma = 0$ consists of a double eigenvalue at $\sigma = 0$.

By a standard perturbation argument there is a neighborhood $\bar{N}$ of $(0, \omega_0, 0)$ in which there are two continuous branches of eigenvalues $\sigma_1(u, \omega, \lambda)$ and $\sigma_2(u, \omega, \lambda)$ which may or may not coincide. We can also find two continuously differentiable branches of corresponding eigenfunctionals $c_1^*(u, \omega, \lambda)$ and $c_2^*(u, \omega, \lambda)$ in the complexification of Π_2^* such that

$$c_i^* \left[F_x(u, \lambda) - \omega K \frac{d}{d\tau} - \sigma_i K \right] = 0, \qquad i = 1, 2. \tag{3.11}$$

We see from the second equation in (3.8) that when $\alpha \neq 0$, $(d/d\tau)z(\alpha, \lambda)$ cannot be identically zero. If, moreover, $\varphi(\alpha, \lambda) = 0$, (3.9) shows that $(1/\alpha)z(\alpha, \lambda; \tau + \theta)$ satisfies (3.7) with $\gamma = 0$, 1 replaced by $\cos\theta$ in the second equation, and 0 replaced by $\sin\theta$ in the third equation. The implicit function theorem then shows that $z(\alpha, \lambda; \tau + \theta)$ is a C^1 function of θ, so that $dz/d\tau$ is in Π_1, even though K may not be invertible. Differentiating (3.8) with respect to τ, we see that when $\varphi(\alpha, \lambda) = 0$,

$$\left(F_x(z(\alpha, \lambda), \lambda) - \omega(\alpha, \lambda)K \frac{d}{d\tau}\right)\frac{dz}{d\tau} = 0.$$

Thus $\sigma = 0$ is an eigenvalue with the corresponding eigenvector $(dz/d\tau)$. We let

$$\sigma_1(z(\alpha, \lambda)\omega(\alpha, \lambda), \lambda) = 0 \qquad \text{when} \quad \varphi(\alpha, \lambda) = 0.$$

This eigenvalue is ignored in the Floquet theory.

We choose $c_2{}^*$ so that

$$\left(c_2{}^*(z(\alpha, \lambda), \omega(\alpha, \lambda), \lambda), K\frac{dz}{d\tau}\right) = 0. \tag{3.12}$$

Since complex eigenvalues occur in pairs, we find that σ_2 is real.

Now $c_2{}^*(0, \omega_0, 0)$ is a linear combination of $e^{-i\tau}b^*$ and $e^{i\tau}\overline{b^*}$. Thus

$$c_2{}^*(z(\alpha, \lambda), \omega(\alpha, \lambda), \lambda)$$
$$= e^{-i\tau}b^* + e^{i\tau}\overline{b^*} + vi(e^{-i\tau}b^* - e^{i\tau}\overline{b^*}) + O(|\alpha| + |\lambda|).$$

for some real v. It follows from the construction of z in terms of the solution of (3.7) that

$$z = \alpha(e^{i\tau}a + e^{-i\tau}\bar{a} + O(|\alpha| + |\lambda|)).$$

Substituting this and the expression for $c_2{}^*$ in (3.12) and using (3.2), we find that

$$-2v\alpha = O(|\alpha|[|\alpha| + |\lambda|]).$$

Thus $v = 0$, and

$$c_2{}^*(z(\alpha, \lambda), \omega(\alpha, \lambda), \lambda) = e^{-i\tau}b^* + e^{i\tau}\overline{b^*} + O(|\alpha| + |\lambda|) \tag{3.13}$$

when $\varphi(\alpha, \lambda) = 0$.

We note that the operator (3.10) is invariant under the transformation $p \to e^{-in\tau}p$, $\sigma \to \sigma + in\omega$. Therefore its spectrum is invariant under translation by integral multiples of $i\omega$. Thus we may confine our attention to the strip $|\text{Im}\,\sigma| \le \frac{1}{2}\omega$.

The Floquet theory states that when the spectrum of (3.10) in the strip $|\text{Im}\,\sigma| \le \frac{1}{2}|\omega|$ that is away from 0 lies in a half-plane $\text{Re}\,\sigma \le \eta < 0$, the stability of the solution $(z(\alpha, \lambda), \omega(\alpha, \lambda), \lambda)$ with $\varphi(\alpha, \lambda) = 0$ is determined by the sign of $\sigma_2(z(\alpha, \lambda), \omega(\alpha, \lambda), \lambda)$. We shall write

$$\sigma(\alpha, \lambda) = \sigma_2(z(\alpha, \lambda), \omega(\alpha, \lambda), \lambda), \qquad c^*(\alpha, \lambda) = c_2{}^*(z(\alpha, \lambda), \omega(\alpha, \lambda), \lambda). \tag{3.14}$$

The linear stability of an equilibrium solution $(0, \lambda)$ is determined by the spectrum of the family of transformations

$$F_x(0, \lambda) - \rho K \tag{3.15}$$

from X_1 to X_2. If the spectrum lies in a half-plane $\operatorname{Re} \rho \leq \eta < 0$, $(0, \lambda)$ is linearly stable while if there is a spectral point with $\operatorname{Re} \rho > 0$, the solution is linearly unstable.

It is easily seen that when $x = 0$, the spectrum of (3.10) is the set of all points of the form $\rho + in\omega$ with ρ in the spectrum of (3.15) and n any integer. Thus, the real parts of ρ are determined by the real parts of the σ in the spectrum of (3.10) with $|\operatorname{Im} \sigma| \leq \frac{1}{2}|\omega|$.

The operator (3.15) has the isolated simple eigenvalue $\rho = i\omega_0$ when $\lambda = 0$. Standard perturbation theory then shows that there is a branch of simple eigenvalues $\rho(\lambda)$ with $\rho(0) = i\omega_0$ near $\lambda = 0$. We make the neighborhood $\bar{N}$ so small that this is the case and that $\frac{1}{2}\omega(0, \lambda) < \operatorname{Im} \rho(\lambda) < \frac{3}{2}\omega(0, \lambda)$. We then put

$$\sigma(0, \lambda) = \sigma_2(0, \omega(0, \lambda), \lambda) = \rho(\lambda) - i\omega(0, \lambda).$$

If we begin with the eigenvalue $-i\omega_0$ of (3.15) we obtain the other eigenvalue

$$\sigma_1(0, \omega(0, \lambda), \lambda) = \overline{\sigma_2(0, \omega(0, \lambda), \lambda)}.$$

There are, of course, corresponding eigenfunctionals $c_2{}^*$ and $c_1{}^*$ which satisfy (3.11).

We let

$$\hat{N}_2 = \{(\alpha, \lambda) \in \hat{N}_1 : (z(\alpha, \lambda), \omega(\alpha, \lambda), \lambda) \in \bar{N}\}.$$

We shall prove the following analog of Proposition 1.

PROPOSITION 2 Let $F \in C^2(N_1, \Pi_2)$ and $K\, d/d\tau \in \mathscr{L}(\Pi_1, \Pi_2)$. Suppose that $F(0, \lambda) = 0$ for $(0, \lambda) \in N_1$ and that

$$F_x(0, 0) - \omega_0 K \frac{d}{d\tau}$$

is a Fredholm operator whose null space consists of the real and imaginary parts of $e^{i\tau}a$ and whose range is the null space of the real and imaginary parts of $e^{-i\tau}b^*$. Assume that the eigenvalue 0 is K-simple and that a, b^*, w, and d^* satisfy the conditions (3.2), (3.5), and (3.6).

Then in a neighborhood $\hat{N}$ of $(0, 0)$ in $R \times \Lambda$

$$\operatorname{sgn} \operatorname{Re} \sigma(\alpha, \lambda) = \operatorname{sgn} \varphi_\alpha(\alpha, \lambda) \qquad \text{when} \quad \varphi(\alpha, \lambda) = 0. \tag{3.16}$$

Proof Differentiate the first equation in (3.8) with respect to α and apply the functional $c^*(\alpha, \lambda)$ to find

$$\sigma(\alpha, \lambda)(c^*(\alpha, \lambda), Kz_\alpha(\alpha, \lambda)) - \omega_\alpha(\alpha, \lambda)\left(c^*, K\frac{dz}{d\tau}\right) = \varphi_\alpha(c^*, w).$$

If $\alpha = 0$, $dz/d\tau = 0$, while if $\alpha \neq 0$, the second term on the left vanishes because of (3.12). Thus

$$\sigma = \varphi_\alpha(c^*, w)/(c^*, Kz_\alpha). \tag{3.17}$$

Now $z(\alpha, \lambda)$ was defined as $\alpha y(\alpha, \lambda)$ and by construction $y(0, 0) = e^{i\tau}a + e^{-i\tau}\bar{a}$. Hence,

$$z_\alpha(0, 0) = e^{i\tau}a + e^{-i\tau}\bar{a}.$$

If $\alpha = 0$, we have $c^*(0, 0) = e^{-i\tau}b^*$. Therefore by (3.2) and (3.5) the real part of the second factor on the right has the limit 1 as $(0, \lambda)$ approaches $(0, 0)$.

We see from (3.13) that this factor has the limit 1 as (α, λ) approaches $(0, 0)$ through values with $\alpha \neq 0$ and $\varphi(\alpha, \lambda) = 0$.

Hence there is a neighborhood $\hat{N}$ of $(0, 0)$ in which

$$\mathrm{Re}(c^*, w)/(c^*, Kz_\alpha) > 0 \qquad \text{when} \quad \varphi(\alpha, \lambda) = 0.$$

Since φ_α is real, (3.16) follows, from (3.17) and the Proposition is proved.

We remark that the obvious analogs of Corollaries 1, 2, and 3 of Proposition 1 are valid as corollaries of Proposition 2.

It is easily seen from the proof that Proposition 2 remains valid when the three conditions (3.2), (3.5), and (3.6) are replaced by the single condition

$$\mathrm{Re}[(e^{-i\tau}b^*, w)d^*a\overline{b^*Ka}] > 0.$$

References

1. M. G. Crandall and P. H. Rabinowitz, Bifurcation, perturbation of simple eigenvalues, and linearized stability. *Arch. Rational Mech. Anal.* **52** (1973), 161–180.
2. M. G. Crandall and P. H. Rabinowitz, The principle of exchange of stability. *Proc. Univ. Florida Internat. Symp. Dynamical Systems.* Academic Press, New York, 1977.
3. J. K. Hale, "Oscillations in Nonlinear Systems." McGraw-Hill, New York, 1963.
4. D. Henry, "Geometric Theory of Semilinear Parabolic Equations," Univ. of Kentucky Lecture Notes. Univ. of Kentucky, Lexington, 1974.
5. E. Hopf. Abzweigung einer periodischen Lösung eines Differentialsystems. *Ber. Verh. Saechs. Akad. Wiss. Leipzig, Math.-Phys. Kl.* **94** (1942), 1–22.
6. D. D. Joseph and D. A. Nield, Stability of birfurcating time-periodic and steady state solutions of arbitrary amplitude. *Arch. Rational Mech. Anal.* **58** (1975), 369–380.
7. D. D. Joseph and D. H. Sattinger, Bifurcating time-periodic solutions and their stability. *Arch. Rational Mech. Anal.* **45** (1972), 79–109.
8. H. Kielhöfer, Stability and semilinear evolution equations in Hilbert space. *Arch. Rational Mech. Anal.* **57** (1974), 150–165.

9. A. M. Lyapounov, Problème de minimum dans une question de stabilité des figures d'équilibre d'une masse fluide en rotation. *Akad. Nauk St. Petersb.*, Mem. Ser. VIII, Vol. 12 (1908) #5.
10. D. H. Sattinger, Stability of bifurcating solutions by Leray–Schauder degree. *Arch. Rational Mech. Anal.* **43** (1971), 154–166.
11. D. H. Sattinger, Stability of solutions of nonlinear equations. *J. Math. Anal. Appl.* **39** (1972), 1–12.
12. E. Schmidt, Zur Theorie der linearen und nichtlinearen Integralgleichungen. 3. Teil: Über die Auflösung der nichtlinearen Integralgleichungen und die Verzweigung ihrer Lösungen. *Math. Ann.* **69** (1908), 370–399.
13. H. F. Weinberger, The stability of solutions bifurcating from steady or periodic solutions. *Univ. Florida Internat. Symp. Dynamical Systems.* Academic Press, New York, 1977.

This work was supported by the National Science Foundation through Grant GP 37660X.

AMS (MOS) 1970 Subject Classification: 34D99, 35B30.

PUBLISHED WORKS OF ERICH H. ROTHE

1925

Associate Author of Volume I of Frank–Mises, "Die Differential- und Integralgleichungen der Mechanik und Physik," 1st ed. 1925; 2nd ed. 1930. Contributed: Einige besondere Probleme partieller Differentialgleichungen, pp. 829–836, 845–880.

1927

Über einige Analogien zwischen linearen partiellen und linearen gewöhnlichen Differentialgleichungen. *Math. Z.* **27**, 76–86.

1928

Ein Beitrag zum Cauchyschen Problem. *Math. Z.* **28**, 48–72.

1929

Über die Approximation stetiger Funktionen durch Eigenfunktionen elliptischer Differentialgleichungen. *Sitzungsber. Berlin. Math. Ges.* **28**, 71–77.

1930

Zweidimensionale parabolische Randwertaufgaben als Grenzfall eindimensionaler Randwertaufgaben. *Math. Ann.* **102**, 650–670.

1931

Über die Wärmeleitungsgleichung mit nicht-konstante Koeffizienten im räumlichen Falle, Erste Mitteilung. *Math. Ann.* **104**, 340–354.

Über die Grundlösung bei parabolischen Gleichungen. *Math. Z.* **33**, 488–504.

Über lineare elliptische Differentialgleichungen zweiter Ordung, deren zugeordnete Massestimmung von Konstanter Krummung ist. *Math. Ann.* **105**, 672–693.

Zweite Mitteilung. *Math. Ann.* **104**, 355–362.

1932

Über eine Verallgemeinerung der Besselschen Funktionen. *Jahresber. Deut. Math. Ver.* **42**, 166–173.

1933

Über asymptotische Entwicklungen bei Randwertaufgaben elliptischer partieller Differentialgleichungen. *Math. Ann.* **108**, 578–594.

Über asymptotische Entwicklungen bei Randwertaufgaben der Gleichung $\Delta\Delta u + \lambda^k u = \lambda^k \psi$. *Math. Ann.* **109**, 167–172.

Zur Theorie des Skin-effekts. *Z. Phys.* **83**, 184–186.

1934

Über die Integralgleichung des Skin-effekts. *J. Reine Angew. Math.* **170**, 218–230.

1936

Über asymptotische Entwicklungen bei gewissen nicht-linearen Randwertaufgaben. *Compositio Math.* **3**, 310–327.

1937

Über Abbildungsklassen von Kugeln des Hilbertschen Raumes. *Compositio Math.* **4**, 294–307.

Über den Abbildungsgrad bei Abbildungen von Kugeln des Hilbertschen Raumes. *Compositio Math.* **5**, 166–176.

Zur Theorie der topologischen Ordnung und der Vektorfelder in Benachschen Raumen. *Compositio Math.* **5**, 177–197.

1939

Asymptotic solution of a boundary value problem. *Iowa State Coll. J. Sci.* **13**, 369–372.

Topological proofs of uniqueness theorems in the theory of differential and integral equations. *Bull. Amer. Math. Soc.* **45**, 606–613.

The theory of the topological order in some linear topological spaces. *Iowa State Coll. J. Sci.* **13**, 373–390.

1941

On topology in function spaces, *in* "Lectures in Topology." *Univ. Mich. Conf., 1940*, pp 303–305. Univ. Mich. Press, Ann Arbor, 1941.

1944

On non-negative functional transformations. *Amer. J. Math.* **66**, 245–254.

1946

Gradient mappings in Hilbert space. *Ann. of Math.* **47**, 580–592.

1948

Completely continuous scalars and variational methods. *Ann. of Math.* **49**, 265–278.

Gradient mappings and extrema in Banach spaces. *Duke Math. J.* **15**, 421–431.

1949

Weak topology and non-linear integral equations. *Trans. Amer. Math. Soc.* **66**, 75–92.

1950

A relation between the type numbers of a critical point and the index of the corresponding field of gradient vectors. *Math. Nachr.* **4**, 12–27.

1951

Critical points and gradient fields of scalars in Hilbert space. *Acta Math.* **85**, 73–98.

1952

A remark on isolated critical points. *Amer. J. Math.* **74**, 253–263.

Leray-Schauder index and morse type numbers in Hilbert space. *Ann. of Math.* **55**, 433–467.

1953

Gradient mappings. *Bull. Amer. Math. Soc.* **59**, 5–19.

A note on the Banach spaces of Calkin and Morrey. *Pacific J. Math.* **3**, 493–499.

Correction to the paper "Leray-Schauder index and Morse type numbers." *Ann. of Math.* **58**, 593–594.

1955

Mapping degree in Banach spaces and spectral theory. *Math. Z.* **63**, 195–218.

1956

Remarks on the application of gradient mappings to the calculus of variations and the connected boundary value problems. *Comm. Pure Appl. Math.* **9**, 551–567. Reprinted in: *Trans. Symp. Partial Diff. Eqs., Univ. of Calif., Berkeley, June 20–July 1, 1955*, pp. 253–270.

1958

Some applications of functional analysis to the calculus of variations. *Proc. Eighth Symp. Appl. Math. Amer. Math. Soc., Univ. of Chicago, April 12–13, 1956*, pp. 143–151. McGraw-Hill, New York, 1958.

Integral equations. *In* "Handbook of Automation, Computation and Control." Vol. I, Chap. 6. Wiley, New York, 1958.

1959

Some remarks on fundamental solutions of parabolic differential equations of second order. *Mich. Math. J.* **6**, 227–245.

A note on gradient mappings. *Proc. Amer. Math. Soc.* **10**, 931–935.

1963

"Some Remarks on Critical Point Theory in Hilbert Space; Nonlinear Problems." (R. E. Langer, ed.) pp. 233–256. Univ. of Wisconsin Press, 1963.

1965

Critical point theory in Hilbert space under general boundary conditions. *J. Math. Anal. Appl.* **11**, 357–409.

1966

An existence theorem in the calculus of variations based on Sobolev's imbedding theorems, *Arch. Rational Mech. Anal.* **21**, 151–162.

Weak/topology and calculus of variations, Centro Internationale Matematico Estovo, course given June 10–18, 1966, Bressance, Italy, 207–237.

1967

Some remarks on critical point theory in Hilbert space (continuation), *J. Math. Anal. Appl.* **20**, 515–520.

1970

Remarks on vector fields in Hilbert space, *Proc. Symp. in Pure Math.* Vol. XVIII, Part 1, 1970, 251–270. A.M.S. Symposium, April, 1968.

1971

Critical point theory in Hilbert space under regular boundary conditions, *J. Math. Anal. Appl.* **36**, 377–431.

1972

On continuity and approximation questions concerning critical Morse groups in Hilbert space, *Symp. Infinite Dimensional Topology, Baton Rouge, 1967* (R. D. Anderson, ed.) *in* Ann. of Math. Studies, Vol. 69.

1973

Morse theory in Hilbert space (Santa Fe Summer Sem. 1971), *Rocky Mountain J. of Math.* **3**, 251–274.

1975

On the theories of Morse and Liusternik–Schnirelman for open bounded sets on Fredholm manifolds, *J. Math. Anal Appl.* **50**, 80–107.

A generalization of the Seifert–Threlfall proof for the Liusternik–Schnirelman inequality, *J. Math. Anal. Appl.* **50**, 243–267.

1976

Expository introduction to some aspects of degree theory, *Proc. Michigan State Univ. Conf. Nonlinear Functional Analysis, Differential equations* (L. Cesari, R. Kannan, and J. Schuur, eds.), pp. 291–317, Dekker, New York.

1977

On the Cesari index and the Browder–Petryshin degree, *in* "Dynamical Systems" (A. R. Bednarek and L. Cesari, eds.), pp. 295–312. Academic Press, New York.

A B C 8 D 9 E 0 F 1 G 2 H 3 I 4 J 5